Lasers in Polymer Science and Technology: Applications

Volume II

Editors

Jean-Pierre Fouassier, Ph.D.
Professor
Laboratory of General Photochemistry
Ecole Nationale Superieure de Chimie
University of Haute-Alsace
Mulhouse, France

Jan F. Rabek, Ph.D.
Department of Polymer Technology
The Royal Institute of Technology
Stockholm, Sweden

CRC Press, Inc.
Boca Raton, Florida

Library of Congress Cataloging-in-Publication Data

Lasers in polymer science and technology:applications / editors, Jean
-Pierre Fouassier and Jan F. Rabek.
   p. cm.
  Bibliography: p.
  Includes index.
  ISBN 0-8493-4844-7 (v. 1)
  1. Polymers--Analysis. 2. Laser spectroscopy. I. Fouassier,
Jean-Pierre, 1947- II. Rabek, J. F.
TP1140.L37 1990
668.9--dc20   89-9822
                                                    CIP

Direct all inquiries to CRC Press, Inc., 2000 Corporate Blvd., N.W., Boca Raton, Florida, 33431.

© 1990 by CRC Press, Inc.

International Standard Book Number 0-8493-4844-7 (v. 1)
International Standard Book Number 0-8493-4845-5 (v. 2)
International Standard Book Number 0-8493-4846-3 (v. 3)
International Standard Book Number 0-8493-4847-1 (v. 4)

Library of Congress Number 89-9822
Printed in the United States

DEDICATION

To our wives, partners through life
Geneviève — Ewelina
and our children
Patrick, Laurence, and Yann — Dominika
for their patience and understanding.

PREFACE

Laser spectroscopy and laser technology have been growing ever since the first laser was developed in 1960 and cover now a wide range of applications. Among them, three groups came into prominence as regards polymer science and technology: molecular gas lasers (notably CO_2 lasers) in the IR region, gas, solid, and dye lasers in the visible and near IR region, and the relatively new group of UV excimer lasers. Lasers are unique sources of light. Many recent advances in science are dependent on the application of their uniqueness to specific problems. Lasers can produce the most spectrally pure light available, enabling atomic and molecular energy levels to be studied in greater detail than ever before. Certain types of laser can give rise to the shortest pulses of light available from any light source, thus providing a means for measuring some of the fastest processes in nature.

Measurements of luminescence (fluorescence and phosphorescence) provide some of the most sensitive and selective methods of spectroscopy. In addition, luminescence measurements provide important information about the properties of excited states, because the emitted light originates from electronically excited states. The measurement of luminescence intensities makes it possible to monitor the changes in concentration of the emitting chemical species as a function of time, whereas the wavelength distribution of the luminescence provides information on the nature and energy of the emitting species.

Such areas as laser luminescence spectroscopy, pico- and nanosecond absorption spectroscopy, CIDNP and CIDEP laser flash photolysis, holographic spectroscopy, and time-resolved diffuse reflectance laser spectroscopy, have evolved from esoteric research specialities into standard procedures, and in some cases routinely applied in a number of laboratories all over the world.

Application of Rayleigh, Brillouin, and Raman laser spectroscopy in polymer science gives information about local polymer chain motion, large-scale diffusion, relaxation behavior, phase transitions, and ordered states of macromolecules.

During the last decade the photochemistry and photophysics of polymers have grown into an important and pervasive branch of polymer science. Great strides have been made in the theory of photoreactions, energy transfer processes, the utilization of photoreactions in polymerization, grafting, curing, degradation, and stabilization of polymers. The progress of powerful laser techniques has not been limited to spectroscopical studies in polymer matrix, colloids, dyed fabrics, photoinitiators, photosensitizers, photoresists, materials for solar energy conversion, or biological molecules and macromolecules; it has also found a number of practical and even industrial applications.

One of the most important applications of lasers is the use of a high intensity beam for material processing in polymers. In these materials, the laser beam can be employed for drilling, cutting, and welding. Lasers can produce holes at very high speeds and dimensions, unobtainable by other processing methods.

Lasers can be successfully used to study surface processes and surface modification of polymeric materials, such as molecular beam scattering, oxidation, etching, annealing, phase transitions, surface mobility, and thin films and vapor phase deposition.

UV laser radiation causes the breakup and spontaneous removal of material from the surface of organic polymers (ablative photodecomposition). The surface of the solid is etched away to a depth of a few tenths of a micron, and the products are expelled at supersonic velocity. This method has found practical applications in photolithography, optics, electronics, and the aerospace industries.

The newest process includes stereolithography, which involves building three-dimensional plastic prototypes (models) from computer-aided designs. Stereolithography is actually a combination of four technologies: photochemistry, computer-aided, laser light, and laser-image formation. The device (which consists in a mechanically scanned, computer driven three-dimensional solid pattern generator) builds parts by creating, under the laser exposure, cross sections of the part out of a liquid photopolymer, then "fusing" the sections together until a complete model is formed.

Another new development is technology of micromachines such as gears, turbines, and motors which are 100 to 200 μm in diameter which can be used in a space technology, microrobots, or missile-guidance systems. These micromachines are made by a process of etching patterns on silicon chips. Beside making such micromachines, microscopic tools on a catheter, inserted through a blood vessel, would enable surgeons to do "closed heart" surgery. Developing of micromachine technology would not be possible without photopolymers and UV lasers.

The editors went to great lengths in order to secure the cooperation of the most outstanding specialists to complete this monography. A number of invited authorities were not able to accept our invitation, due to other commitments, but all authors who presented their contributions "poured their hearts out" in this endeavour. We would like to thank them for their efforts and cooperation. This monography strongly favors the inclusion of experimental details, apparatus, and techniques, thus allowing the neophyte to learn the "tricks of the trade" from the experts. This is an effort to show, in compact form, the bulk of information available on applications of lasers to polymer science and technology. The editors are pleased to submit to the readers the state-of-the art in this field.

J.-P. Fouassier and J. R. Rabek

THE EDITORS

Dr. Jean-Pierre Fouassier is head of the Laboratoire de Photochimie Générale, Ecole Nationale Supérieure de Chimie de Mulhouse, and Centre National de la Recherche Scientifique, and Professor of Physical Chemistry at the University of Haute Alsace.

Prof. J. P. Fouassier graduated in 1970 from The National School of Chemistry, Mulhouse, with an Engineer degree and obtained his Ph.D. in 1975 at the University of Strasbourg. After doing postdoctoral work at the Institüt für Makromolekulare Chemie, Freiburg, (West Germany), he was appointed as lecturer. It was in 1980 that he assumed his present position.

Prof. J. P. Fouassier is a member of the Société Française de Chimie, the Groupe Français des Polymères, the European Photochemistry Association, the ACS Polymer Division, and Radtech Europe.

Prof. J. P. Fouassier has been the recipient of research grants from the Centre National de la Recherche Scientifique, the Ministère de la Recherche, the Association Nationale pour la Valorisation et l'Aide à la Recherche, and French and European private industries. He has published more than 100 research papers. His current major research interests include time-resolved laser spectroscopies, excited state processes in photoinitiators and photosensitizers, laser-induced photopolymerization reactions, development of photosensitive systems for holographic recording, and light radiation curing.

Dr. Jan F. Rabek is Professor of Polymer Chemistry in the Department of Polymer Technology, The Royal Institute of Technology, Stockholm, working in the field of polymer photochemistry and photophysics since 1960. His research interests lie in the photodegradation, photooxidation, and photostabilization of polymers, singlet oxygen photooxidation, spectroscopy of molecular complexes in polymers, and recently photoconducting polymers and polymeric photosensors.

Dr. Rabek obtained his D.Sc. in Polymer Technology at the Department of Polymer Technology, Technical University, Wroclaw, Poland (1965) and his Ph.D. in Polymer Photochemistry at the Department of Chemistry, Sileasian Technical University, Gliwice, Poland (1968). He has published more than 120 research papers, review papers, and books on the photochemistry of polymers.

CONTRIBUTORS

Paritosh K. Das, Ph.D.
Research Chemist
Phillips Petroleum Company
Bartlesville, Oklahoma

Jacques Delaire, Ph.D.
Professor
Department of Physical Chemistry
University of Paris South
Orsay, France

Herbert Dreeskamp, Ph.D.
Professor
Institute of Physical and Theoretical
  Chemistry
Technical University of Braunschweig
Braunschweig, West Germany

Jean Faure, Ph.D.
Professor
Department of Physical Chemistry
University of Paris South
Orsay, France

Jean-Pierre Fouassier, Ph.D.
Professor
Laboratory of General Photochemistry
Ecole Nationale Superieure de Chimie
University of Haute-Alsace
Mulhouse, France

Marye Anne Fox, Ph.D.
Rowland Pettit Centennial Professor
Department of Chemistry
University of Texas
Austin, Texas

Hisaharu Hayashi, Ph.D.
Head
Physical Organic Chemistry Laboratory
Institute of Physical and Chemical
  Research
Wako, Saitama, Japan

Akira Itaya, Ph.D.
Associate Professor
Department of Polymer Science and
  Engineering
Kyoto Institute of Technology
Kyoto, Japan

Prashant V. Kamat, Ph.D.
Associate Professional Specialist
Radiation Laboratory
University of Notre Dame
Notre Dame, Indiana

Daniel-Joseph Lougnot, Ph.D.
Director of Research
Department of General Photochemistry
ENSCMu/CNRS
Mulhouse, France

Hiroshi Masuhara, Ph.D.
Professor
Department of Polymer Science and
  Engineering
Kyoto Institute of Technology
Kyoto, Japan

William Grant McGimpsey, Ph.D.
Assistant Professor
Department of Chemistry
Worcester Polytechnic Institute
Worcester, Massachusetts

Wolf-Ulrich Palm, Dipl. Chem.
Institute of Physical and Theoretical
  Chemistry
Technical University of Braunschweig
Braunschweig, West Germany

Yoshio Sakaguchi, Ph.D.
Researcher
Physical Organic Chemistry Laboratory
Institute of Physical and Chemical
  Research
Wako, Saitama, Japan

Wolfram Schnabel, Dr. rer. nat.
Strahlenchemie
Hahn-Meitner-Institut Berlin
Berlin, West Germany

Francis Wilkinson, Ph.D.
Professor
Department of Chemistry
Loughborough University of Technology
Loughborough, Leicestershire, England

Charles John Willsher, Ph.D.
Department of Chemistry
Loughborough University of Technology
Loughborough, Leicestershire, England

SERIES TABLE OF CONTENTS

Application of Lasers in the Scattering for the Study of Solid Polymers
Laser Spectroscopy in Life Sciences
Emission and Laser Raman Spectroscopy of Nucleic Acid Complexes
Picosecond Laser Spectroscopy and Optically Detected Magnetic Resonance on Model
  Photosynthetic Systems in Biopolymers

TABLE OF CONTENTS

Chapter 1

MAGNETIC FIELD EFFECTS AND LASER FLASH PHOTOLYSIS — ESR OF RADICAL REACTIONS IN POLYMERS AND MODEL COMPOUNDS

Hisaharu Hayashi and Yoshio Sakaguchi

TABLE OF CONTENTS

I. INTRODUCTION

Recently, many chemical reactions which occur through radical pairs and biradicals in solution have been found to show anomalous intensities in ESR and NMR spectra and also to be influenced by external and internal magnetic fields.[1] The anomalous ESR and NMR intensities arise from radicals and reaction products which are formed with nonequilibrium spin populations. These phenomena are called chemically induced electron and nuclear polarizations (CIDEP and CIDNP, respectively).

Usual electromagnets can supply magnetic fields less than about 2.5 T. In such an ordinary magnetic field, Zeeman energies of electron and nuclei are much smaller than the activation energies for chemical reactions. An external magnetic field of such an ordinary intensity, however, has been proved to change not only the reaction yields but also the reaction rates. The internal magnetic field effects can be brought about by magnetic species such as magnetic nuclei and paramagnetic ions.

Most of the above-mentioned phenomena of CIDEP, CIDNP, and magnetic field effects have been explained by the singlet-triplet conversion of radical pairs and biradicals (radical pair mechanism, RPM)[2,3] and some of them by the spin polarization of precedent triplet molecules (triplet mechanism, TM).[4] These phenomena have opened a new research field in chemistry and biology, which can be called "spin chemistry". Moreover, new techniques have been expected to be developed with new principles which have been obtained in the course of the studies of these phenomena.

Most of the transient radicals produced by laser flash photolysis-ESR show CIDEP in their initial stages and the reactions of such radicals are expected to be influenced by magnetic fields. In this chapter, therefore, we will explain theoretical and experimental aspects of the magnetic field effects and CIDEP of radical reactions in polymers and model compounds, taking mainly the results obtained by our group as examples without giving a comprehensive review of these phenomena. Since some magnetic field effects and CIDEP found recently in micellar solutions at room temperature cannot be explained by ordinary RPM and TM, new mechanisms for these phenomena (relaxation mechanism, RM, and nonzero exchange integral mechanism, JM, respectively) will also be explained. Those who prefer to read only magnetic field effects should choose Sections II, IV, and V and those who would like to read only CIDEP should consult Sections II, III, VI, and VII. CIDNP is dealt with in other chapters.

II. MECHANISM OF MAGNETIC FIELD EFFECTS

A. SINGLET-TRIPLET (S-T) MIXING IN RADICAL PAIRS

In many photo- and radiation-induced decomposition, electron transfer, and hydrogen transfer reactions, radical pairs and biradicals are produced in their initial stages. Kurita[5] was the first to observe ESR spectra of radical pairs in X-irradiated organic crystals. The central three lines of the nine hyperfine (HF) ones of the pair "K" of iminoxy radicals trapped in a single crystal of dimethylglyoxime at 77 K are not equally spaced. The authors[6] noticed the importance of this anomalous HF structure and interpreted this anomaly in terms of the S-T mixing in the radical pair, calculating generally the angular dependence of this S-T matrix elements.

According to this analysis, the radical pair, where the $=N-O\cdot$ groups of two iminoxy radicals are separated by 0.61 nm with each other, was proved to have an exchange integral (J) of -0.10 ± 0.01 cm^{-1}. Here the Hamiltonian of the exchange interaction (H_{ex}) of a radical pair is written as

$$H_{ex} = -J(2S_1 S_2 + 1/2) \tag{1}$$

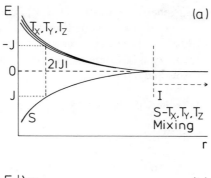

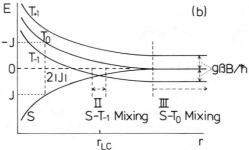

FIGURE 1.   Schematic diagrams showing the distance dependence of the energy levels of the singlet (S) and triplet (T) states of a radical pair with a negative J (a) in the absence and (b) in the presence of an external magnetic field. Here, r, g, and B are the distance between the radical centers of the pair, the mean value of the g-values of the component radicals (g_1 and g_2), and the strength of the applied magnetic field (in Tesla), respectively.

This means that the energy (E(S)) of the singlet state (S) of the pair is J and that the energy (E(T_n)) of the triplet state (T_n, n = X, Y, and Z) at B = 0 is $-J$. Here, X, Y, and Z are the directions of the principal axes of the zero-field splittings (ZFS) of the pair and B is the strength of an applied magnetic field.[*1]

The energy relation, E(S) < E(T_n), is usually realized for most radical pairs and biradicals where the coupling between the component radicals is relatively weak. The arguments about weakly coupled biradicals are quite parallel to those about weakly coupled radical pairs. Therefore, the word ''biradical(s)'' will be dropped hereafter in the arguments about weakly coupled radical pairs and biradicals for simplicity.

The distance (r) dependence of E(S) and E(T_n) at B = 0 is illustrated schematically in Figure 1a. Here, r is the distance between two radical centers and T_n represents each of three sublevels of a triplet radical pair in the molecular coordinate. As shown in this figure, S and T_n become nearly degenerate with one another at large r (region I). In Figure 1b, the r-dependence of E(S) and E(T_i) in the presence of a magnetic field of B is shown. Here T_i (i = 0, ±1) represents each of three Zeeman sublevels of a triplet radical pair in the laboratory coordinate, g is the mean value of the g-factors of the component radicals (g_1 and g_2, respectively), and β is the Bohr magneton. At a certain distance (r = r_{LC}), a level crossing occurs between S and T_{-1}. Around r = r_{LC} (region II), S and T_{-1} become nearly degenerate with each other. If J is positive, a level crossing occurs between S and T_{+1}. At large r (region III), S becomes nearly degenerate with only T_0 among the three Zeeman sublevels.

[*1]   Since the values of ZFS of usual radical pairs are much smaller than those of triplet molecules, the former are not considered in this chapter.

When S and $T_{n(i)}$ are nearly degenerate, an efficient S-$T_{n(i)}$ conversion is expected to occur though the S-$T_{n(i)}$ matrix element of the magnetic Hamiltonian (H_{mag}) of a radical pair. The authors[6] obtained the general forms of the S-T_i matrix element of H_{mag} in the presence of a magnetic field, but the elements of a fast tumbling radical pair in solution*[2] can only be represented by the following isotropic terms:

$$H_{mag} = \beta(g_1 S_1 + g_2 S_2)\hbar^{-1}B + \Sigma_j a_{1j} I_j S_1 + \Sigma_k a_{2k} I_k S_2 \tag{2}$$

$$Q_{NM} = \langle S, N, M|H_{mag}|T_0, N, M,\rangle$$

$$= \Delta g \beta \hbar^{-1} B/2 + (\Sigma_j a_{1j} m_{1j} - \Sigma_k a_{2k} m_{2k})/2 \tag{3}$$

$$V_{\pm} = \langle S, N, M|H_{mag}|T_{\pm 1}, N', M'\rangle$$

$$= \mp(a_r 8^{-1/2})\{I_r(I_r + 1) - m_r(m_r \mp 1)\}^{1/2} \tag{4}$$

Here, a_{1j} and a_{2k} are the HF coupling constants of magnetic nuclei in two component radicals, I_j and I_k are the nuclear spin operators, m_{1j} and m_{2k} are the nuclear magnetic quantum numbers, and Δg is $g_1 - g_2$. Sets of nuclear spin states are represented by N ($= \{m_{1j}\}$), M ($= \{m_{2k}\}$), N' ($= \{m'_{1j}\}$), and M' ($= \{m'_{2k}\}$). The S-T_i conversion must obey the following selection rule:

$$\Delta m_S + \Delta m_I = 0 \tag{5}$$

Here, Δm_S (Δm_I) is the change of one electron (nuclear) magnetic quantum number. For the S-T_0 conversion,

$$\Delta m_S = \Delta m_I = 0 \tag{6}$$

and for the S-$T_{\pm 1}$ conversion,

$$\Delta m_S = \mp 1 \quad \text{and} \quad \Delta m_I = \pm 1 \tag{7}$$

In the absence of a magnetic field (B = 0), all three S-T_n (n = X, Y, and Z) conversions occur at region I of Figure 1a through the HF interactions.

$$V_n = \langle S|H_{mag}|T_n\rangle$$

$$= V_n(a_{1j}, m_{1j}, a_{2k}, m_{2k}) \tag{8}$$

B. S-T CONVERSIONS IN THE RADICAL PAIR MECHANISM (RPM)

Let us consider the mechanism of the S-T_0 conversion of radical pairs in solution as a typical example of the RPM. The reaction scheme of radical pairs is illustrated in Figure 2. A radical pair is produced as an intermediate through a decomposition, electron transfer, or hydrogen transfer reaction from a singlet or triplet excited state. These reaction precursors are called "S- and T-precursors".*[3]

The generated pair is surrounded by solvent molecules (solvent cage) and retains the

---

*[2] In usual solvents the rotationl correlation time (τ_c) is $10^{-12} \sim 10^{-11}$s.

*[3] Radical pairs are also produced through the encounter of free radicals, which are called "F-precursors". Since the dynamic behavior of radical pairs produced from F-precursors are similar to that from T-precursors, the results obtained for the reactions from T-precursors can also be applied to those from F-precursors.

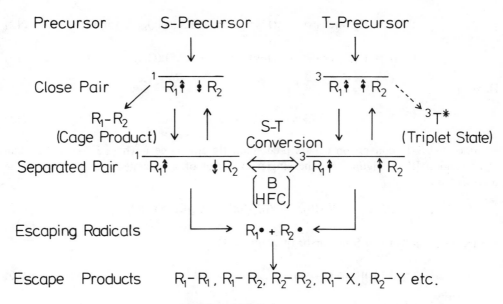

FIGURE 2. Reaction scheme of radical pairs generated from singlet(S)- and triplet(T)-precursors.

spin multiplicity of its precursor. Two radicals are produced as nearest neighbors. Such a pair is called "a close pair". In some of singlet (S)-close pairs,[*4] a recombination (or disproportionation) reaction takes place immediately after the formation of the pairs. This process and the products of this process are called "primary recombination" and "cage products", respectively. In this chapter, such products will be hereafter called "S-cage products." During the primary recombination, no S-T_0 conversion is expected to occur because the time scale for this process is too short ($\sim 10^{-11}$ s) and because the exchange interaction is too large (r is too small to locate in region III of Figure 1b).

The remaining radicals in a close pair start to diffuse from each other (r becomes larger). At a sufficiently large r (region III), $|J|$ becomes very small. This radical pair is called "a separated pair". During this period, the S-T_0 conversion occurs through the S-T_0 mixing. The time dependence of the wave function [ψ (t)] of a radical pair can be represented by the Schrödinger equation,[7]

$$i\delta\psi(t)/\delta t = H\psi(t) \tag{9}$$

$$H = H_{ex} + H_{mag} \tag{10}$$

The wave function can be expanded with a coupled basis set of singlet ($|S> = \{|\alpha\beta> - |\beta\alpha>\}/2^{1/2}$ and triplet ($|T_0> = \{|\alpha\beta> + |\beta\alpha>\}/2^{1/2}$) states combined with the nuclear spin state (ϕ_{NM}).

$$\psi(t) = \{C_S(t)|S\rangle + C_T(t)|T_0\rangle\}\phi_{NM} \tag{11}$$

S and T_0 states are not stationary states because they are coupled through Q_{NM} (see Equation 3).

Equation 11 can be solved as

[*4] Usually, no recombination occurs from a triplet (T)-close pair because usual ground state molecules have singlet spin multiplicity. In some cases, however, triplet excited states can also be produced from T-close pairs but such recombinations will not be considered, for simplicity, in this section. The results obtained in this section can easily be extended to such cases. In this chapter, such products are called "T-cage products".

$$C_S(t) = C_S(0)(\cos wt - i(J/w)\sin wt) - iC_T(0)(Q_{NM}/w)\sin wt \qquad (12)$$

$$C_T(t) = C_T(0)(\cos wt + i(J/w)\sin wt) - iC_S(0)(Q_{NM}/w)\sin wt \qquad (13)$$

Here

$$w = (Q_{NM}^2 + J^2)^{1/2} \qquad (14)$$

Since the S-T_0 conversion occurs predominantly at a large r where J^2 is smaller than Q_{NM}^2 (region III of Figure 1b), the singlet character of a separated radical pair can be represented as

$$|C_S(t)|^2 = |C_S(0)|^2 + (|C_T(0)|^2 - |C_S(0)|^2)\sin^2|Q_{NM}|t \qquad (15)$$

For the reaction from an S-precursor ($|C_S(0)|^2 = 1$),

$$|C_S(t)|^2 = 1 - \sin^2|Q_{NM}|t \qquad (16)$$

and for the reaction from a T-precursor ($|C_T(0)|^2 = 1/3$)[*5]

$$|C_S(t)|^2 = (1/3)\sin^2|Q_{NM}|t \qquad (17)$$

Some radicals in separated pairs reencounter to their initial partners but others escape from the solvent cages becoming "escaping radicals". Noyes[8] used the following probability for the first reencounter [f(t)]:

$$f(t) = mt^{-3/2}\exp(-\pi m^2/p^2 t) \qquad (18)$$

Here, p ($= \int_0^\infty f(t)\,dt$) is the total probability of at least one reencounter and m $\sim \tau_D^{1/2}$, where τ_D is the average time between diffusive steps with a typical value of 10^{-12}s

Some of the reencountered radicals react with their partners giving also cage products. This process is called "secondary recombination". The probability of product formation during the first reencounter at t [$P_1(t)$] is given as[7]

$$P_1(t) = \lambda|C_S(t)|^2 f(t) \qquad (19)$$

Here, λ is the probability for recombination reaction. Since the time scales for the secondary recombination and the S-T_0 conversion are $10^{-10} \sim 10^{-7}$ and $10^{-9} \sim 10^{-8}$ s, respectively, $P_1(t)$ is possible to be influenced by a magnetic field.

Radicals in a close pair that fail to react during the first reencounter start again their random walk. They have a new chance of meeting their initial partners or escaping from the solvent cage. The above-mentioned processes of the generated radical pairs occur successively, giving cage products and escaping radicals. Each of the escaping radicals reacts at last with another escaping radical from a different pair and/or with a scavenger, giving so-called "escape products".

[*5] Here, the spin polarization within T_i is not considered for simplicity. When the spin polarization of a T-precursor is transferred to the generated radical pair, $|C_T(0)|^2$ is possible to be different from 1/3 and to be dependent on B as shown by Equation 46.

From Equations 16 to 19, Kaptein[7] obtained the total probability of giving an S-cage product from a Γ-precursor ($P_{\Gamma\text{-}S}$, Γ = S or T) as follows:*6

$$P_{S\text{-}S} = \lambda\{p - m|\pi Q_{NM}|^{1/2}\} \tag{20}$$

$$P_{T\text{-}S} = (\lambda m/3)|\pi Q_{NM}|^{1/2}/(1 - p) \tag{21}$$

These are the key equations for the magnetic field effects on chemical reactions as well as CIDNP. The yield of giving the jth escape product ($P_{\Gamma\text{-}j}$) is

$$P_{\Gamma-j} = c_j(1 - P_{\Gamma-S}) \tag{22}$$

Here, c_j is the fraction of forming the jth escape product from the escaping radicals.

C. CLASSIFICATION OF MAGNETIC FIELD EFFECTS THROUGH THE RPM
1. The Δg Mechanism

In this section, the magnetic field effects on chemical reactions through the RPM are classified into some typical cases.[9,10] At first, let us consider a case where $a_{1j} = a_{2k} = 0$ and $\Delta g \neq 0$. In this case, Equations 3, 4, and 8 can be simplified as

$$Q_{NM} = \Delta g \beta \hbar^{-1} B/2 = Q \tag{23}$$

$$V_{\pm} = V_n = 0 \tag{24}$$

Thus, the S-T_0 conversion rate is expected to increase with increasing ($\Delta g \cdot B$). Therefore, the magnetic field effects classified in the present case are called the effects due to the Δg mechanism (ΔgM).

From Equations 20 to 23, $P_{S\text{-}S}$ and $P_{T\text{-}S}$ are simplified as follows:

$$P_{S\text{-}S} = \lambda(p - x_B) \tag{25}$$

$$P_{T\text{-}S} = \lambda x_B/3(1 - p) \tag{26}$$

Here $\tag{27}$

$$x_B = m(\pi \Delta g \beta \hbar^{-1} B/2)^{1/2}$$

Thus, the yield of an S-cage product from an S-precursor is proved to decrease with increasing B. On the other hand, the yield of an S-cage product from a T-precursor is proved to increase with increasing B.*7 The magnetically induced changes in both cases are proportional to ($\Delta g \cdot B$)$^{1/2}$. From Equation 22, the yield of escape products from an S(T)-precursor is shown to increase (decrease) with increasing B. The B-dependence of the yields of various products are schematically illustrated in Figure 3a.

2. The Hyperfine Coupling Mechanism (HFCM)

The magnetic field effects on chemical reactions through the RPM are explained for a

---

*6 In this calculation, $\int_0^\infty f(t)\sin^2|Q_{NM}|dt$ is approximated to be $\int_0^\infty m\, t^{-3/2}\sin^2|Q_{NM}|dt$ and a definite integral, $\int_0^\infty x^{-1/2}\sin ax\, dx = (\pi/2a)^{1/2}$, is used.

*7 Similarly, the yield of a triplet state from an S(T)-precursor can also be shown to increase (decrease) with increasing B.

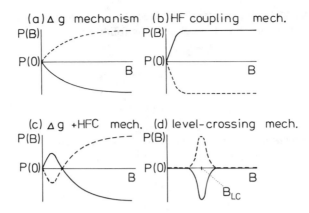

FIGURE 3.   Theoretical classification of magnetic field dependence of
product yields through the radical pair mechanism. The full curves indicate
the dependence of singlet-cage (escape) products produced from a sin-
glet(triplet or free radical)-precursor; the broken curves indicate the dependence of escape (singlet-cage) products produced from a singlet(triplet
or free radical) precursor.

case where $\Delta g = 0$, $a_{1j} \neq 0$, and $a_{2k} \neq 0$ in Equation 3. The effects classified in this case
are called those due to the HFCM. In contrast to the effects due to the ΔgM, it is difficult
to derive an analytical formulation for the present case. Thus, the theoretical studies of the
HFCM have been made so far through numerical calculation with some ideal models.[11]

In this section, let us consider qualitatively the magnetic field effects due to the HFCM
using the RPM shown in Section II.B. In the absence of a magnetic field, the S-T conversion
of a separated radical pair occurs through all of the three S-T_n (n = X, Y, Z) mixings (see
Equation 8) in region I of Figure 1a. In the presence of a large enough field where the
electronic Zeeman energy of a radical pair ($g\beta\hbar^{-1}B$) is much larger than the HF interactions;
the T_{+1} and T_{-1} states of a radical pair are splitted from its T_0 one as shown in Figure 1b.
Therefore, only the S-T_0 conversion occurs in region III of this figure. Usually, the S-T_{-1}
conversion cannot be considered because the time spent in region II of Figure 1b is too short
for usual solutions at room temperature.

Thus, in the case of the HFCM, the total rate of the S-T conversion for a separated
radical pair is decreased by a magnetic field and attains a saturated value at a relatively
weak field. It is also noteworthy that the S-T conversion rate due to the HFCM increases
with increasing of the amount of the HFC interactions. For example, the rate at a fixed
magnetic field (or even at zero field) is increased by a substitution of a nonmagnetic ^{12}C
atom in a component radical by a magnetic ^{13}C. On the other hand, the rate is decreased
by a substitution of protons by deuteriums because the HFC constant of H is about six times
larger than that of D. This is the origin of magnetic isotope effects.

With a similar consideration as shown in Section II.B, the magnetic field effects on
product yields through the HFCM can be obtained and the results are illustrated schematically
in Figure 3b. As shown in this figure, the yield of an S-cage product from an S(or T)-
precursor is increased (decreased) by a magnetic field and attains its saturated value at a
relatively weak field.[*8] The yield of an escape product from an S(T)-precursor is decreased
(increased) by a field and also attains its saturated value at a relatively weak field. Interestingly, the directions of the magnetically induced changes due to the ΔgM are proved to
be opposite to those due to the HFCM as shown in Figure 3a and b.

*8   The yield of a triplet state from an S(T)-precursor is also shown to be decreased (increased) by a weak magnetic
     field.

Werner et al.[12] defined the field strength ($B_{1/2}$) at which the magnetic field effect takes half the saturated value as follows:

$$P(B_{1/2}) - P(0) = \{P(B \to \infty) - P(0)\}/2 \qquad (28)$$

They obtained the $B_{1/2}$ value as

$$B_{1/2} \sim \Sigma_r\{a_r^2 I_r(I_r + 1)\}^{1/2} \qquad (29)$$

Here, r represents every magnetic nuclei contained in a radical pair. Since $B_{1/2}$ of a usual radical pair of organic radicals is less than 10 mT, the magnetic field effects due to the HFCM can be induced by such a weak field and attain their saturated values at a lower field than 50 mT. Low field saturation of magnetic field effects is characteristic of the HFCM, but the effects due to the ΔgM have been shown to appear clearly at a much larger field.

3. The Mixed Mechanism of the ΔgM and HFCM

For a general case where all of Δg, a_{1j}, and a_{2k} are not zero, a combination of the magnetic field effects due to the ΔgM and HFCM appears. At a relatively weak magnetic field, the effects due to the HFCM are predominant. On the other hand, at a relatively strong field, the effects due to the ΔgM become predominant. Therefore, the magnetic field effects on product yields through the mixed mechanism of the ΔgM and HFCM can be illustrated schematically as shown in Figure 3c. Here, the yield of an S-cage product from an S(T)-precursor is increased (decreased) by a weak magnetic field but is decreased (increased) by a strong field. The yield of an escape product from an S(T)-precursor is decreased (increased) by a weak field but is increased (decreased) by a strong field.

4. The Level-Crossing Mechanism (LCM)

In usual radical pairs in solution at room temperature, the time spent in the level-crossing region between S and T_{-1} (region II of Figure 1b) is so short that the S-T_{-1} conversion is not expected to occur. On the other hand, when two radical centers of a biradical are linked by a chain of a finite length, the $|J|$ value of the biradical cannot become zero. In such a case, the S-T_{-1}[*9] conversion is expected to occur when the following relation is fulfilled:

$$g\beta\hbar^{-1}B_{LC} = |2J| \qquad (30)$$

Here, B_{LC} is called "a level-crossing field".

For radical pairs in solution, the S-T_{-1} conversion is also considered to occur in the following cases:

1. When an employed solvent is very viscous even at room temperature or when the solvent becomes very viscous at low temperature, the time spent in region II of Figure 1b becomes so long that the S-T_{-1} conversion occurs.
2. When generated radicals are fixed by such external forces as the hydrogen bonding and the confinement in a fixed space, the J value is fixed to a constant one.
3. Because the S-T_{-1} conversion occurs through the HF interactions, large sizes of the HFC due to such atoms as proton and phosphorous induce considerably the S-T_{-1} conversion even in nonviscous solvents.

[*9] Because J is usually negative, the S-T_{-1} conversion usually occurs. When J is positive, the S-T_{+1} conversion occurs. In both cases, the expected magnetic field dependencies of product yields are very similar to each other.

At a level-crossing field, a sudden S-T$_{-1}$ conversion is expected to occur in these reactions. The magnetic field effects brought about in this case are called the effects due to the LCM. The magnetic field dependence of product yields through the LCM is illustrated schematically in Figure 3d. As shown in this figure, the yield of an S-cage product from an S(or T)-precursor shows a sudden decrease (increase) at B = B$_{LC}$.*10 The yield of an escape product from an S(T)-precursor shows a sudden increase (decrease) at B = B$_{LC}$.

III. MECHANISMS OF CIDEP

A. TWO MECHANISMS OF CIDEP

There are several origins to produce radical pair CIDEP. The processes giving CIDEP can be classified into two cases:

1. When a radical pair is generated, the electron spins of the component radicals are not polarized. The spin polarization is brought about afterward through the S-T$_i$ (i = 0 and ±1) mixing of the pair. This mechanism is also called the RPM as in the case of magnetic field effects. Although CIDEP can be generated through the S-T$_0$ mixing as well as the S-T$_{\pm 1}$ one (the S-T$_0$ mechanism (M) and S-T$_{\pm 1}$M, respectively), the former is more important than the latter in usual reactions at room temperature.
2. When the spin polarization of a triplet molecule (T-precursor) is transferred to a radical pair, the electron spins of the component radicals are polarized at the beginning of their formation. Thus, this mechanism is called the TM.[4] For the appearance of CIDEP through the TM, the precursor triplet state should be polarized and the reaction producing the radical pair should occur fast enough to retain the original spin polarization.

Since the relaxation time (τ_{RL}) of the electron spin in triplet molecules is $10^{-10} \sim 10^{-8}$ s at room temperature, the reactions which occur more slowly than this time scale cannot give any CIDEP through the TM. On the other hand, CIDEP through the RPM is possible to be induced for such slow reactions. Carbonyl molecules are famous for their selective intersystem crossing (ISC) to triplet sublevels and high reactivity in their triplet states. Therefore, many reactions from triplet carbonyls have been found to show strong CIDEP through the TM.

In this section, the theoretical foundation of the S-T$_0$M, some typical examples of CIDEP through the S-T$_0$M, and brief explanations of the S-T$_{\pm 1}$M and TM will be given.

B. THE S-T$_0$ MECHANISM OF CIDEP

The S-T$_0$ mixing of the radical pair wave function shown in Equations 9 to 14 can also bring about radical pair CIDEP, but the process giving CIDEP is more complicated than those giving magnetic field effects and CIDNP. Adrian[13] obtained the spin polarization of two radicals (ρ_1 and ρ_2) in a radical pair as follows:

$$\rho_1(t) = -\rho_2(t) = \langle \psi^*(t)|S_{1z} - S_{2z}|\psi(t)\rangle \tag{31}$$

$$= C_T(t)C_S^*(t) + C_T^*(t)C_S(t) \tag{32}$$

Here $\rho > 0$ means emission (E) and $\rho < 0$ means enhanced absorption (A).

From Equations 12 and 13, the spin polarization induced at the first reencounter (t = t') is given as

---

*10 The yield of a triplet state from an S(T)-precursor shows a sudden increase (decrease) at B = B$_{LC}$.

$$\rho_1(t') = \{C_T(0)C_S^*(0) + C_T^*(0)C_S(0)\}\{\cos 2wt' + 2(Q_{NM}/w)^2\sin^2 wt'\}$$

$$+ (iJ/w)\{C_T(0)C_S^*(0) - C_T^*(0)C_S(0)\}\sin 2wt'$$

$$+ (2Q_{NM}J/w^2)\{|C_S(0)|^2 - |C_T(0)|^2\}\sin^2 wt' \tag{33}$$

It is shown that $|\rho_1(t')|$ is too small to give appreciable CIDEP because $C_T(0) \cdot C_S(0) = 0$ for the reactions from S- and T-precursors and because J is zero at almost all of the time region between $t = 0 \sim t'$. Thus, CIDEP is not expected to be induced at the time of the first reencounter although magnetic field effects and CIDNP are expected to occur at this time.

CIDEP, however, is expected to be developed when $|J|$ keeps a much larger value than $|Q_{NM}|$ for $t - t'$ as follows:

$$\rho_1(t - t') = \{C_T(t')C_S^*(t') + C_T^*(t')C_S(t')\}\{\cos 2w(t - t')$$

$$+ 2(Q_{NM}/w)^2\sin^2 w(t - t')\}$$

$$+ (iJ/w)\{C_T(t')C_S^*(t') - C_T^*(t')C_S(t')\}\sin 2w(t - t')$$

$$+ (2Q_{NM}J/w^2)\{|C_S(t')|^2 - |C_T(t')|^2\}\sin^2 w(t - t') \tag{34}$$

The first term of Equation 34 becomes zero because there is no polarization at $t = t'$.

$$\rho_1(t') = C_T(t')C_S^*(t') + C_T^*(t')C_S(t') = 0 \tag{35}$$

The third term of Equation 34 is also negligible because $|Q_{NM}J/w^2| << 1$ during $t - t'$.
The second term of Equation 34, however, can induce CIDEP because

$$C_T(t')C_S^*(t') - C_T^*(t')C_S(t') = -(iQ_{NM}/|Q_{NM}|)\{|C_S(0)|^2 - |C_T(0)|^2\}\sin 2|Q_{NM}|t' \tag{36}$$

Thus, CIDEP induced between t and t' is given as

$$\rho_1(t - t') = (Q_{NM}J/|Q_{NM}J|)\{|C_S(0)|^2 - |C_T(0)|^2\} \times \sin 2|Q_{NM}|t'\sin 2|J|(t - t') \tag{37}$$

When τ_J represents the time range of $|J| >> |Q_{NM}|$, the spin polarization can be given as

$$\rho_1(\tau_J) = \int_0^t f(t')\rho_1(t - t')dt' \sim \int_0^\infty mt'^{-3/2}\rho_1(t - t')dt'$$

$$= 2\pi^{1/2}m\{|C_S(0)|^2 - |C_T(0)|^2\}\sin 2|J|\tau_J \times (Q_{NM}J/|Q_{NM}J|)|Q_{NM}|^{1/2} \tag{38}$$

More generally, Monchick and Adrian[14] obtained an asymptonic solution for CIDEP assuming J(r) as

$$J(r) = J_0\exp(-\xi r) \tag{39}$$

The spin polarization of one radical in the Nth nuclear state can be expressed as

$$\rho_N = (3^{1/2}\pi/\xi d)\{|C_S(0)|^2 - |C_T(0)|^2\} \times \Sigma_M(Q_{NM}J_0/|Q_{NM}J_0|)|Q_{NM}\tau_D|^{1/2} \tag{40}$$

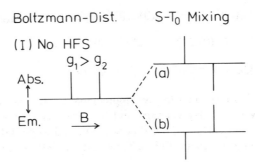

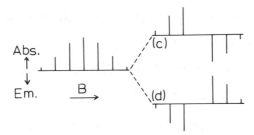

FIGURE 4.   ESR spectra of radicals in the Boltzmann distribution and CIDEP spectra predicted by the S-T_0 mechanism (I) for a radical pair where $g_1 > g_2$, $a_{1j} = 0$, and $a_{2k} = 0$ and (II) for a radical pair where $g_1 = g_2$, $a_{1j} = a$ (for $j = 1 - 6$), $a_{1j'} = 0$ (for $j' > 6$), and $a_{2k} = 0$. (In the Boltzmann distribution ESR of (II), the signal due to radical 2 is omitted for simplicity.) The polarization patterns of CIDEP spectra are shown for the cases of (a) $\mu J > 0$, (b) $\mu J < 0$, (c) $\mu J > 0$, and (d) $\mu J < 0$. Here, μJ is given in Equations 41 and 43, respectively.

Here, d is the distance of closest approach. The observed CIDEP spectra through the S-T_0M have been well interpreted by this equation. Thus, CIDEP through the S-T_0M is shown to be proportional to $|Q_{NM}|^{1/2}$ and its phase (E or A) to depend on the signs of J, Q_{NM}, and $\{|C_S(0)|^2 - |C_T(0)|^2\}$,

C. EXAMPLES OF CIDEP THROUGH THE S-T_0 MECHANISM

In this section, we will consider some typical examples of CIDEP spectra induced through the S-T_0M using Equation 40. The first example deals with the case where there is no nuclear spin in a radical pair. Therefore, only two ESR signals are expected to appear according to their g-values as shown in Figure 4I.

In this case, Q_{NM} of Equation 40 can be exchanged by Q of Equation 23. Thus, the phase pattern (Λ) of CIDEP spectra can be represented by the following sign rule:

$$\Lambda_{ne} = \mu J \Delta g \text{ (positive A, negative E)} \tag{41}$$

Here μ is + for T-precursors and − for S-precursors. When $g_1 > g_2$ and $J < 0$. Equation 41 predicts that the CIDEP signals of radicals 1 and 2 which are produced from an S(or T)-precursor show A(E)- and E(A)-phases, respectively. Figure 4I shows the ESR spectrum of the present system with the Boltzmann distribution and the CIDEP spectra predicted by the S-T_0M using Equation 40.

The second example deals with the case where $\Delta g = 0$, $a_{2k} = 0$, but $a_{1j} \neq 0$. In this case, CIDEP through the S-T$_0$M gives signals due to radical 1 only. Q$_{NM}$ of Equation 40 can be exchanged by

$$Q_{NM} = \Sigma_j a_{1j} m_{1j}/2 \tag{42}$$

The phase pattern of CIDEP spectra in the present case can be represented by the following sign rule:

$$\Lambda_{me} = \mu J \text{ (positive A/E, negative E/A)} \tag{43}$$

Here A/E implies that the low-field half of the spectrum is in absorption and that the high-field half in emission, and conversely for E/A.

As a typical example of the present case, the predicted CIDEP spectra by the S-T$_0$M for a radical having six equivalent protons are shown in Figure 4II. As shown in this figure, the intensities of the CIDEP signals are different from those of the usual ESR ones. Especially, the central line is missing in the CIDEP spectra. Since such disappearances often occur in CIDEP spectra, attention should be paid in their analyses. It is also noteworthy that the character of the reaction precursor can be supposed from observed phase patterns of CIDEP spectra. Because J is usually negative, an E/A(A/E)-phase shows that the reaction occurs from a T(S)-precursor.

The third example deals with the case where $\Delta g \neq 0$, $a_{1j} \neq 0$, and $a_{2k} \neq 0$. As a simple model of the present case, let us consider the following radical pair: (1) the pair is assumed to be produced from a T-precursor.[*11] (2) J is taken to be negative as usual. (3) Radicals 1 and 2 are assumed to have two equivalent protons and one proton, respectively, with $a_{11} = a_{12} = 2a_{21}$. Figure 5 shows the calculated CIDEP spectra of the present model having some Δg values together with the ESR spectra in the Boltzmann distribution. As shown in this figure, all of the CIDEP spectra in this figure have E/A phase patterns although their intensity distributions are complex and some lines are missing.

D. THE S-T$_{\pm 1}$ Mechanism of CIDEP

In this section, CIDEP through the S-T$_{\pm 1}$M is explained qualitatively. Since J is usually negative, the S-T$_{-1}$ mixing is more important than the S-T$_{+1}$M as shown in Figure 1b. Therefore, let us consider first the effect of the S-T$_{-1}$ mixing. For appearance of the S-T$_{-1}$M, the same conditions that are required for the LCM (see Section II.C.4) should also be satisfied.

There are two parts in the S-T$_{-1}$ polarization, the HF-dependent and -independent ones.[15] The former is generated through the flip-flop of an electron spin and a nuclear spin within a common radical of a radical pair and the latter through the flip-flop of an electron spin in one radical and of a nuclear spin in the counterradical in the pair. As shown in Equations 5 and 7, the S-T$_{-1}$ conversion needs the following selection rule:

$$\Delta m_S = 1 \quad \text{and} \quad \Delta m_I = -1 \tag{44}$$

Therefore, when the S-T$_{-1}$ conversion occurs through the flip-flop within one radical in a radical pair produced from a T-precursor, the radical with β-electron spin and upper nuclear spin recombines with its counterradical more efficiently than the radical with other spins.[15] Since α-electron spin and lower nuclear spin become excessive in the remaining radical, an E-phase pattern is expected for the CIDEP spectrum observed in the reaction

---

[*11] When the reaction occurs from an S-precursor, the phases of the CIDEP spectra shown in Figure 5 should be inverted.

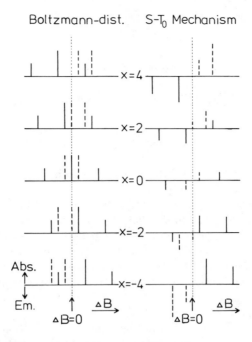

FIGURE 5. ESR spectra of a radical pair in the Boltzmann distribution and its CIDEP spectra predicted by the S-T_0 mechanism (T-precursor and $J < 0$) for a model system where $g_1 \neq g_2$, $a_{11}/2 = a_{12}/2 = a_{21} = a_2$. The full lines show the signals of radical 1 and the broken lines show the signals of radical 2. When the microwave energy is given by E_0 ($= g\beta\hbar^{-1}B_0$) and when $(g_1 + g_2)/2$ and $g_1 - g_2$ are written by g and Δg, respectively, the signal positions ($B = B_0 + \Delta B$) can be given as follows:

$$E_0 = (g + \Delta g/2)\beta\hbar^{-1}B + 2a_2(m_{11} + m_{12}) \qquad \text{for radical 1}$$

$$E_0 = (g - \Delta g/2)\beta\hbar^{-1}B + a_2 m_{21} \qquad \text{for radical 2}$$

Thus, ΔB can be obtained as

$$\Delta B = -x \cdot C/2 - 4C \cdot (m_{11} + m_{12}) \qquad \text{for radical 1}$$

$$\Delta B = x \cdot C/2 - 2C \cdot m_{21} \qquad \text{for radical 2}$$

Here, $C = a_2/(2g\beta\hbar^{-1})$ and $x = \Delta g B_0/(gC)$. From Equation 3, the corresponding Q_N for radical 1 and Q_M for radical 2 can be given

$$Q_N \propto x \cdot C + 4C \cdot (m_{11} + m_{12}) - \Sigma_M 2C \cdot m_{21} \qquad \text{for radical 1}$$

$$Q_M \propto x \cdot C - 2C \cdot m_{21} + \Sigma_N 4C \cdot (m_{11} + m_{12}) \qquad \text{for radical 2}$$

Using these Q values, the signal intensities of the CIDEP spectra can be calculated from Equation 40.

from a T-precursor, and conversely for that from an S-precursor. The nuclear spin selection rule brings about an HF-dependent emissive or absorptive phase pattern in this case.

When the S-T_{-1} conversion occurs between the two component radicals in a radical pair, the electron spin polarization occurs in one radical and the nuclear spin polarization occurs in the counterradical. Thus, an HF-independent emissive or absorptive phase pattern is induced in this case.

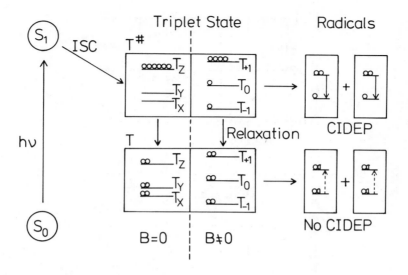

FIGURE 6. The triplet mechanism (TM) for electron spin polarization in free radicals. The spin polarization of a triplet state (T^*) arises in an anisotropic intersystem crossing (ISC). The spin relaxation in T^* is so fast ($10^{-10} \sim 10^{-8}$ s) at room temperature that the spin polarization of T^* can be transferred to generated radicals only when the radical formation is faster than or comparable to the spin relaxation of T^*.

From a similar consideration, the S-$T_{+1}M$ is also proved to give an E(or A)-phase pattern for the reaction from an S(or T)-precursor. Thus, the sign rule for the S-$T_{\pm 1}M$ can be expressed as

$$\Lambda_\pm = \mu J \quad \text{(positive A, negative E)} \qquad (45)$$

E. THE TRIPLET MECHANISM OF CIDEP

CIDEP through the TM requires that the generated radicals must be formed from a triplet molecule.[4] In some molecules, three sublevels (T_n, n = X, Y, and Z) of their lowest triplet states are not equally populated from their excited singlet states at zero magnetic field.[16] For example, in some carbonyl molecules, the ISC process to the top sublevel (T_Z) has been shown to be much more predominant than others (T_X and T_Y)[*12] Such a situation is schematically illustrated in Figure 6.

As a typical example of CIDEP through the TM, let us first consider the radical pair produced from such a triplet carbonyl where the initial population of its top sublevel (T_Z) is much larger than those of other sublevels in the absence of a magnetic field. Since ESR is measured in the presence of a magnetic field, three Zeeman sublevels (T_i, i = 0 or ± 1) in solution can be expanded by T_n

$$T_i = \Sigma_n C_{in}(B)T_n \qquad (46)$$

Since the triplet molecule in solution rotates very fast at room temperature ($\tau_C = 10^{-12} \sim 10^{-11}$ s), the rotational averaged value of C_{in}^2 becomes 1/3 at a sufficiently large field. In this case, there appears no spin polarization in T_i. In a usual ESR measurement with X-band microwave, however, the electronic Zeeman splitting (~ 0.3 cm^{-1}) is not sufficiently larger than the zero-field splittings of triplet molecules (0.1 $\sim$ 1.0 cm^{-1}). Therefore, C_{1Z}^2

[*12] Here, the molecular axes Z, X, and Y are taken to be along the C=O bond of the carbonyl group, perpendicular to the molecular plane, and completing the right-handed Cartesian frame, respectively.

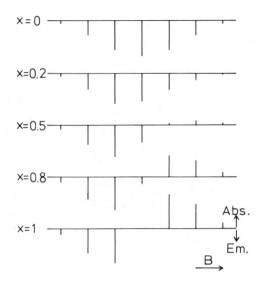

FIGURE 7.   CIDEP spectra due to the combination of an emissive spectrum generated through the TM and an E/A one generated through the S-
T_0M for a radical having six equivalent protons. The portion of the S-T_0M
is represented by x. When x = 0 (1), the CIDEP spectrum is explained
only by the TM (S-T_0M). When $0 < x < 1$, the spectrum shows an E*/
A phase pattern.

at B $\sim$ 0.33 T is possible to be larger than C_{1X}^2 and C_{1Y}^2 for such a triplet carbonyl as
shown in Figure 6.

The spin polarization of a triplet molecule in the presence and absence of a magnetic
field decays very fast ($\tau_{RL} = 10^{-10} \sim 10^{-8}$ s) at room temperature and a polarized triplet
state ($T^\#$) becomes an unpolarized one (T) with the Boltzmann distribution as shown in
Figure 6. If the time of the radical pair formation is faster than or comparable to τ_{RL} of $T^\#$,
the spin polarization of $T^\#$ can be transferred to the generated radicals. As shown in Figure
6, the radicals have an excess of α-electron spin.

Thus E-phase patterns are expected to appear for both radicals without any distortion
in their HF structures. On the other hand, if the formation time is much longer than τ_{RL} of
$T^\#$, no CIDEP through TM can be expected. For the reactions from polarized triplet states,
similar results can be obtained generally for CIDEP through the TM; if an upper (or lower)
sublevel of a triplet state is overpopulated, the generated radicals from this precursor show
E(or A)-phase patterns.

Actually, CIDEP through the TM is often superimposed on that through the S-T_0M.
Indeed, the former occurs immediately after the formation of radical pairs but the latter is
developed after the radical pair formation as explained in Section III.B. As a typical example
of the combination of the TM and S-T_0M, CIDEP spectra are calculated for a radical having
six equivalent protons. For such a model radical, some combinations of the E-phase pattern
through the TM and the E/A one through the S-T_0M are illustrated in Figure 7. This figure
shows characteristic features of mixing of E and E/A phase patterns (E*/A). Similarly,
mixtures of A and E/A (E/A*), E and A/E (A/E*), and A and A/E (A*/E) can also be
considered.

IV. EXAMPLES OF MAGNETIC FIELD EFFECTS

A. MAGNETIC FIELD EFFECTS ON PRODUCT YIELDS

To control chemical and biological reactions by magnetic fields has long been one of the dreams of scientists. Although there have been some papers reporting magnetic field effects on the reactions, the study of the effects had been considered to be "a romping ground for charlatans"[17] before 1976. Exceptionally, the effects on the predissociation of gaseous iodine[18,19] and ortho/para-conversion of hydrogen molecule[20,21] had only been established at that time. On the other hand, so many reactions through radical pairs and biradicals have recently been proved to be influenced by relatively weak magnetic fields.

The early studies of such magnetic field effects were concerned with product yields. In 1972, Sagdeev et al.[22] studied a magnetic field effect on the thermal reaction of pentafluorobenzyl chloride (ACl) with *n*-butyl lithium (BLi) in solution measuring the NMR signal intensities of reaction products.

$$
C_6F_5CH_2Cl \ (ACl) \ + \ C_4H_9Li \ (BLi) \xrightarrow{\Delta} {}^1\overline{A\cdot \ \cdot B}
$$

$$
{}^1\overline{A\cdot \ \cdot B} \rightarrow {}^3\overline{A\cdot \ \cdot B}
$$

$$
\downarrow \qquad\qquad \downarrow
$$

$$
A\text{-}B \qquad \searrow A\cdot \ + \ B\cdot \rightarrow A\text{-}A, \ A\text{-}B, \ B\text{-}B
$$

$$
\text{(cage product)} \qquad\qquad \text{(escape products)} \qquad\qquad (47)
$$

They found the ratio of the intensity of a cage product (AB) to that of an escape one (AA) to be increased by magnetic fields. This effect can be explained by the HFCM as shown in Figure 3b, since most thermal reactions are considered to occur through S-precursors.

In 1974, Brocklehurst[23] carried out a pulse radiolysis study of ion recombination in the squalane (S) solution of fluorene (F) at room temperature.

$$
S \rightsquigarrow S^{\ddagger} + e^-
$$

$$
S^{\ddagger} + F \rightarrow S + F^{\ddagger}
$$

$$
e^- + F \rightarrow F^{\div}
$$

$$
(48)
$$

$$
{}^1\overline{F^{\ddagger} \cdot F^-} \rightarrow {}^1F^* + F
$$

The fluorescence of fluorene, which should be proportional to the yield of the lowest excited singlet state ($^1F^*$), was found to be increased in intensity by magnetic fields. Since this reaction is also considered to occur from S-precursors, this effect can also be explained by the HCFM.[24]

In 1976, four groups (two German and two Japanese) separately showed some photochemical reactions in solution to be influenced by magnetic fields at room temperature. Since this year, the fact that reactions through radical pairs in solution can be influenced by magnetic fields has been accepted widely to the world of science. Michel-Beyerle et al.[25] and Shulten et al.[26] independently studied magnetic field effects on the quenching reactions of singlet pyrene ($^1P^*$) with amines (D).

$$
P \xrightarrow{h\nu} {}^1P^*
$$

$$
{}^1P^* + D \rightarrow {}^1\overline{P^{\div} \cdot D^+}
$$

$$
{}^1\overline{P^{\div} \cdot D^+} \rightarrow {}^3\overline{P^{\div} \cdot D^+} \rightarrow {}^3P^* + D
$$

$$
\searrow \qquad\qquad \swarrow
$$

$$
P^{\div} + D^{\ddagger} \qquad\qquad\qquad (49)
$$

With the aid of nanosecond-laser flash photolysis-optical absorption (LFP-OA) techniques, they showed the yield of triplet pyrene ($^3P^*$) to be reduced by weak magnetic fields (less than 0.1 T) and their magnetic field effects can be explained by the HFCM because their reactions occur from the lowest excited singlet state of pyrene (S-precursor) and because triplet pyrene is produced through cage recombination within triplet radical pairs (T-cage-product).

The authors[27] studied the singlet-sensitized ($^1sens^*$) photodecomposition of dibenzoyl peroxide in toluene at room temperature using a superconducting magnet.

$$(C_6H_5CO_2)_2 \xrightarrow{\,^1Sen^*\,} {}^1\overline{C_6H_5CO_2\cdot\ \cdot C_6H_5} + CO_2$$

$$^1\overline{C_6H_5CO_2\cdot\ \cdot C_6H_5} \rightarrow {}^3\overline{C_6H_5CO_2\cdot\ \cdot C_6H_5}$$

$$C_6H_5CO_2C_6H_5 \qquad C_6H_5CO_2\cdot\ +\ C_6H_5\cdot \rightarrow \text{escape products}$$

(cage product)   (50)

The yields of some reaction products were directly measured by a gas chromatograph. With increasing B, the yield of a cage product (phenyl benzoate) was proved to decrease and those of escape ones (methylbiphenyls and 1,2-diphenylethane) to increase. The magnetically induced changes in the yield of phenyl benzoate were confirmed to be proportional to $B^{1/2}$ and could be quantitatively reproduced by Equation 25. Since this reaction occurs from a singlet sensitizer (S-precursor), these effects can be explained by the ΔgM as shown in Figure 3a.

Hata[28] studied the photoisomerization of isoquinoline N-oxide (QNO) in ethanol.

$$QNO \text{ --- } HOR \xrightarrow{\ h\nu\ } {}^1QNO^* \text{ --- } HOR \rightarrow {}^1\overline{QN^+O\cdot \text{ --- } \cdot HOR}$$

$$^1\overline{QN^+O\cdot \text{ --- } \cdot HOR} \rightarrow {}^3\overline{QN^+O\cdot \text{ --- } \cdot HOR} \rightarrow QNO \text{ --- } HOR$$

$$\downarrow$$

Lactam (cage product)   (51)

He found the yield of the lactam (an S-cage product) to decrease sharply around B = 1 T. This effect can be explained by the LCM as shown in Figure 3d because this reaction occurs from the lowest excited singlet state of QNO which is considered to form a hydrogen bond with alcohol.

Recently, some magnetic field effects on the reactions through biradicals, which are produced upon photolysis of polymethylene-linked compounds, have been observed.

$$A\text{---}D \xrightarrow{\ h\nu\ } {}^1A^*\text{---}D \rightarrow {}^1\overline{\cdot A^-\text{---}D^{\ddagger}}$$

$$^1\overline{\cdot A^-\text{---}D^{\ddagger}} \rightarrow {}^3\overline{\cdot A^-\text{---}D^{\ddagger}} \rightarrow {}^3A^*\text{---}D$$

$$\downarrow$$

Cage product   (52)

Weller et al.,[29] Nakagaki et al.,[30] and Tanimoto et al.[31] found the effects on the yields of triplet states and reaction products as well as the effects on lifetimes and fluorescence intensities of exciplexes. When the number of the methylene group is small (usually less than 8), no magnetic field effect can be observed with electromagnets because $|J|$ is too large. When the number is intermediate (usually $8 \sim 10$), the observed effects can be explained by the LCM because B_{LC} (see Equation 30) becomes available with electromagnets. When the number is large (usually more than 10), the observed effects can be explained by the HFCM because J becomes almost zero.

One of the most important applications of the RPM is isotope separation with magnetic isotope effects. As shown in Section II.C.2, this is a new technique developed during the studies of magnetic field effects. In 1976, Buchachenco et al.[32] discovered a magnetic isotope effect of ^{13}C in the photodecomposition of dibenzyl ketone (DBK) in benzene and hexane solutions.

$$C_6H_5CH_2COCH_2C_6H_5 \xrightarrow{h\nu} {}^3DBK^* \rightarrow {}^3\overline{C_6H_5CH_2CO\cdot \ \cdot CH_2C_6H_5}$$
$$(DBK)$$

$$^3\overline{C_6H_5CH_2CO\cdot \ \cdot CH_2C_6H_5} \xrightarrow{{}^{13}CO} {}^1\overline{C_6H_5CH_2CO\cdot \ \cdot CH_2C_6H_5} \rightarrow DBK$$

$$\downarrow {}^{12}CO$$

$$^3\overline{C_6H_5CH_2\cdot \ \cdot CH_2C_6H_5} + CO \rightarrow 2C_6H_5CH_2\cdot + CO \qquad (53)$$

Then, Turro et al.[33] found that this isotope effect can be enhanced very much by the confinement of radical pairs in micelles. Here, the escape of radicals from a radical pair is so suppressed by the confinement in a micelle that the lifetime of the pair increases very much.

Another application of magnetic field effects has been carried out for polymerization. Although Tabata et al.[34] reported the effect on the radiation-induced polymerization of formaldehyde in the solid state at 77K in as early as 1964, the mechanism of this effect has not yet been clarified. Turro et al.[35] studied the effect on an emulsion polymerization of styrene at room temperature.

$$n \times \text{(styrene)} \xrightarrow[DBK/SDS]{h\nu} \text{(polystyrene)}_n \qquad (54)$$

This reaction was initiated photochemically with DBK in a micelle. They found the average molecular weight of polymerized styrene to be increased by a factor of 5 if the reaction was performed in the presence of a magnetic field above 0.1 T.

They interpreted this effect as follows: (1) DBK is solubilized in the micelle together with monomer styrene molecules. (2) The polymerization is initiated within the micelle by the benzyl radical generated through the photodecomposition of triplet DBK (T-precursor). (3) Termination is carried out through the cage recombination within the micelle (S-cage product). (4) Thus, the recombination rate can be reduced by a magnetic field through the HFCM as shown in Figure 3b. More recently, Morita et al.[36] studied the magnetic field effects on the photocrosslinking efficiency of bromo- and chloromethylated polystyrene sensitized by a triplet state in thin films at room temperature and found the gel fraction of the polymers to be enhanced by magnetic fields. These effects can also be explained by the HFCM.

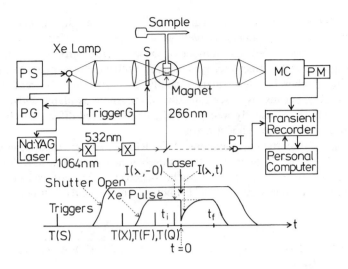

FIGURE 8.   A schematic diagram of our nanosecond-LFP-OA apparatus[38] for measuring magnetic field effects on reaction rates. A trigger generator (G) controls a cooperative sequence of components with some triggers (T). T(S) opens an electromagnetic shutter (S), T(X) initiates the action of a pulse generator (PG) to a Xe lamp which is continuously lighted by a power supply (PS). When the Xe lamp is intensified by PG, T(F) fires the flash lamps of the Nd:YAG laser then T(Q) operates the Q-switch of the laser. After the generation of the fundamental oscillation of the laser (1064 nm), its second (532 nm) and fourth (266 nm) harmonics are obtained by crystals (X). The fourth harmonic excites the sample which is put in an electromagnet and the second harmonic is detected by a phototransistor (PT), which triggers the action of a transient recorder. The monitor light passes through the sample where the laser excitation takes place; hence it is absorbed by transient species generated by the laser excitation. Next, the monitor light is monochromatized by a monochromator (MC) and is detected by a photomultiplier (PM). The time dependence of the signal intensity [$I(\lambda, t)$] is recorded from t = t_i (<0) to t = t_f (>0) by the transient recorder. The time origin (t = 0) is taken at the laser excitation and $I(\lambda, -0)$ is the averaged intensity of the flat part of the monitor light before the laser excitation (t < 0). The time dependence of the absorbance [$A(\lambda, t)$] is calculated from Equation 55 by a personal computer.

B. APPARATUS FOR MEASURING MAGNETIC FIELD EFFECTS ON REACTION RATES

Since the S-T conversion of radical pairs was found to be the key step of their magnetic field effects, the authors tried to measure the magnetic field effects on the S-T conversion rates by directly monitoring the component radicals in radical pairs with the aid of a LFP-OA technique in 1980.[37,38] Our apparatus is schematically shown in Figure 8.

The degassed sample solution in a rectangular quartz cell (10 × 10 mm) with a pyrex branch is set in the gap of an electromagnet. The gap width is 15 mm. The maximum and residual fields of this magnet are 1.34 T and 18 mT, respectively. The residual field can be cancelled by a countercurrent and the lowest field is less than 0.2 mT. Hereafter, the experiments under the lowest field are denoted as those in the absence of a magnetic field.

The sample is irradiated with the fourth harmonic (266 nm) of a Nd:YAG laser from the bottom of the cell. The pulse width is 5 ns and the pulse energy is less than 10 mJ per pulse. Since this laser is designed to be most efficient at 10-Hz repetition rate, the flash lamps are fired at 10 Hz but the Q-switch is usually operated at 0.625 Hz. The second or third harmonic (532 or 355 nm) of this laser, a nitrogen gas laser (337.1 nm), or an excimer laser (KrF 249 nm or XeCl 308 nm) can also be used for the excitation light source of the present LFP-OA apparatus.

The generated reaction intermediates with the laser excitation are monitored by a usual

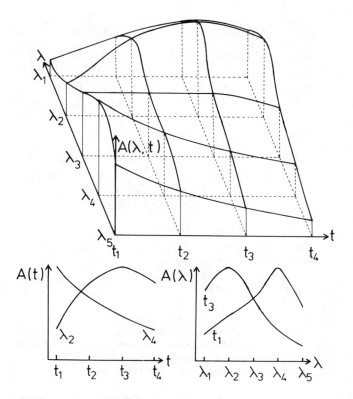

FIGURE 9. From the data for the time dependence of the absorbances measured at various λ [A(λ, t)], the time dependence [A(t)] of the absorbance at a fixed λ and the transient optical absorption spectrum [A(λ)] at a fixed t can be calculated by a personal computer.

Xe-lamp (300 W). In prior to the laser excitation, the light intensity of the Xe-lamp is increased for about 1 ms by a pulse current and an electromagnetic shutter is opened for about 10 ms. This shutter is used for preventing photochemical reactions by the Xe-lamp.

The monitor light passes through the sample and is partially absorbed by the reaction intermediates. The remaining light is monochromatized (wavelength = λ) by a monochromator (f = 250 mm) then detected by a photomultiplier. The time dependence of the signal intensity $I(\lambda,t)$ detected by the photomultiplier is recorded by a transient recorder (10 ns per point and ±50 mV per full scale) or a storage oscilloscope (100 MHz and 5 mV per div). The recorded data are transferred to a personal computer and their logarithms are calculated. The time sequence of triggers and operations is schematically shown in Figure 8.

If necessary, the above-mentioned procedure is repeated, then the obtained $\log I_i(\lambda,t)$ data are accumulated by the computer. The accumulated data $[\Sigma_i^n \log I_i(\lambda,t)]$ are stored on a floppy disk. Thus the time dependence of the absorbance $[A(\lambda,t)]$ of the generated intermediates measured at a fixed wavelength (λ) can be calculated by the computer as shown in Figure 9:

$$A(\lambda,t) = \epsilon(\lambda)c(t)L$$
$$= (1/n)\Sigma_i^n \log\{I_i(\lambda, -0)/I_i(\lambda,t)\} \qquad (55)$$

Here, $\epsilon(\lambda)$, $c(t)$, and L are the molar extinction coefficient, concentration of transient species, and optical length, respectively. The time origin (t = 0) is taken at the time of the laser

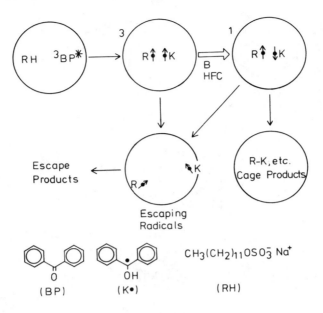

FIGURE 10.   Reaction scheme of the hydrogen abstraction reaction of triplet benzophenone (^{3}BP*) from an SDS molecule (RH) in an SDS micelle. The rate of the T-S conversion in the generated radical pair of the ketyl (K·) and alkyl (R·) radicals can be influenced by an external magnetic field (B) and the hyperfine couplings (HFC) of the pair.

excitation and I(λ, -0) is the average light intensity of the flat part before the laser excitation. I(λ, -0) can also be measured by the transient recorder or storage oscilloscope.

Next, the same measurements are repeated at different wavelengths. If the A(λ,t_1) data at a fixed delay time (t_1) are plotted against λ as is shown in Figure 9, the transient optical absorption spectrum [A(λ)] at t = t_1 can be obtained. If the A(λ_1,t) data at a fixed wavelength (λ_1) are plotted against t, the time dependence of the absorbance [A(t)] at λ = λ_1 can be obtained.

C. MAGNETIC FIELD EFFECTS ON REACTION RATES

The authors studied the magnetic field effects on the reaction rates of radical pairs with the aid of the LFP-OA technique, taking the photoreduction reactions of benzophenone (BP), naphthoquinone (NQ), and their derivatives in micelles as examples.[37-43] We found the primary processes of the reaction of BP in a sodium dodecyl sulfate (SDS = RH) micelle to proceed as shown in Figure 10.[37,41] The processes of other carbonyl compounds in micelles have also been proved to occur similarly.[38-40]

As shown in Figure 10, the excited triplet state of BP (^{3}BP*) abstracts a hydrogen atom from a micellar molecule (RH), forming a triplet radical pair of the ketyl (K·) and aklyl (R·) radicals. Figure 11 shows the time-resolved absorption spectra [A(λ)] observed for this reaction. Here, ^{3}BP* and K· were detected but R· was not. However, R· was confirmed to be produced with the aid of a LFP-ESR technique as discussed in Section VI.

As shown in Figure 10, the triplet pair of K· and R· is converted into a singlet radical pair and then disappears by cage recombination (and disproportionation). The rate of this triplet-singlet (T-S) conversion is expected to show magnetic field and magnetic isotope effects as shown in Section II. The remaining radicals escape from the triplet and singlet pairs and become the escaping radicals.

Thus, if the time dependence of a transient absorbance [A(t)] is measured for at least one of the component radicals, the above-mentioned dynamic behavior of a radical pair can

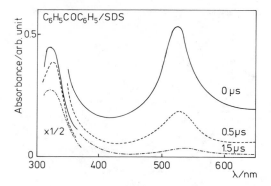

FIGURE 11.   Time-resolved absorption spectra [A(λ)] observed for the photoreduction reaction of benzophenone (1 m*M*) in a micellar SDS (80 m*M*) solution at room temperature. (From Sakaguchi, Y., Hayashi, H., and Nagakura, S., *J. Phys. Chem.*, 86, 3177, 1982. With permission.)

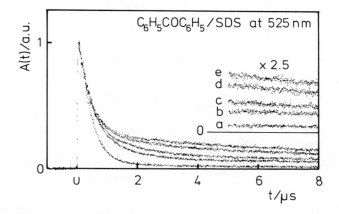

FIGURE 12.   Time dependence of the transient absorption [A(t)] observed at 525 nm for the photoreduction reaction of benzophenone (1 m*M*) in a micellar SDS (80 m*M*) solution at room temperature: (a) in the absence of a magnetic field; in the presence of a magnetic field of (b) 0.1, (c) 0.2, (d) 0.5, and (e) 1.34 T. (From Hayashi, H., *Sci. Pap. Inst. Phys. Chem. Res.*, 80, 87, 1986. With permission.)

be clarified. The A(t) curve observed at 525 nm for this reaction is shown in Figure 12. The absorbance at 525 nm is due to ^{3}BP* and K·. Decay of the A(t) curve mainly shows the disappearance of K· by recombination of K· and R· in a radical pair and an invariant component of A(t) is ascribed to the escaping K· from the pair because the lifetime of the escaping radical was proved to be much longer than the time scale of Figure 12 (10 μs).

As clearly shown in Figure 12, magnetic fields of 0.1 ~ 1.34 T were found to reduce the recombination rate of the radical pair and to enhance the yield of the escaping radical. Such magnetic field effects were also observable with a magnetic field of as low as 40 mT. Since no magnetic field effect was observed in 2-propanol solution, the confinement of the generated pairs in micelles is shown to be important for the appearance of the effects.

In the absence of a magnetic field, the decay of the A(t) curve at 525 nm can be represented by an exponential function except the initial fast decay of ^{3}BP*. The rate constant for the decay of K· at B = 0 T was calculated to be 2.8×10^6 s^{-1}. In the presence of a magnetic field, the decay of the A(t) curve due to the decay of K· can be expressed by a combination of two exponential functions. Although the constant of the faster decay could

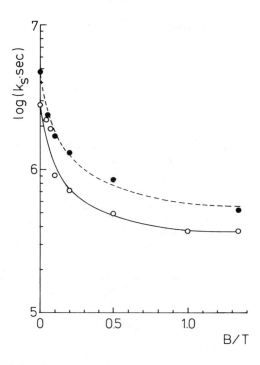

FIGURE 13.   Rate constant for the decay of a radical observed at B = 0 T and the constant for the slower decay of the radical observed in the presence of a magnetic field of B: (○) the rate constants of the ketyl radical generated in the photoreduction reaction of benzophenone (1 m*M*) in a micellar SDS (80 m*M*) solution [42] and (●)the constants of the semiquinone radical generated in the reaction of naphthoquinone (0.5 m*M*) in the micellar SDS solution. [38]

not be decided precisely, that of the slower one could be calculated as shown in Figure 13. This figure shows that the observed constant decreases with increasing B and that no saturation at fields lower than 0.1 T is noticed in the B-dependence of the decay constant. The constant at B = 1.34 T was observed to be as small as $0.3_7 \times 10^6$ s^{-1}. Such a large magnetic field effect had not been observed before our study. [40]

The B-dependence of the yield of the escaping radical can be obtained by A(t = 5 μs)/A(t = 0 μs) of Figure 12. We also carried out similar experiments for benzophenone-d$_{10}$ (BP-d$_{10}$) and benzophenone-*carbonyl*-^{13}C (BP-^{13}C) and compared their results with those for BP. Figure 14 shows the B-dependence of A(t = 5 μs)/A(t = 0 μs) for these compounds. As shown in this figure, the yields of the escaping ketyl radicals increases quickly below 0.5 T and then slowly above it but no saturation is noticed in these yields.

Although the magnetic isotope effect on the yields of the escaping radicals was almost indistinguishable in the absence of a magnetic field, the effect became clear in the presence of magnetic fields. The isotope effect at B = 0.2 T especially was found to be very large for a single reaction. In general, the yield of the BP-^{13}C ketyl radical was smaller than that of the BP one and the yield of the BP-d$_{10}$ one was larger than that of the BP one. This isotope effect can be explained by the T-S conversion of the radical pairs, where the conversion rate is expected to be enhanced by an increase of the hyperfine interaction in the pair.

Similar isotope effects on the decay rates of the ketyl radicals were also observed in the presence of magnetic fields for the reactions of BP, BP-^{13}C, and BP-d$_{10}$ in SDS micelles. [42] Similar magnetic field effects on the decay rates of generated radicals and the yields of

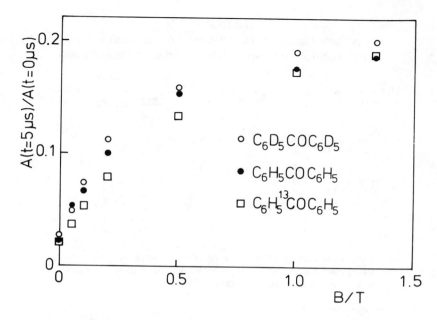

FIGURE 14.   Magnetic field dependence of the ratio [A(t = 5 μ s)/A(t = 0 μ s)] observed at 525 nm for the micellar SDS (80 m*M*) solutions of (●) benzophenone (1 m*M*), (○) benzo-phenone-d$_{10}$ (1 m*M*), and (□) benzophenone-*carbonyl*-^{13}C (1 m*M*). The ratio shows the relative yield of the escaping ketyl radical. (From Hayashi, H., *Sci. Pap. Inst. Phys. Chem. Res.*, 80, 87, 1986. With permission.)

escaping radicals were also found for the photoreduction reactions of BP derivatives,[40] NQ, and an NQ derivative[38] in micelles. For example, the decay rate constants of the NQ semiquinone radical in an SDS micelle observed in the absence and presence of magnetic fields are also plotted against B in Figure 13. The rate constants observed for the photo-reduction reactions of some carbonyl compounds in the presence and absence of magnetic fields are listed in Table 1. Recently, Tanimoto et al.[44] and Nakamura et al.[45] observed similar magnetic field effects on the reactions of other carbonyl compounds and porphyrin-viologen linked systems, respectively.

The magnetic field and magnetic isotope effects described in this section cannot be explained by the ordinary ΔgM, HFCM, LCM, and their mixed mechanisms (see Section II) because

1.   The decays of the generated radical pairs in the absence of a magnetic field can be expressed by single exponential functions but those in the presence of magnetic fields by combinations of two exponential ones. This effect cannot be derived from the ordinary mechanisms.
2.   With increasing B, the decay rate constants of the pairs decrease but the yields of the escaping radicals increase. This effect cannot be explained by the ΔgM because the inverse effects are expected for this mechanism. This effect cannot be fully explained by the HFCM because the magnetically induced changes of the present reactions are not saturated below 0.1 T.
3.   The magnetic isotope effects are almost indistinguishable in the absence of a magnetic field but the effects become most appreciable at B ~ 0.2 T. This effect cannot be simply explained by the HFCM.

Thus, the authors proposed a new mechanism for explaining the above-mentioned prob-lems, taking the relaxation of electron spins in triplet radical pairs into consideration.[42] This

TABLE 1
Decay Rate Constants (k) Observed for Radical Pairs in
Micellar SDS Solutions of Some Carbonyl Compounds

Magnetic field B/T	$k/10^6 \ s^{-1}$					
	BP[a]	BPD[b]	BP13[c]	BPF[d]	NQ[e]	MNQ[f]
0	2.8	2.8	2.8	2.9	4.7	4.4
0.04	2.2	2.1	2.4	—	—	—
0.05	—	—	—	—	2.4	1.9
0.07	1.9	1.9	2.2	—	—	—
0.10	0.9_1	—	1.1	1.3	1.7	1.4
0.20	0.7_1	—	0.8_9	0.7_5	1.3	0.9_5
0.50	0.4_9	—	0.5_3	0.4_5	0.8_5	0.7_7
1.00	0.3_7	—	0.3_8	—	—	—
1.34	0.3_7	—	0.4_2	0.3_0	0.5_2	0.6_5

Note: Each rate constant observed for the decay of the radical pair in the absence of a magnetic field and each one observed for the slower decay of the radical pair in the presence of a magnetic field are listed.

[a]   Benzophenone.
[b]   Benzophenone-d_{10}.
[c]   Benzophenone-*carbonyl*-^{13}C.
[d]   4,4'-Difluorobenzophenone.
[e]   1,4-Naphthoquinone.
[f]   2-Methyl-1,4-naphthoquinone.

From Hayashi, H. and Nagakura, S., *Bull. Chem. Soc. Jpn.*, 57, 322, 1984. With permission.

mechanism is called the relaxation mechanism (RM), the details of which will be shown in the next section.

V. RELAXATION MECHANISM

A. THEORY

In 1976 Brocklehurst proposed that the relaxation of electron spins in radical pairs might affect the evolution of their singlet characters.[46] In usual solvents, however, the time scales of the secondary recombination and S-T conversion are estimated to be $10^{-10} \sim 10^{-7}$ s and $10^{-9} \sim 10^{-8}$ s,[1] respectively, at room temperature. The escape from a usual solvent cage is also considered to have a similar time scale at room temperature. On the other hand, the spin-lattice and spin-spin relaxation times (τ_1 and τ_2)[*13] of organic radicals in usual solvents are estimated to be $10^{-6} \sim 10^{-4}$ s and $10^{-6} \sim 10^{-5}$ s,[1] respectively, at room temperature. Thus, the effect of the spin relaxation does not usually appear for the magnetic field effects observed in usual solvents at room temperature.

On the other hand, in some confined systems such as in micelles, polymers, and linked systems, the time scales of the cage recombination and the escape from a cage are possible to become much longer, even at room temperature, than those in usual solvents and close to τ_1 and τ_2. In such cases, the spin relaxation may play an important role in the reactions of radical pairs and biradicals (the RM). The authors considered dynamic behavior of a radical pair (or biradical) in such a case.[42] Here the energy levels of a radical pair and the rate constants (k) concerning the levels are shown in Figure 15.

[*13] Usually, the spin-lattice and spin-spin relaxation times are called "T_1 and T_2", respectively. In this chapter, however, they are written as τ_1 and τ_2 in order to avoid the confusion with the symbols concerning triplet states.

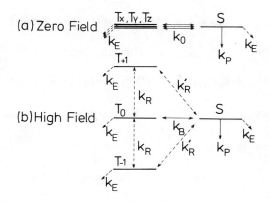

FIGURE 15. Energy levels and conversion processes of a radical pair (a) at zero magnetic field and (b) at a sufficiently high field, respectively. See text for the meanings of the rate constants (k's) of the conversion processes.

We introduced rate equations for the populations of the singlet state ([S]) and triplet sublevels ([T_n] n = X, Y, and Z or [T_i] n = 0, ±1) of a radical pair. At B = 0 T, they were represented by

$$d[T]/dt = -(k_0 + k_E)[T] + 3k_0[S] \tag{56}$$

$$d[S]/dt = k_0[T] - (k_P + 3k_0 + k_E)[S] \tag{57}$$

Here, [T] = [T_X] + [T_Y] + [T_Z], k_0 is the rate constant for the S-T conversion at B = 0 T, k_E for the escape from a solvent cage, and k_P for the recombination in a singlet radical pair.

At a sufficiently large field, where the S-T_0 conversion becomes much faster than the S-$T_{\pm 1}$ ones, the following equations were introduced:

$$d[T_i]/dt = -(\Sigma_{i \neq j}k_{ij} + k_{iS} + k_E)[T_i] + \Sigma_{i \neq j}k_{ji}[T_j] + k_{Si}[S] \tag{58}$$

$$d[S]/dt = \Sigma_i k_{iS}[T_i] - (k_P + \Sigma_i k_{Si} + k_E)[S] \tag{59}$$

where i = j = 0 or ±1, k_{ij} is the rate constant for the T_i-T_j conversion, and k_{iS} for the T_i-S one. Since $k_{ij} = k_{ji}$ and $k_{Si} = k_{iS}$, Equations 58 and 59 were summarized as

$$d[T']/dt = -(k_R + k_R' + k_E)[T'] + 2k_R[T_0] + 2k_R'[S] \tag{60}$$

$$d[T_0]/dt = k_R[T'] - (k_B + 2k_R + k_E)[T_0] + k_B[S] \tag{61}$$

$$d[S]/dt = k_R'[T'] + k_B[T_0] - (k_P + k_B + 2k_R' + k_E)[S] \tag{62}$$

Here, [T'] = [T_{+1}] + [T_{-1}], $k_B = k_{S0}$, $k_R = k_{\pm 10}$, and $K_R' \cdot = k_{\pm 1S}$ as shown in Figure 14.

For the reaction from a T-precursor, the approximated solutions were obtained as, at zero magnetic field:

$$\text{Case a } (k_P \gg k_0): \quad [R] = \exp\{-(k_0 + k_E)t\} \tag{63}$$

$$\text{Case b } (k_0 \gg k_P): \quad [R] = \exp\{-(k_P/4 + k_E)t\} \tag{64}$$

where $[R] = [S] + [T] + [T']$, and at a sufficiently large field:*[14]

$$\text{Case a } (k_P \gg k_B): \quad [R] = (1/3)\exp(-k_f t) + (2/3)\exp(-k_s t) \tag{65}$$

$$\text{Case b } (k_B \gg k_P): \quad [R] = (1/3)\exp(-k_f' t) + (2/3)\exp(-k_s t) \tag{66}$$

where k_f, k_f', and k_s are given as follows:

$$k_f = k_B + 2k_R + k_E \tag{67}$$

$$k_f' = k_P/2 + k_s \tag{68}$$

$$k_s = k_R + k_R' + k_E \tag{69}$$

Similar solutions were also obtained for the reaction from an S-precursor.[42]

From Equations 63 to 66, the following predictions could be given for the reaction from a T-precursors:

1. In the absence of a magnetic field, the decay of the generated radical pair can be expressed by a single exponential function with a decay rate constant of (Case a) $k_0 + k_E$ or (Case b) $k_P/4 + k_E$.
2. In the presence of a sufficiently large field, the decay of the pair can be expressed by a combination of two exponential functions. The rate constant of the fast component can be given by (Case a) k_f or (Case b) k_f' and that of the slow one by k_s for both cases.
3. No magnetic isotope effect can be expected for the rate constant observed in the absence of a magnetic field for Case b, because k_P and K_E should show no kinetic isotope effect. On the other hand, a magnetic isotope effect can be expected for the constant observed in the absence of the field for Case a, because k_0 is possible to depend on the HF interactions in the generated radical pair.
4. A magnetic isotope effect can be expected for the rate constant of the slow decay component observed in the presence of a magnetic field for both cases, because k_R and k_R' are possible to depend on the HF interactions. Little isotope effect, however, can be expected for k_f and k_f', because k_P is usually much larger than k_R and k_R'.

B. COMPARISON WITH EXPERIMENTS

In this section, we will demonstrate that the magnetic field and magnetic isotope effects shown in Section IV.C can well be explained by the above-mentioned RM. Here these effects have been proved to concern the reactions from T-precursors. The above conclusion can be derived for the following reasons:[42]

1. The decays of the generated radical pairs in the absence of a magnetic field can be expressed by single exponential functions but those in the presence of magnetic fields above 40mT by combinations of two exponential ones. This result can be explained by Equations 63 to 66. It is noteworthy that a magnetic field of as small as 40mT can be considered as a sufficiently large field.

---

*[14] The initial population of each triplet sublevel (T_i) was assumed to be equal (1/3) for simplicity. For a general case, the initial population may depend on i and B. For such a case, similar results can be obtained by changing the coefficients of Equations 65 and 66.

2. The decay rate constants observed at B = 0 (see Figure 12 and Table 1) are considered to correspond to $k_P/4 + k_E$ of Equation 64 (Case b) because they are too small to be explained by $k_0 + k_E$ of Equation 63 (Case a) and because they show little magnetic isotope effect. Since k_P is much larger than k_E, the k_P values are, for example, obtained to be 1.1×10^7 and 1.9×10^7 s^{-1} for the photoreduction reactions of BP and NQ in SDS micelles, respectively.

3. The slow rate constants observed for the decays of the radical pairs in the presence of magnetic fields decrease with increasing B as shown in Figure 13 and Table 1. This can be explained by $k_R + k_R'$ in r_S (see Equation 69). Although k_E is considered to be independent of B, k_R and k_R' are possible to be reduced by B. Indeed, the transition probability (P_{nm}) between the n and m spin states of a radical pair can be expressed as

$$P_{nm} = \hbar^2 |V_{nm}|^2 2\tau_C/(1 + \omega_{nm}^2 \tau_C^2) \qquad (70)$$

Here, V_{nm} is the off-diagonal matrix element between the n and m states, τ_C is the rotational correlation time, and ω_{nm} is the energy difference between the n and m states.

From Equation 70, the transition probability between the $T_{\pm 1}$-T_0 and $T_{\pm 1}$-S states which correspond to k_R and k_R', respectively, should decrease with increasing the Zeeman splitting ($\omega = (g_1 + g_2) \beta \hbar^{-1} B/2$) and increase with increasing the HF interactions because V_{nm} consists of the anisotropic Zeeman and HF interactions. Thus, the magnetic field and magnetic isotope effects observed for the photoreduction reactions of carbonyl compounds in micelles in the presence of magnetic fields can be proved to be explained well by the RM.

C. FURTHER EVIDENCES FOR THE RELAXATION MECHANISM

Recently, some further evidences for proving the applicability of the RM to the photoreduction reactions of carbonyl compounds in micelles have been obtained. The authors tried to enhance the relaxation rates by adding some paramagnetic ions to the sample solutions.[47] Since the lanthanoid ions are chemically inactive to triplet states and organic radicals, we carried out a LFP study of the photoreduction of NQ in an SDS micelle at room temperature using these ions. Here, ^{3}NQ* abstracts a hydrogen atom from an SDS molecule (RH), forming a triplet radical pair between the NQ semiquinone (SQ·) and alkyl (R·) radicals.

(71)

For example, the A(t) curves observed for the NQ semiquinone radical at B = 0 and 1.0 T in the absence and presence of Gd^{3+} ion are shown in Figure 16. As shown in this figure, the rate constant of the slow decay component in the A(t) curve at B = 1.0 T was found to increase with increasing the concentration of Gd^{3+}. On the other hand, no Gd^{3+} effect was observed at B = 0 T. The above-mentioned effect can be called ''an internal magnetic field effect of Gd^{3+}'' because the paramagnetic ion can enhance the relaxation rates of the generated radical pair as shown in Figure 17.

Thus, k_R and k_R' were found to be increased by a paramagnetic Gd^{3+} ion through the RM. Some other paramagnetic ions such as Nd^{3+}, Sm^{3+}, Dy^{3+}, Ho^{3+}, and Er^{3+} were also proved to show similar internal magnetic field effects due to the RM but diamagnetic ions such as La^{3+} and Lu^{3+} were not. From the present study, a further evidence for the applicability of the RM has been obtained.

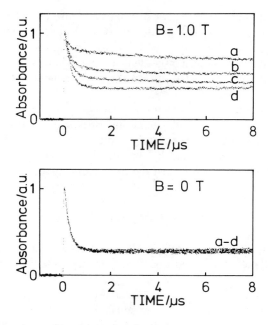

FIGURE 16. Time dependence of transient absorbances [A(t)] observed at 380 nm in the presence and absence of a magnetic field of 1.0 T for the micellar SDS (80 mM) solutions containing naphthoquinone (0.5 mM) and Gd^{3+} of (a) 0, (b) 0.47, (c) 0.93, and (d) 1.9 mM, respectively. (From Sakaguchi, Y. and Hayashi, H., *Chem. Phys. Lett.*, 106, 420, 1984. With permission.)

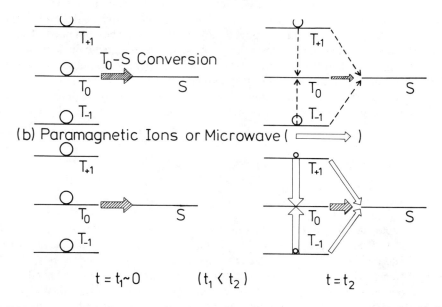

FIGURE 17. Relaxation mechanism of a radical pair (a) in the absence and (b) presence of a paramagnetic ion or microwave. A paramagnetic ion dissolved in solution and microwave resonant with the energy differences of a radical pair can enhance the relaxation of electron spins in the pair. The size of the circle on each level indicates the population of the level.

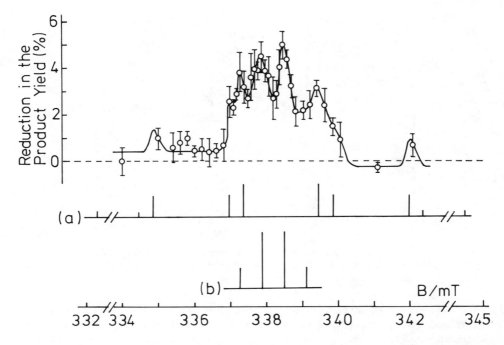

FIGURE 18.   Reduction in the yield of the spin adduct during photoreduction of 2-methyl-1,4-naphthoquinone (0.3 mM) in a micellar SDS (0.4 M) solution as a function of the applied magnetic field (B). Here, photoirradiation (500 W ultra-high pressure mercury lamp) and microwave irradiation (9.488 GHz, 160 mW) were performed simultaneously for 30 s at 300 K. ESR observation of the spin adduct produced from phenyl-*tert*-butylnitrene (17 mM) was made 1.0 min after irradiation. Stick diagrams show the signal positions and intensities (in the Boltzmann distribution) (a) for the alkyl radical from SDS (g = 2.0034, $a_H(\alpha)$ = 2.1 mT, and $a_H(\beta)$ = 2.5 mT) and (b) for the semiquinone radical (g = 2.0046 and a_H(methyl) = 0.63 mT). (The original data were kindly supplied by Dr. Masaharu Okazaki of Government Industrial Research Institute, Nagoya, Japan.)

More evidence has been given by Okazaki et al.[48] At room temperature, they irradiated a micellar SDS solution containing 2-methyl-1,4-naphthoquinine and a spin trap (TNO) with a mercury lamp in the absence and presence of a magnetic field. They used two spin traps, phenyl-*tert*-butylnitron and perdeutero-2,4-dimethyl-3-nitrosobenzenesulfonate. Immediately after the irradiation, they measured the ESR signals of each spin adduct which is formed by trapping of the alkyl radical (R·)

$$R\cdot \;+\; TNO \rightarrow TRNO\cdot \tag{72}$$

They found the signal intensity of each spin adduct to increase with increasing the strength (B) of the applied field. This magnetic field effect agrees very well with the effect on the yield of the escaping semiquinone radical observed by the authors[38] with the aid of the LFP-OA technique.

Furthermore, Okazaki et al. carried out the photoreduction of 2-methyl-1,4-naphthoquinone in the presence of microwave (9.488 GHz and 160 mW). Surprisingly, the yield of a spin adduct was found to be decreased by microwave at the positions of the ESR transition of the generated radical pair as shown in Figure 18. The spectrum of this figure could be explained by a radical pair of the semiquinone and alkyl radicals which had been observed by us with the aid of a LFP-ESR technique[49] as shown in the next section. Such a product-yield-detected ESR spectrum can be explained by the enhancement of the relaxation processes ($T_{\pm1}$-T_0 and $T_{\pm1}$-S) by microwave as shown in Figure 17.

This is a very skillful method to detect short-lived radical pairs with a conventional ESR

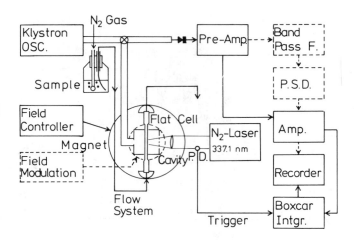

FIGURE 19.   Block diagram of the setup of an apparatus for LFP-ESR. The broken lines show the parts removed for the time-resolved measurement.

spectrometer. Thus, they have proved that the relaxation rate of radical pairs can be enhanced by microwave through the RM. They also analyzed carefully their product-yield-detected ESR spectrum and found the $|J|$ value of the generated radical pair to be less than 0.3 mT. From this estimation, the LCM is not considered to play an important role in the magnetic field effects of the photoreduction reactions of carbonyl compounds in micelles at room temperature.

VI. EXAMPLES OF CIDEP

A. APPARATUS FOR LASER FLASH PHOTOLYSIS-ESR

Since the signals of a conventional ESR spectrometer are detected with a 100-kHz field-modulation amplifier in order to improve S/N ratio, the time resolution of the spectrometer is about 30 μs. This is often unsatisfactory for studying initial stages of physical and chemical processes. To achieve faster time resolution, some kinds of apparatus for LFP-ESR have been developed. Among them, a spectrometer used in the superheterodyne mode with a 2-MHz modulation,[50] a time-resolved electron spin-echo spectrometer,[51] and an unmodulated broad-band spectrometer[52] are specially famous. These apparatus, however, are too sophisticated and too expensive for those who are not specialists of ESR.

On the other hand, a much simpler and fairly satisfactory apparatus for LFP-ESR can be made with a conventional ESR spectrometer, a nanosecond-pulsed laser, and a boxcar integrator. The block diagram of the setup of such an apparatus[53] is shown in Figure 19. A conventional X-band ESR spectrometer (Varian E-109) is used without field modulation. A pulsed nitrogen laser with a pulse duration of 10 ns and output power of 9 mJ per pulse is used for excitation of the sample inside a rectangular cavity resonant in the TE_{102} mode. An excimer laser or a Q-switched Nd:YAG laser can also be used for laser excitation.

The signal from the fast response preamplifier is further amplified by the main amplifier of this spectrometer. Since the maximum frequencies of these amplifiers are extended to a few megahertz, a submicrosecond time resolution can easily be obtained. The time resolution can be improved to 50 ~ 100 ns with a better amplifier system.[54]

A time-resolved ESR spectrum is measured with a boxcar integrator by fitting the time delay after the laser excitation of the sample and by changing the strength (B) of a magnetic field. The ESR spectrum thus measured has a pure absorption/emission mode instead of a conventional first derivative one observed with field modulation. The time profile of the ESR signal at a fixed magnetic field is obtained with a boxcar integrator by changing the time delay or with a conventional oscilloscope or transient recorder.

In order to carry out an LFP study of excited triplet states with the above-mentioned apparatus, convention quartz sample tube and quartz dewar for liquid nitrogen (or temperature control system) can be used. For the study of a liquid sample, however, the solution should be kept on flowing through a flat quartz cell (50 × 10 mm) during a series of measurements. Especially for polar solvents such as water and alcohol, the gap of the cell should be designed to be as thin as 0.3 mm in order to avoid lowering of the Q value of the cavity.

Although the handling system of sample solutions is somewhat complicated at temperatures other than room temperature,[55] the authors made a much simpler flow system[49] for the studies at room temperature as shown in Figure 19. Here, the solution is bubbled with nitrogen gas and then is introduced into the cell with a teflon tube system. A constant flow of the solution can easily be performed by gravity without any pump.

B. TYPICAL CIDEP SPECTRA

In this section, some typical ESR spectra measured at room temperature with the above-mentioned apparatus for LFP-ESR are shown. The photoreduction of xanthone (XO) in 2-propanol was taken as the first example.[56] From the study of this reaction with a LFP-OA technique, Scaiano[57] measured the time profile [A(t)] of the peak at 610 nm, which is assigned as $^3XO^*$, and obtained the triplet lifetime to be 370 ns. Thus, the present reaction was confirmed to proceed as

$$(73)$$

Figure 20a shows the time-resolved ESR spectrum observed at a time delay of 1.2 μs after the laser excitation of the 2-propanol solution of XO. From the ESR spectra reported in the literature,[55,58] the signals denoted by spectra I and II in Figure 20a can be assigned as the 2-propanol and ketyl radicals, respectively. As clearly seen in this figure, the signals of the 2-propanol radical show a typical E/A phase pattern. This pattern can be explained by Equation 43 of the S-T_0M because μ is positive (T-precursor) and because J is considered to be negative as usual.

Although an extra HF spitting due to the O-H proton (0.07 mT) is observed in the CIDEP spectrum of the 2-propanol radical, the spectrum has the HF structure due to six equivalent protons (1.97 mT) and its signal intensities correspond well with spectrum (d) of Figure 4. Since the triplet lifetime was observed to be 370 ns for this reaction, it may be longer than the typical spin relaxation time of usual triplet molecules at room temperature ($\tau_{RL} = 10^{-10} \sim 10^{-8}$ s). Thus, the CIDEP through the TM is considered to be less important than that through the S-T_0M for the present reaction. Indeed, the CIDEP signals of the 2-propanol radical shown in Figure 20a show only a little distortion by an A-phase pattern due to the TM.

Although the HF structures of the ketyl radical was not completely resolved in its observed spectrum, its signal positions can be reproduced by the reported HF coupling constants[58] as shown in Figure 20a. These signals have an E/A* phase pattern, which can also be explained by a combination of the S-T_0M and the TM. A larger contribution of the TM in the ketyl radical compared with the 2-propanol one may be due to the smaller HFC constants of the former than the latter.

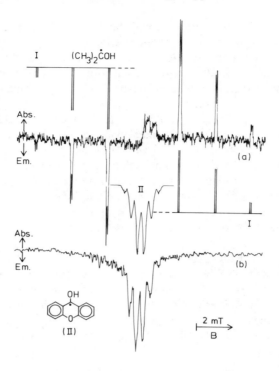

FIGURE 20.   Typical CIDEP spectra observed at room temperature at a
time delay of 1.2 μs (gate width, 200 ns) after laser excitation of the 2-
propanol solutions containing (a) xanthone (2 mM) and (b) xanthone (1
mM) and N-methylaniline (1.54 M). Stick diagram I shows the signal
positions of the 2-propanol radical[55] and spectrum II is the calculated one
from the observed HFC constants of the ketyl radical of xanthone.[58]

As the second example, we measured the CIDEP spectrum at a time delay of 1.2 μs
after laser excitation of a ternary solution of 2-propanol containing XO and N-methylaniline.[59]
The observed spectrum is shown in Figure 20b. This spectrum clearly shows the signals
due to the ketyl radical produced by the following reaction:

$$^3 \text{XO}^* + C_7NH_9 \longrightarrow \text{XOH} \cdot + C_7NH_8 \cdot$$

$$(^3XO^*) \qquad\qquad\qquad (XOH\cdot) \qquad\qquad\qquad\qquad (74)$$

The signals due to the radical ($C_7NH_8\cdot$) generated from N-methylaniline were not clearly be
obtained. This may be explained by the loss of the initial polarization through the following
exchange reaction:

$$C_7NH_8 \cdot \text{ (spin polarized)} + C_7NH_9 \rightarrow C_7NH_9 + C_7NH_8 \cdot \text{ (unpolarized)} \qquad (75)$$

Since no signals due to the 2-propanol radical were observed in Figure 20b, 2-propanol is
not considered to act as the hydrogen donor in Reaction 74.

The observed CIDEP spectrum of the ketyl radical has an almost E-phase pattern as
shown in Figure 20b. This E one indicates that Reaction 74 is fast enough to transfer the
spin polarization of ^{3}XO* to the generated radical pair through the TM. Indeed, the abstraction
reactions of hydrogen by triplet carbonyls from aromatic amines have been confirmed to

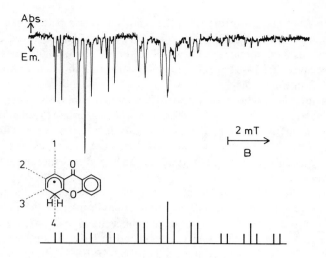

FIGURE 21. CIDEP spectrum observed at room temperature at a time delay of 1.2 μs (gate width, 200 ns) after laser excitation of the 2-propanol solution containing xanthone (2 m*M*) and tri-*n*-butyltin hydride (0.2 *M*). The stick diagram shows the signal positions of a cyclohexadienyl-type radical where the hydrogen abstraction occurs at C(4) of xanthone. The weak extra signals in the observed spectrum corresponds to another cyclohexadienyl-type radical where the hydrogen abstraction occurs at C(1) of xanthone.

occur very fast in polar solvents because of the creation of charge-transfer complexes.[60,61] The spectrum of the ketyl radical is slightly distorted from a wholly E-phase pattern as shown in Figure 20b. This distortion can be explained by a slight contribution of an E/A one due to the S-T$_0$M.

As the third example, we added tri-*n*-butylstannyl hydride (Bu$_3$SnH) to the 2-proponal solution of XO.[56] To our great surprise, as shown in Figure 21, the CIDEP spectrum showed a drastic change from the spectrum of Figure 20(a). From the stick diagram of Figure 21, the observed signals can be explained by the cyclohexadienyl-type radical (XHO·) produced as follows:

$$(^3XO^*) + Bu_3SnH \longrightarrow (XHO·) + Bu_3Sn· \tag{76}$$

No signal due to the stannyl radical was observed. This may be due to the fast exchange reaction which destroy the spin polarization of an initially prepared radical

$$Bu_3Sn· \text{ (spin polarized)} + Bu_3SnH \rightarrow Bu_3SnH + Bu_3Sn· \text{ (unpolarized)} \tag{77}$$

As shown in Figure 21, the CIDEP signals of XHO· have an overall E-phase pattern, but the observed CIDEP spectrum seems to deviate more and more from an ideal one having a wholly E-phase pattern as B increases. Since the authors measured the rate constant of the reaction indicated by Reaction 76 to be $5.2 \times 10^8 \, M^{-1} \, s^{-1}$ with the LFP-OA technique,[59] the triplet lifetime in the presence of Bu$_3$SnH (0.2 *M*) becomes as short as 10 ns. This lifetime may be shorter than or comparable to the spin relaxation time of triplet xanthone at room temperature. Thus, the wholly E-component of the observed spectrum can be explained by the TM. The distortion from the wholly E-phase pattern can be explained by the E/A one induced from the $S-T_0M$. As shown in Figure 7, an observed CIDEP spectrum becomes, in general, a combination of a wholly E(or A)-phase pattern and an E/A(or A/E) one when the TM and $S-T_0M$ are both effective. The spectrum shown in Figure 21 corresponds to that of x ~ 0.2 in Figure 7.

The photoreduction reactions of carbonyl compounds are very typical photochemical processes. The high reactivity of triplet carbonyls has been ascribed to either the nπ* character of the carbonyl group[60] or the creation of charge-transfer complexes with amines.[61] The formation of the ketyl or semiquinone radicals*[15] in these two types of reactions has been well established by analysis of reaction products and by direct mesurement with LFP-OA and -ESR techniques.

On the other hand, such a new reaction as hydrogen addition to the benzene ring had never been found for the photoreduction of any carbonyl molecule before our discovery of the cyclohexadienyl-type radical of xanthone.[62] Since its signal was not observed with a LFP-OA technique,[59] an LFP-ESR one is considered to have an advantage of discovering new intermediate radicals. Indeed, with the aid or the LFP-ESR technique, we found similar cyclohexadienyl-type radicals in the photoreduction reactions of acetophenone and 2-acetonaphthone[63] and the benzyl-type radicals in the reactions of flavones.[64]

We have studied the reaction mechanism for such new reactions changing solvents and hydrogen donors and obtained the following results:[56,59,62-65]

1. Some special hydrogen donors with B-H, Ge-H, and Sn-H bonds can give rise to such new reactions but usual donors with C-H and N-H ones cannot. This is due to the polarization of these bonds. The hydrogen atom of the former donors is negatively polarized but that of the latter ones is positively polarized.
2. The photoreduction through the nπ* triplet states of carbonyl molecules with any hydrogen donor occurs at the carbonyl oxygen and gives the corresponding ketyl or semiquinone radicals as has been believed for many years.[60,61]
3. The photoreduction through the ππ* triplet states of carbonyl molecules with the former donors gives such new radicals as cyclohexadienyl- and benzyl-type radicals but no photoreduction occurs with the latter donors as has been believed.

As the fourth example, the CIDEP spectrum observed for the photoreduction of XO with triethylgermanium hydride (Et_3GeH) in an SDS micelle is shown in Figure 22.[62] As shown in this figure, the signal positions of this spectrum coincide with those of the cyclohexadienyl-type radical of xanthone. Thus, the present reaction was also proved to proceed as

$$^3XO^* + Et_3GeH \rightarrow XHO\cdot + Et_3Ge\cdot \qquad (78)$$

Here, no signal due to the germyl radical could be observed because of the similar exchange reaction as Reaction 77.

It is noteworthy that the spectrum of Figure 22 has an alternating E/A phase pattern. This pattern looks like a usual ESR spectrum measured with field modulation but was observed actually in a pure absorption/emission mode. Since such a peculiar phase pattern could not be explained by the ordinary CIDEP mechanisms ($S-T_0M$, $S-T_{\pm 1}M$, and TM), our finding of this peculiarity[49] attracted considerable attention. Recently, its mechanism has been clarified with a new mechanism, which will be explained in Section VII.

---

*[15] Sometimes, proton is dissociated from the O-H bond of these radicals and then they become anion radicals.

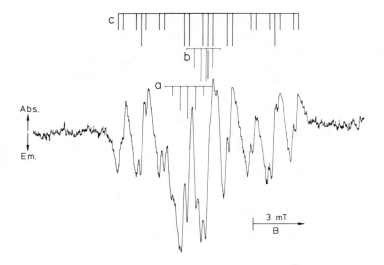

FIGURE 22. CIDEP spectrum observed at room temperature at a time delay of 1.5 μs (gate width, 200 ns) after laser excitation of a micellar SDS (80 mM) solution containing xanthone (1 mM) and triethylgermanium hydride (20 mM). Diagrams a, b, and c represent the signal positions of the germyl, ketyl, and cyclohexadienyl-type radicals, respectively. (From Sakaguchi, Y., Hayashi, H., Murai, H., I'Haya, Y. J., and Mochida, K., *Chem. Phys. Lett.*, 120, 401, 1985. With permission.)

C. CIDEP STUDIES OF THE PHOTOREDUCTION REACTIONS OF CARBONYL COMPOUNDS IN MICELLES

As shown in Section IV.C, the authors found large magnetic field and magnetic isotope effects on the reaction rates as well as the yields of escaping radicals in the photoreduction reactions of carbonyl compounds in micelles. These new types of effects have been explained by the RM (see Section V). Since the above-mentioned effects, however, were studied with a LFP-OA technique, no direct information on the structures and relaxations of electron spins in radical pairs could be obtained. Therefore, we performed CIDEP studies with a LFP-ESR technique.

Figure 23 shows the CIDEP spectra observed after the excitation of the micellar SDS solutions of naphthoquinone (NQ) and benzophenone (BP) at room temperature.[49] As shown in this figure, the CIDEP signals due to the two component radicals of each radical pair were detected in the same time: for the reaction of NQ, the semiquinone (SQ·) and alkyl (R·) radicals were observed as had been anticipated by Equation 71, and for the reaction of BP, the ketyl (K·) and alkyl radicals as had been anticipated in Figure 10.

No optical absorption signal due to R· is expected to appear in the usual wavelength region (300 ~ 800 nm). Our CIDEP studies,[49] however, directly confirmed that a micellar molecule in a micelle acts as a good hydrogen donor in the photoreduction reactions of carbonyl compounds. Thus, ESR measurement is shown to play a complementary role to optical one, which is not always able to detect all intermediates in the studies of fast reactions.

The time profiles of the CIDEP signals of SQ·, K·, and R· were also measured.[49] The decays of these signals were proved to correspond well with those of the transient absorbances [A(t)] of SQ· and K· measured in the presence of the same magnetic field (about 0.34 T) that had been applied for the CIDEP studies. Moreover, the decays of the CIDEP signals were found to be enhanced by the addition of some paramagnetic lanthanoid ions to the present micellar solutions. Thus, the applicability of the RM to the photoreduction reactions of carbonyl molecules in micelles is also proved by the present CIDEP studies.

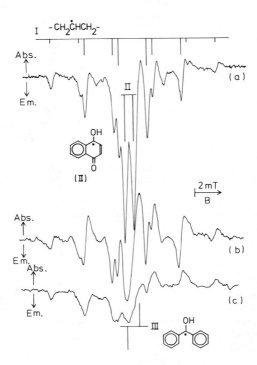

FIGURE 23.   CIDEP spectrum observed at room temperature at a time delay of 1.2 μs (gate width, 200 ns) after laser excitation of (a) a micellar SDS (80 mM) of naphthoquinone (1 mM) and the spectra observed at time delays of (b) 1.2 μs and (c) 3.0 μs after excitation of a micellar SDS (80 mM) solution of benzophenone (2 mM). Diagrams I, II, and III show the signal positions of the alkyl ($a_H(\alpha)$ = 2.1 mT and $a_H(\beta)$ = 2.5 mT), naphthosemiquinone ($a_H(3)$ = 0.7 mT), and benzophenone ketyl radicals, respectively. (From Sakaguchi, Y., Hayashi, H., Murai, H., and I'Haya, Y. J., *Chem. Phys. Lett.*, 110, 275, 1984. With permission.)

The phase patterns of CIDEP spectra also present some important information. The spectrum observed at a time delay of 1.2 μs after excitation of the NQ solution has an almost E-phase pattern as shown in Figure 23a. Since the E one can be explained by the TM, the reaction of ^{3}NQ* with an SDS molecule is considered to be faster or comparable to the spin relaxation of ^{3}NQ*. As the time delay was increased from 1.2 μs, a little distortion due to an E/A pattern appeared through the S-T_0M.

On the other hand, an alternating E/A phase pattern was observed for R· at a time delay of 1.2 μs after excitation of the BP solution as shown in Figure 23b. Although the signals due to K· also showed an E/A pattern at this time delay, it cannot be determined whether this pattern is attributed to a usual E/A one through the S-T_0M or an alternating E/A one. As the time delay was increased from 1.2 μs, the phase pattern of this reaction changed into a usual E/A one as shown in Figure 23c. The spectrum observed at a time delay of 3.0 μs can be explained by Equation 43, where μ is positive (T-precursor) and J is negative.

From a careful examination of the CIDEP spectrum observed for the reaction of NQ (see Figure 23a), the spectrum was also shown to have a little component due to a similar alternating E/A phase pattern. As shown in Figure 22, we also found a similar one in the CIDEP spectrum observed for the cyclohexadienyl-type radical of xanthone in an SDS micelle. On the other hand, the same radical did not show such a peculiar pattern in 2-propanol as shown in Figure 21. Thus, it is probable that the alternating phase patterns can be observed for reactions in confined systems such as micelles and viscous solvents. The mechanism for these peculiar phase patterns will be explained in the next section.

VII. J MECHANISM IN CIDEP

In 1977, Trifunac and Nelson[66] carried out a pulse radiolysis study of some aqueous micelle systems with the aid of a time-resolved ESR technique. They observed the CIDEP signals due to the alkyl radicals produced through hydrogen elimination from the employed micellar molecules. At all surfactant concentrations, the phase patterns of the observed CIDEP spectra could be explained mainly by E/A ones which were produced through the S-T_0M. In pulse radiolysis, two independently generated radicals can encounter to give a radical pair (free radical precursor, μ is positive) and J is negative as usual. Thus, Λ of Equation 43 becomes negative.

At concentration greater than the critical micelle concentration (CMC), each of the six hyperfine (HF) lines of the observed alkyl radicals seems to be more or less distorted by an E/A phase pattern (see Figures 1 to 3 of Reference 66). Trifunac and Nelson, however, interpreted these distortions in terms of excess emission polarization caused by the S-T_{-1}M.

In 1984, the authors studied the photoreduction of benzophenone (BP) in an SDS micelle with the aid of an LFP-ESR technique.[49] As shown in Figure 23b, however, each of the HF lines observed for the alkyl radical at a time delay of 1.2 μs after laser excitation was found to show a much clearer E/A phase pattern. We also found similar alternating E/A ones for the alkyl radicals produced from other micellar molecules[67] as well as for the cyclohexadienyl-type radical produced in an SDS micelle from xanthone and triethyl-germanium hydride (see Figure 22).[62] As mentioned in Section VI.C, these peculiar phase patterns cannot be explained by the ordinary CIDEP mechanisms (the S-T_0M, S-$T_{\pm 1}$M, and TM).

Our finding of such peculiar phase patterns has attracted considerable attention. At first, McLauchlan and Stevens[68] proposed that the alternating E/A phase pattern of the alkyl radical produced from the photoreduction of BP with an SDS molecule (RH) could be explained by two radicals generated from different types of precursors: one is the ordinary alkyl radical generated from the excited triplet state of BP (^{3}BP*) and the other is a similar but somewhat different alkyl one generated from the excited singlet state of BP (^{1}BP*).

$$^3BP^* + RH \rightarrow K\cdot + R\cdot \tag{79}$$

$$^1BP^* + RH \rightarrow K\cdot + R'\cdot \tag{80}$$

Indeed, from Equation 43, R· (or R'·) should give an ordinary E/A (A/E) phase pattern. If these alkyl radicals have somewhat different HF coupling constants, the observed alternating E/A phase pattern can be reproduced by a combination of the above-mentioned E/A (from R·) and A/E (from R'·) ones.

One of the difficulties in this explanation is the reality of Reaction 80. The lifetime of ^{1}BP* is so short that all the photochemical reactions of BP have been believed to occur from ^{3}BP*. Actually, we added naphthalene (NP) to a micellar SDS solution of BP and observed its CIDEP spectrum with the aid of a LFP-ESR technique.[67,69] Here, NP acts as a triplet quencher.

$$^3BP^* + NP \rightarrow BP + ^3NP^* \tag{81}$$

Although the signal intensities of the alkyl radical were reduced by the addition of NP, the phase pattern observed for the radical at a time delay of 1.2 μs after laser excitation of the solution with NP was almost the same as that observed without NP. Since the employed laser (nitrogen laser) could not excite NP directly, the alternating E/A phase pattern observed for the alkyl radical was proved to be induced through the reaction of ^{3}BP*. Thus, no contribution from ^{1}BP* can be considered for explaining this peculiar phase pattern.

Second, Murai et al.[67] proposed intramolecular hydrogen migration within the alkyl radicals for their peculiar phase patterns.

$$-CH_2-\overset{\centerdot}{\underline{C}}\underline{H}-CH_2-CH_2- \rightarrow CH_2-CH_2-\overset{\centerdot}{\underline{C}}\underline{H}-CH_2- \tag{82}$$

Here, the hydrogens which are initially polarized through the S-T$_0$M are represented by "$\underline{H}$" and unpolarized ones by "H". If one of the β-protons in the initially prepared radical moves to the next carbon atom as shown in Reaction 82, the new radical should give a different phase pattern from that of the initial radical. The alternating phase patterns observed for the aklyl radicals are possible to be explained by a series of such hydrogen transfers. Although an ordinary ESR study gives the same spectra for the alkyl radicals before and after Reaction 82, such migration reactions can be studied with the aid of an LFP-ESR technique.

The above-mentioned explanation with fast intramolecular hydrogen migration, however, requires drastic change in conventional thinkings of free radical chemistry because fast 1,2-shifts of hydrogen in straight chain alkyl radicals have no precedent. Therefore, a different explanation for the peculiar phase patterns has recently been proposed independently by Buckley et al.[70] and Closs et al.[71] Let us consider their mechanism, taking a simple model as explained by Buckley et al. This model corresponds to the case that has been considered for the ΔgM in Section II.C. Here, $a_{1j} = a_{2k} = 0$ and $\Delta g \neq 0$.

In this case, the S-T$_0$ mixing occurs through Q of Equation 23 but no S-T$_{\pm1}$ one. The eigenstates ($|i>$) and eigenvalues (E$_i$) of the radical pair for this case (see Figure 24) are

$$|1\rangle = |T_{+1}\rangle \qquad\qquad E_1 = -J + E_0 \tag{83}$$

$$|2\rangle = \cos x|S\rangle + \sin x|T_0\rangle \qquad\qquad E_2 = w \tag{84}$$

$$|3\rangle = -\sin x|S\rangle + \cos x|T_0\rangle \qquad\qquad E_3 = -w \tag{85}$$

$$|4\rangle = |T_{-1}\rangle \qquad\qquad E_4 = -J - E_0 \tag{86}$$

where $E_0 = (g_1 + g_2)\beta\hbar^{-1}B/2$, $\tan 2x = Q/J$, and w is given by Equation 14.

The four possible ESR transitions have energy differences ($E_{ij} = E_i - E_j$) and transition probability (P_{ij}) given by

$$E_{12} = E_0 - w - J \qquad\qquad P_{12} = \sin^2 x \tag{87}$$

$$E_{34} = E_0 - w + J \qquad\qquad P_{34} = \cos^2 x \tag{88}$$

$$E_{13} = E_0 + w - J \qquad\qquad P_{13} = \cos^2 x \tag{89}$$

$$E_{24} = E_0 + w + J \qquad\qquad P_{24} = \sin^2 x \tag{90}$$

When the triplet sublevels are equally populated, the triplet probabilities of states $|1>$ and $|4>$ are both 1/3 while those of $|2>$ and $|3>$ become $(1/3)\sin^2 x$ and $(1/3)\cos^2 x$, respectively. With these occupation probabilities (N$_i$) the intensities (I$_{ij}$) of the four transitions become

$$I_{ij} = C(N_j - N_i)P_{ij} \tag{91}$$

$$I_{24} = I_{34} = -I_{12} = -I_{13} = (C/12)\sin^2 2x = CQ^2/(12w^2) \tag{92}$$

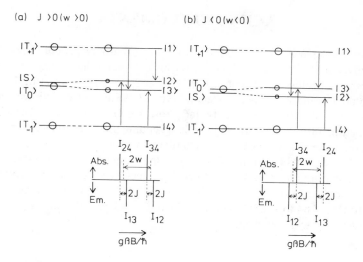

FIGURE 24. S-T$_0$ mixing of a radical pair ($g_1 > g_2$, $a_{1j} = 0$, $a_{2k} = 0$, through Q (see Equation (23) and its possible ESR transitions (I_{ij}) whose signal intensities can be calculated from Equation 92. Here, $g = (g_1 + g_2)/2$ and the radical pair is assumed to be produced from a T-precursor. The initial population of each sublevel is represented schematically by a circle.

Here, C is a positive constant. The calculated spectra for $J < 0$ and $J > 0$ are shown in Figure 24, where each polarized spectrum consists of two antiphase doublets.

As shown by this figure and Equations 87 to 90, the doublet splitting corresponds to 2 $|J|$ and the separation between the centers of the doublets to 2w. When $|J|$ is much larger than Q, little polarization is observed because $|I_{ij}|$ becomes very small. When $|J|$ is small compared to the line width, the CIDEP signals also vanish because the antiphase components of each doublet cancel one another. For intermediate $|J|$, the polarized spectrum becomes much more intense than the equilibrium one.

At present, the third mechanism, which can be called "the non zero J mechanism (JM), is considered to be the most probable mechanism for explaining the alternating phase patterns that have been observed in the CIDEP spectra of the radicals produced from the reactions in micelles at room temperature. From the observed splittings of the antiphase doublets,[49,62,67] the $|J|$ values can be estimated to be less than 0.3 mT. This estimation agrees well with that made by Okazaki et al.[48] with the aid of a product-yield-detected ESR technique (see Section V.C). Thus the $|J|$ values of usual radical pairs in micelles have been proved to be so small that the S-T$_{-1}$M as proposed by Tanimoto et al.[44b] can again be considered to be not so important in interpreting the magnetic field and magnetic isotope effects observed for the photoreduction reactions of carbonyl compounds in micellar solutions at room temperature.

It is noteworthy that the JM is also applicable in explaining CIDEP spectra observed in other confined systems, such as in viscous solvents and linked systems. Recently, Buckley et al. found the photoreduction of acetone in 2-propanol to give such a peculiar CIDEP spectrum at low temperature.[70]

$$(CH_3)_2C{=}O \ + \ (CH_3)_2CH{-}OH \rightarrow 2(CH_3)_2\overset{\bullet}{C}{-}OH \qquad (93)$$

Indeed, the spectrum observed at 199 K for the 2-propanol radical can be explained by a combination of an E*/A phase pattern due to the S-T$_0$M and S-T$_{-1}$M with an alternating E/A one due to the JM. More recently, Closs and Forbes found similar alternating E/A phase patterns in the CIDEP spectra observed for some polymethylene-linked biradicals at room temperature.[72]

ACKNOWLEDGMENTS

The authors are sincerely grateful to Professor Saburo Nagakura, President of The Graduated University for Advanced Studies, for his continuing interest and encouragement. We are indebted to Professor Yasumasa J. I'Haya and Dr. Hisao Murai of The University of Electro-Communications for their collaboration on CIDEP studies. We wish to thank Dr. K. A. McLauchlan of Oxford University and Professor G. L. Closs of The University of Chicago for their helpful discussions and for their sending preprints to us before publication.

The support from the Science and Technology Agency (Special Project Research on New Reaction Fields, 1984 to) and those from the Ministry of Education, Science, and Culture (to H. Hayashi: Grant-in-Aid for Special Project Research on Molecular Assembly, 1983 to 1986; Grant-in-Aid No. 59430002, 1984 to 1987; and to Y. Sakaguchi: Grant-in-Aid No. 59740226, 1984 to 1985; Grant-in-Aid No. 61740267, 1986 to 1987) for the present project are gratefully acknowledged.

REFERENCES

1. **Muus, L. T., Atkins, P. W., McLauchlan, K. A., and Pedersen, J. B.,** *Chemically Induced Magnetic Polarization,* D. Reidel, Dordrecht, Netherlands, 1977.
2. **Kaptein, R. and Oosterhoff, J. L.,** Chemically induced dynamic nuclear polarization. II, *Chem. Phys. Lett.,* 4, 195, 1969.
3. **Closs, G. L.,** A mechanism explaining nuclear spin polarizations in radical combination reactions, *J. Am. Chem. Soc.,* 91, 4552, 1969.
4. **Wong, S. K., Hutchinson, D. A., and Wan, J. K. S.,** Chemically induced dynamic electron polarization. II. A general theory for radicals produced by photochemical reactions of excited triplet carbonyl compounds, *J. Chem. Phys.,* 58, 985, 1973.
5. **Kurita, Y.,** Electron spin resonance study of radical pairs trapped in irradiated single crystals of dimethylglyoxime at liquid nitrogen temperatures, *Nippon Kagaku Zasshi,* 85, 833, 1964.
6a. **Hayashi, H., Itoh, K., and Nagakura, S.,** The determination of singlet-triplet separation from the anomalous hyperfine structure observed with a radical pair, *Bull. Chem. Soc. Jpn.,* 39, 199, 1966.
6b. **Itoh, K., Hayashi, H., and Nagakura, S.,** Determination of the singlet-triplet separation of a weakly interacting radical pair from the E.S.R. spectrum, *Mol. Phys.,* 17, 561, 1969.
7. **Kaptein, R.,** Chemically induced dynamic nuclear polarization. VIII. Spin dynamics and diffusion of radical pairs, *J. Am. Chem. Soc.,* 94, 6251, 1972.
8. **Noyes, R. M.,** Models relating molecular reactivity and diffusion in liquids, *J. Am. Chem. Soc.,* 78, 5486, 1956.
9. **Hayashi, H. and Nagakura, S.,** The theoretical study of external magnetic field effect on chemical reactions in solution, *Bull. Chem. Soc. Jpn.,* 51, 2862, 1978.
10. **Sakaguchi, Y., Hayashi, H., and Nagakura, S.,** Classification of the external magnetic field effects on the photodecomposition reaction of dibenzoyl peroxide, *Bull. Chem. Soc. Jpn.,* 53, 39, 1980.
11. **Schulten, Z. and Schulten, K.,** The generation, diffusion, spin motion, and recombination of radical pairs in solution in the nanosecond time domain, *J. Chem. Phys.,* 66, 4616, 1977.
12. **Werner, H. -J., Staerk, H., and Weller, A.,** Solvent, isotope, and magnetic field effects in the geminate recombination of radical ion pairs, *J. Chem. Phys.,* 68, 2419, 1978.
13. **Adrian, F. J.,** Theory of anomalous electron spin resonance spectra of free radicals in solution. Role of diffusion-controlled separation and reencounter of radical pairs, *J. Chem. Phys.,* 54, 3918, 1971.
14. **Monchick, L. and Adrian, F. J.,** On the theory of chemically induced electron polarization (CIDEP): vector model and an asymptotic solution, *J. Chem. Phys.,* 68, 4376, 1978.
15. **Buckley, C. D. and McLauchlan, K. A.,** The influence of ST_{-1} mixing in spectra which exhibit electron spin polarization (CIDEP) from the radical pair mechanism, *Chem. Phys. Lett.,* 137, 86, 1987.
16. **Kinoshita, M., Iwasaki, N., and Nishi, N.,** Molecular spectroscopy of the triplet state through optical detection of zero-field magnetic resonance, *Appl. Spectrosc. Rev.,* 17, 1, 1981.
17. **Atkins, P. W.,** Magnetic field effects, *Chem. Br.,* 127, 214, 1976.
18. **Steubling, W.,** Action of a magnetic field on fluorescence intensity, *Verh. Dtsch. Physik. Ges.,* 15, 1181, 1913.

19. **Turner, L. A.,** The magnetic quenching of iodine fluorescence and its connection with predissociation phenomena, *Z. Phys.,* 65, 464, 1930.
20. **Misono, M. and Selwood, P. W.,** Extrinsic field acceleration of the magnetic parahydrogen conversion, *J. Am. Chem. Soc.,* 90, 2977, 1968.
21. **Selwood, P. W.,** The effect of a magnetic field on the catalyzed nondissociative parahydrogen conversion rate, *Adv. Catal.,* 27, 23, 1978.
22. **Sagdeev, R. Z., Molin, Yu. N., Salikhov, K. M., Leshina, T. V., Kamha, M. A., and Shein, S. M.,** Effects of magnetic field on chemical reactions, *Org. Magn. Reson.,* 5, 603, 1973.
23. **Brocklehurst, B., Dixon, R. S., Gardy, E. M., Lopata, V. J., Quinn, M. J., Singh, A., and Sargent, F. P.,** The effect of a magnetic field on the singlet/triplet ratio in geminate ion recombination, *Chem. Phys. Lett.,* 28, 361, 1974.
24. **Sargent, F. P., Brocklehurst, B., Dixon, R. S., Gardy, E. M., Lopata, V. J., and Singh, A.,** Pulse radiolysis in an applied magnetic field. The time dependence of the magnetic field enhancement of the fluorescence from solutions of fluorene in squalane, *J. Phys. Chem.,* 81, 815, 1977.
25. **Michel-Beyerle, M. E., Haberkorn, R., Bube, W., Steffens, E., Schröder, H., Neusser, H. J., Schlag, E. W., and Seidlitz, H.,** Magnetic field modulation of geminate recombination of radical ions in a polar solvent, *Chem. Phys.,* 17, 139, 1976.
26. **Schulten, K., Staerk, H., Weller, A., Werner, H. -J., and Nickel, B.,** Magnetic field dependence of the geminate recombination of radical ion pairs in polar solvents, *Z. Phys. Chem. N. F.,* 101, S371, 1976.
27. **Tanimoto, Y., Hayashi, H., Nagakura, S., Sakuragi, H., and Tokumaru, K.,** The external magnetic field effect on the singlet sensitized photolysis of dibenzoyl peroxide, *Chem. Phys. Lett.,* 41, 267, 1976.
28. **Hata, N.,** The effect of external magnetic field on the photochemical reaction of isoquinoline N-oxide, *Chem. Lett.,* 547, 1976.
29. **Weller, A., Staerk, H., and Treichel, R.,** Magnetic-field effects on geminate radical-pair recombination, *Faraday Discuss. Chem. Soc.,* 78, 271, 1984.
30. **Nakagaki, R., Hiramatsu, M., Mutai, K., Tanimoto, Y., and Nagakura, S.,** Photochemistry of bi-chromophoric chain molecules containing electron donor and acceptor moieties. External magnetic field effects upon the photochemistry of N-[ω-(p-nitrophenoxy)alkyl] anilines, *Chem. Phys. Lett.,* 134, 171, 1987.
31. **Tanimoto, Y., Okada, N., Itoh, M., Iwai, K., Sugioka, K., Takemura, F., Nakagaki, R., and Nagakura, S.,** Magnetic field effects on the fluorescence of intramolecular electron-donor-acceptor systems, *Chem. Phys. Lett.,* 136, 42, 1987.
32. **Buchachenko, A. L., Galimov, E. M., Ershow, V. V., Nikiforov, G. A., and Pershin, A. D.,** Isotope enrichment induced by magnetic interactions in chemical reactions, *Dokl. Acad. Nauk SSSR,* 228, 379, 1976.
33. **Turro, N. J. and Kraeutler, B.,** Magnetic isotope and magnetic field effects on chemical reactions. Sunlight and soap for the efficient separation of ^{13}C and ^{12}C isotopes, *J. Am. Chem. Soc.,* 100, 7432, 1978.
34. **Tabata, Y., Kimura, H., Sobue, H., and Oshima, K.,** The effect of magnetic field on the radiation-induced polymerization of formaldehyde in the solid state at a low temperature, *Bull. Chem. Soc. Jpn.,* 37, 1713, 1964.
35. **Turro, N. J., Chow, M. -F., Chung, C. -J., and Tung, C. -H.,** An efficient, high conversion photoinduced emulsion polymerization. Magnetic field effects on polymerization efficiency and polymer molecular weight, *J. Am. Chem. Soc.,* 102, 7391, 1980.
36. **Morita, H., Higasayama, I., and Yamaoka, T.,** Magnetic field effect on photocrosslinking reaction of bromo- and chloro-methylated polystyrene, *Chem. Lett.,* 963, 1986.
37. **Sakaguchi, Y., Nagakura, S., and Hayashi, H.,** External magnetic field effect on the decay rate of benzophenone ketyl radical in a micelle, *Chem. Phys. Lett.,* 72, 420, 1980.
38. **Sakaguchi, Y. and Hayashi, H.,** Laser-photolysis study of the photochemical reactions of naphthoquinones in a sodium dodecyl sulfate micelle under high magnetic fields, *J. Phys. Chem.,* 88, 1437, 1984.
39. **Sakaguchi, Y., Nagakura, S., Minoh, A., and Hayashi, H.,** Magnetic isotope effect upon the decay rate of the benzophenone ketyl radical in a micelle, *Chem. Phys. Lett.,* 82, 213, 1981.
40. **Sakaguchi, Y. and Hayashi, H.,** Laser-photolysis study of the photochemical processes of carbonyl compounds in micelles under high magnetic fields, *Chem. Phys. Lett.,* 87, 539, 1982.
41. **Sakaguchi, Y., Hayashi, H., and Nagakura, S.,** Laser-photolysis study of the external magnetic field effect upon the photochemical processes of carbonyl compounds in micelles, *J. Phys. Chem.,* 86, 3177, 1982.
42. **Hayashi, H. and Nagakura, S.,** The theoretical study of relaxation mechanism in magnetic field effects on chemical reactions, *Bull. Chem. Soc. Jpn.,* 57, 322, 1984.
43. **Hayashi, H.,** Research at physical organic chemistry laboratory. Recent studies of excited molecules and reaction intermediates, *Sci. Pap. Inst. Phys. Chem. Res.,* 80, 87, 1986.
44a. **Tanimoto, Y. and Itoh, M.,** Nanosecond laser photolysis study of the magnetic field effect on p-xylo-semiquinone radical in sodium dodecylsulfate micelle, *Chem. Phys. Lett.,* 83, 626, 1981.

44b. **Tanimoto, Y., Udagawa, H., and Itoh, M.,** Magnetic field effects on the primary photochemical processes of anthraquinones in SDS micelles, *J. Phys. Chem.,* 87, 724, 1983.

44c. **Tanimoto, Y., Takashima, M., and Itoh, M.,** Magnetic field effect on the hydrogen abstraction reactions of aromatic carbonyls in SDS micellar solution, *Chem. Lett.,* 1981, 1984.

44d. **Tanimoto, Y., Takashima, M., Hasegawa, K., and Itoh, M.,** Magnetic field effects on the intramolecular hydrogen abstraction reaction of polymethylene-linked systems, *Chem. Phys. Lett.,* 137, 330, 1987.

45. **Nakamura, H., Uehata, A., Motonaga, A., Ogata, T., and Matsuo, T.,** Reaction control of photoinduced electron transfer in porphyrin-viologen linked system by the use of external magnetic fields, *Chem. Lett.,* 543, 1987.

46. **Brocklehurst, B.,** Spin correlation in the geminate recombination of radical ions in hydrocarbons, *J. Chem. Soc. Faraday 2,* 72, 1869, 1976.

47. **Sakaguchi, Y. and Hayashi, H.,** Internal magnetic field effect of lanthanoid ions on the photochemical reaction of naphthoquinone in a micelle, *Chem. Phys. Lett.,* 106, 420, 1984.

48. **Okazaki, M. and Shiga, T.,** Product yield of a magnetic-field-dependent photochemical reaction modulated by electron spin resonanse, *Nature,* 323, 240, 1986.

49. **Sakaguchi, Y., Hayashi, H., Murai, H., and I'Haya, Y. J.,** CIDEP study of the photochemical reactions of carbonyl compounds showing the external magnetic field effect in a micelle, *Chem. Phys. Lett.,* 110, 275, 1984.

50. **Atkins, P. W., McLauchlan, K. A., and Simpson, A. F.,** A flash-correlated 1 μs response electron spin resonance spectrometer for flash photolysis studies, *J. Phys. E.,* 3, 547, 1970.

51. **Trifunac, A. D., Norris, J. R., and Lawler, R. G.,** Nanosecond time-resolved EPR in pulse radiolysis via the spin echo method, *J. Chem. Phys.* 71, 4380, 1979.

52. **Basu, S., McLauchlan, K. A., and Sealy, G. R.,** A novel time-resolved electron spin resonance spectrometer, *J. Phys. E.,* 16, 767, 1983.

53. **Murai, H., Hayashi, T., and I'Haya, Y. J.,** Time Resolved ESR Study of the Lowest Excited Triplet States of Para-Quinones in Glassy Matrices at 77 K, Annu. Rep. The University of Electro-Communications, Chofu, Tokyo, 1983, 24.

54. **Trifunac, A. D., Thurnauer, M. C., and Norris, J. R.,** Submicrosecond time-resolved EPR in laser photolysis, *Chem Phys. Lett.,* 57, 471, 1978.

55. **Livingston, R. and Zeldes, H.,** Paramagnetic resonance study of liquids during photolysis: hydrogen peroxide and alcohols, *J. Chem. Phys.,* 44, 1245, 1966.

56. **Sakaguchi, Y. and Hayashi, H.,** Solvent dependence of the formation of a cyclohexadienyl-type radical in the reaction of triplet xanthone with tri-*n*-butyltin hydride, *J. Phys. Chem.,* 90, 550, 1986.

57. **Scaiano, J. C.,** Solvent effects in the photochemistry of xanthone, *J. Am. Chem. Soc.,* 102, 7747, 1980.

58. **Wilson, R.,** Electron spin resonance study of radicals photolytically generated from aromatic carbonyl compounds. II. Hydroxy-radicals from acetophenone, xanthone, thioxanthone, *p*-hydroxybenzophenone, perinaphthenone and *p*-benzoquinone, *J. Chem. Soc. B,* 1581, 1968.

59. **Sakaguchi, Y., Hayashi, H., Murai, H., and I'Haya, M. J.,** Formation of cyclohexadienyl-type radicals in the photoreduction of xanthone as studied by a laser photolysis-ESR technique, *J. Am. Chem. Soc.,* 110, 7479, 1988.

60. **Scaiano, J. C.,** Intermolecular photoreductions of ketones, *J. Photochem.,* 2, 81, 1973/74.

61. **Cohen, S. G., Parola, A., and Parsons, G. H., Jr.,** Photoreduction by amines, *Chem. Rev.,* 73, 141, 1973.

62. **Sakaguchi, Y., Hayashi, H., Murai, H., I'Haya, Y. J., and Mochida, K.,** CIDEP study of the formation of a cyclohexadienyl-type radical in the hydrogen abstraction reactions of triplet xanthone, *Chem. Phys. Lett.,* 120, 401, 1985.

63. **Hayashi, H., Sakaguchi, Y., Murai, H., and I'Haya, Y. J.,** CIDEP studies of the formation of cyclohexadienyl-type radicals in the photoreduction of aromatic ketones, *J. Phys. Chem.,* 90, 4403, 1986.

64. **Sakaguchi, Y., Hayashi, H., Murai, H., and I'Haya, Y. J.,** CIDEP studies of the formation of benzyl-type radicals in the photoreduction of flavones, *J. Phys. Chem.,* 90, 6416, 1986.

65. **Hayashi, H. and Sakaguchi, Y.,** Direct observation of new intermediate radicals in photoreduction of aromatic ketones, *Studies Org. Chem.,* 31, 107, 1986.

66. **Trifunac, A. D. and Nelson, D. J.,** Chemically induced dynamic electron polarization. Pulse radiolysis of aqueous solutions of micelles, *Chem. Phys. Lett.,* 46, 346, 1977.

67. **Murai, H., Sakaguchi, Y., Hayashi, H., and I'Haya, Y. J.,** An anomalous phase effect in the individual hyperfine lines of the CIDEP spectra observed in the photochemical reactions of benzophenone in micelles, *J. Phys. Chem.,* 90, 113, 1986.

68. **McLauchlan, K. A. and Stevens, D. G.,** Comment "on alternating phase effects in individual spin multiplets in electron spin-polarized (CIDEP) free-radical spectra", *Chem. Phys. Lett.,* 115, 108, 1985.

69. **Hayashi, H., Sakaguchi, Y., Murai, H., and I'Haya, Y. J.,** Reply to comment "on alternating phase effects in individual spin multiplets in electron spin-polarized (CIDEP) free-radical spectra", *Chem. Phys. Lett.,* 115, 111, 1985.

70. **Buckley, C. D., Hunter, D. A., Hore, P. J., and McLauchlan, K. A.,** Electron spin resonance of spin-correlated radical pairs, *Chem. Phys. Lett.*, 135, 307, 1987.
71. **Closs, G. L., Forbes, M. D. E., and Norris, J. R., Jr.,** Spin polarized EPR spectra of radical pairs in micelles. Observation of electron spin spin interaction, *J. Phys. Chem.*, 91, 3592, 1987.
72. **Closs, G. L. and Forbes, M. D. E.,** Observation of medium chain length polymethylene biradicals in liquid solutions by time resolved EPR spectroscopy, *J. Am. Chem. Soc.*, 109, 6185, 1987.

Chapter 2

APPLICATION OF THE CIDNP-DETECTED LASER FLASH PHOTOLYSIS IN STUDIES OF PHOTOINITIATORS

Herbert Dreeskamp and Wolf-Ulrich Palm

TABLE OF CONTENTS

I. INTRODUCTION

While studying polymerization reactions initiated by thermal decomposition of diben-zoylperoxide, Bargon et al.[1] discovered that NMR signals may be polarized, that is, the population of the Zeeman levels of a diamagnetic reaction product deviate strongly from the thermal equilibrium. Independently Ward and Lawler[2] made a similar observation while running the reaction of *n*-butylbromide with *t*-butyllithium in the field of an NMR spectrometer. In both cases free radicals are involved and a first interpretation of this observation was a dynamic interaction between electron and nuclear spins, hence the term chemical induced dynamic nuclear polarization (CIDNP) was coined.

It soon became aparent, however, that a correct interpretation relies on a spin sorting mechanism.[3,4] Since many properties of nuclear spins in an external magnetic field can adequately be described by a vector model, a simple description of this mechanism is possible.[5] Due to the weak hyperfine interaction between the magnetic moments of electron and nuclear spins, the precession frequencies of electrons in an external magnetic field depend not only on the g-value but also on the orientation of the nuclear spins in the radicals. Thus the rate of interconversion between singlet and triplet states of radical pairs depends on the orientation of nuclear spins. A reaction of radical pairs to diamagnetic products occurs in singlet states only, excepting a few very special cases.[6] Through this selection the initial population of nuclear spin states in the products may strongly deviate from the thermal Boltzmann population and will persist for the nuclear spin lattice relaxation time T_1. By NMR methods using either continuous wave (CW) or Fourier transform (FT) techniques, this polarization may be detected.

For the sign of the net polarization of nuclei i Kaptein derived the simple rule:[7]

$$\Gamma_{Ne}(i) = \Delta g \cdot \epsilon \cdot \mu \cdot a_i$$

where Δg is the difference in g-values of the two radicals, a_i is the hyperfine constant, and by definition μ is positive for precursors in a triplet state or free radical pairs, negative for singlet state precursors, and ϵ is positive for cage products and negative for escape products. An enhanced absorption for the nuclei i is then predicted for positive $\Gamma_{Ne}(i)$ and an emission for negative $\Gamma_{Ne}(i)$. Very often the signs of three of the factors determining the sign of $\Gamma_{Ne}(i)$ are known and thus from the experimentally observed polarization the fourth can be deduced. In particular, photoreactions can be studied by irradiating the sample inside the probe of the NMR magnet.[8,9] Hence from the polarization the multiplicity of the precursor — singlet vs. triplet — can be determined. An alternative to the irradiation inside the field of the NMR spectrometer is the irradiation in an external magnetic field and fast transfer of the sample to the spectrometer.[10] In these experiments the influence of the magnetic field strength on the CIDNP effect can be studied.

The interpretation of the intensity of polarizations is a much more difficult task, since it involves (1) the dynamics of the radical pair, (2) the relaxation of electron and nuclear spins, and (3) compensation of polarization produced in separate reaction channels. A theory given by Freed and Pedersen deals with the first problem.[11]

The field of CIDNP was most recently reviewed by Muus et al.[12] and Molin[13] while the application of photo-CIDNP to study the reaction mechanism of photoinitiators was recently reviewed by Hageman.[14] Up to now mostly continuous irradiation was used.

It was soon realized that much more information might be obtained by studying time-resolved photoinitiated CIDNP signals. The chronology of processes leading to a polarization of nuclear spins is schematically depicted in Figure 1. After photoexcitation to a singlet state, intersystem crossing (ISC) into the triplet state may occur within nanoseconds. From either state dissociation into a radical pair may occur. The time of change between singlet

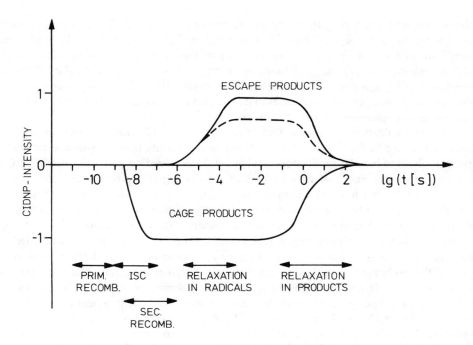

FIGURE 1.   Chronology of the processes leading to the CIDNP effect.

and triplet character of radical pairs depends on the difference in the precession frequency of the electron spins, which in turn is affected by nuclear spins; it is in the order of tens of nanoseconds. The nuclear polarization due to secondary recombination in the cage will appear during this time while that due to reaction of radicals escaping from the cage appears later. With the advent of FT NMR spectroscopy[15] it became possible to detect populations in a nuclear spin system with the time resolution of the duration of the electronic radio frequency pulse, i.e., some 5 μs for ^{1}H and some 25 μs for ^{13}C in modern spectrometers. In a fundamental paper Ernst and co-workers[16] initiated this technique and analyzed the effect of different pulse lengths on the CIDNP signals observed. In their experiments the time resolution was limited to about 1 ms by the use of a commercial pulsed photo flash lamp. They further showed that with pulsed photoexcitation and pulsed detection of the spin system it became possible to presaturate the spin system and hence to observe the pure polarization spectra due to the CIDNP effect undisturbed by the signals of the system in thermal equilibrium.

II. LASERS APPLIED TO TIME-RESOLVED PHOTO-CIDNP

An increase in time resolution by three orders of magnitude from milliseconds to microseconds was readily achieved with pulsed lasers as photoexcitation source.[17-19] Particularly useful appear to be excimer lasers with several emission wavelengths in the UV region where photoinitiators generally absorb. Since the pulse length of these lasers is of the order of 10 ns, the time resolution of the experiment will completely be determined by the detection pulse of the FT-NMR spectrometer. Furthermore, the highly collimated beam of a laser compared to that of a conventional flash lamp simplifies the problem of focusing the light beam into the probe which is situated in the magnet of the NMR spectrometer.

To increase the time resolution even further two possibilities have been realized: (1) Miller and Closs,[20] by adding a power amplifier in the transmitter line, increased the field acting on the nuclei and thus reduced the width of a 10° pulse to 0.12 μs. A diode switching of the resonator was necessary to reduce a distortion of the pulse. (2) By numerical methods, a deconvolution of the signal observed with the NMR pulse applied, Closs et al.[21] and Läufer[22] showed that a significant improvement in time resolution may be obtained. By applying both methods (1) and (2) a time resolution of 25 ns has been reached by Closs and co-workers.[21] With this nanosecond time-resolved photo-CIDNP it is even possible to detect the lifetime of primary excited molecules in their triplet states as shown for cyclic ketones[23,24] or anthraquinone quenched by dimethylaniline.[22]

In a first application of microsecond time-resolved CIDNP using a laser for photoexcitation, Closs and Miller[17] were able to differentiate geminate from free radical processes. In the photolysis of benzylphenylketone the benzyl/ benzoyl radical pair is formed in a triplet state. From the appearance of polarizations of the diamagnetic products as a function of time it was established that, for example, the starting material is formed promptly (<1 μs) while dibenzyl is formed in a recombination of freely diffusing benzyl radicals on the time scale of some 10 μs. The same technique was subsequently applied by Läufer and Dreeskamp[25] to study the kinetics of photoinitiated radicals of several benzylketones. In a detailed experimental and theoretical investigation Fischer and co-workers[26] used the time-resolved photo-CIDNP technique to study the photolysis of t-butylketones. They clarified details of the reaction mechanism and determined values of the polarization and the efficiency of spin relaxation in excellent agreement with the Freed-Pedersen high field radical theory. As first shown by Turro et al.[27] with time-resolved CIDNP, the kinetics of benzyl radicals formed by photolysis of bibenzylketone in micelles is strongly influenced by the supercage of the micelle and the exit rate of benzyl radicals into the surrounding aqueous phase can be determined.

Perhaps the most interesting application of time-resolved CIDNP is to cyclic radical reactions with no net chemical change.[28,29] This occurs, for example, in the quenching of an excited molecule by electron transfer to produce an intermediate radical ion pair which reacts by reverse electron transfer back to the starting material. Except for some relaxation of nuclear spins in the intermediate radicals, a complete cancellation of polarization is expected and no CIDNP effect in a continuous illumination is observed. If, however, self-exchange of the electron occurs, the polarization is transfered to a diamagnetic molecule. This polarization can then be measured and thus rate constants of self-exchange can be measured directly. Photoexcitation of chlorophylles by quinones[28] or photoreduction of anthraquinone by amines[22] were studied in this way. The topic of time-resolved photo-CIDNP was reviewed by Closs et al.[21]

III. THE EXPERIMENT

To illustrate the technique of time-resolved photo-CIDNP a detailed description of our experimental setup will be given and the photolysis of benzoin taken as an example. The equipment consists essentially of an XeCl-Excimerlaser (Lambda Physics, type EMG 101, emission at 308 nm with a pulse energy of 100 to 150 mJ) and of a superconducting FT-NMR spectrometer (Bruker, WM 250, ^{1}H at 250 MHz). The laser beam is directed to the bottom of the magnet by three mirrors and three quartz lenses focusing it on the end of a suprasil quartz rod, fixed in a modified ^{1}H-probe. The whole light pathlength to the quartz rod is about 5.4 m, the length of the quartz rod is 32 cm, and its diameter 5 mm (see Figure 2). At the upper end of the quartz rod a quartz prism reflects the light perpendicular to the NMR tube axis through the receiver coils so that the size of the light spot on the NMR tube is some 5×6 mm. The pulse energy reaching the sample is about 10% of the laser pulse energy. The fluctuations in pulse energies of the laser are $\sim \pm 5\%$. The perpendicular irradiation chosen here allows to use samples of varying optical density. In the alternative design of superconducting magnet systems with axial irradiation either from the bottom[30]

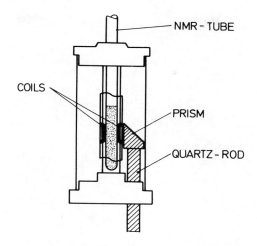

FIGURE 2.   Modified NMR probe for *in situ* irradiation.

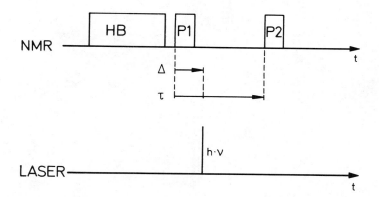

FIGURE 3.   Schematic diagram of laser and NMR pulses.

or from the top,[31] the optical density critically affects the amount of light absorbed in the effective volume of the probe. Three methods can be used for recording the spectra:

1. Steady-state spectra. Irradiation of the sample with a laser frequency of 5 to 10 Hz during the NMR measurement delivers the CIDNP spectrum, superimposed by the spectrum of molecules in thermal equilibrium.
2. Difference spectra. The pure CIDNP lines can be obtained by substraction of spectra with and without illumination.
3. Time-resolved spectra. CIDNP spectra time resolved in the microsecond region are obtained as follows (Figure 3): after saturation of the ^{1}H NMR transitions with a broadband pulse (HB) of typically 3-s length, a triggerpulse (P1) of 0.7-μs length will be given through an external exit of the NMR console to the laser. The laser pulse is emitted with a delay of $\Delta = 1.65$ μs. Between the trigger pulse P1 and the NMR excitation pulse P2 (length 1 μs corresponding to a pulse angle of ~21°) a variable delay is inserted which may be incremented in steps of 0.1 μs. The high-frequency pulse P2 rotates the nuclear magnetization vector out of the direction of the static magnetic field, the free precision of this vector is recorded during an acquisition time T and after FT gives the NMR spectrum. Hence this spectrum reflects the population of nuclear Zeeman levels at the time of the NMR-excitation pulse P2.

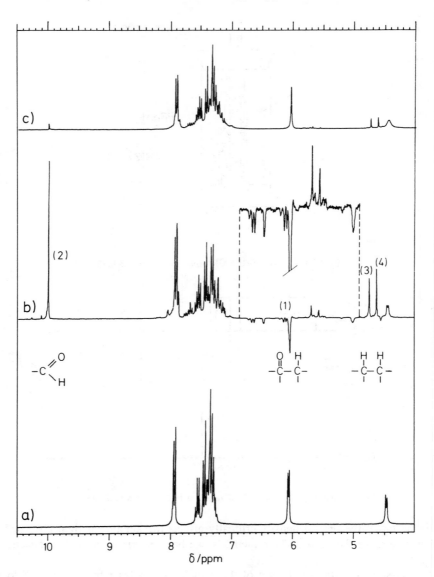

FIGURE 4.   ¹H-NMR spectrum $7 \cdot 10^{-2}$ *M* (deaerated) benzoin in acetonitrile-d3 (δ = 1.93 ppm) solution (a) before irradiation, (b) during irradiation with 8-Hz pulses of a XeCl excimer laser, and (c) after irradiation.

The photolysis of benzoin is taken as an example to illustrate the method.[32-34] Figure 4 shows (a) the spectrum of a $7 \cdot 10^{-2}$ *M* (deaerated) solution of benzoin (1) in acetonitrile-d3, (b) the steady-state CIDNP spectrum, and (c) a spectrum after ~30% conversion of (1). Typical parameters for these spectra are acquisition time T = 1.3 s; number of scans = 32; laser pulse frequency = 8 Hz; pulse width = 2 μs (corresponding to a pulse angle of ~42°). The polarizations and products observed can be explained according to the scheme given in Figure 5. The line at 10.00 ppm is assigned to benzaldehyde (2) and those at 4.75 ppm to meso-1,2-diphenyl-1,2-ethanediol (3) and at 4.62 ppm to d,l-1,2-diphenyl-1,2-etha-nediol (4) (in a proportion 1:1) as shown by addition of authentic substances. Beside these products benzil (5) from the recombination of two radicals (1a) will be formed,[35] as verified by high-performance liquid chromatography spectroscopy. The product distribution is strongly

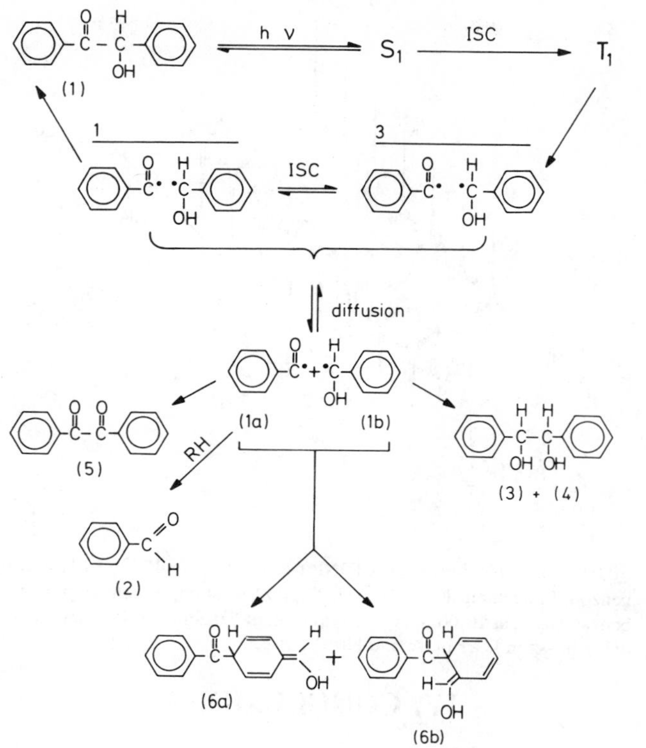

FIGURE 5.   Reaction scheme of the photolysis of benzoin (solvent: acetonitrile-d3).

solvent dependent.[36] Moreover the formation of semibenzenes (6a) and (6b) is possible and from the spectral pattern we assign the minor but well-resolved signals in the region between 6.9 and 4.9 ppm to the semibenzene (6a). The resonances at 4.48 and 6.05 ppm are due to the -OH and benzyl protons of (1), respectively. The loss of spin-spin splitting of these signals with prolonged irradiation is due to an intermolecular proton exchange caused by traces of acid formed in the photolysis.

Formation of the two radicals (1a) and (1b) has been assumed for the first time by Kornis and de Mayo.[37] From the product distribution, ESR,[38] CIDNP,[39] and radical trap[40] investigations there is no doubt of the formation of (1a) and (1b) as a result of a Norrish type I cleavage. From the polarizations in the CIDNP spectrum and the Kaptein rule given above the precursor multiplicity of the radical pair (1a) and (1b) can be obtained. The benzyl proton in (1) is negatively polarized ($\Gamma_{Ne} = \ominus$). The hyperfine constant of this proton in the radical (1b) is negative[41] ($a_i = \ominus$), $g(1b)^{42}$-$g(1a)^{42} = \oplus$ and (1) is a cage product ($\epsilon = \oplus$). Therefore $\mu = \oplus$, i.e., the precursor to the radical pair is a triplet state. As (3) and (4) are escape products ($\epsilon = \ominus$), a positive polarization is observed as expected. Figure 6 shows an example of a time-resolved spectrum ($7 \cdot 10^{-3}$ M (deaerated) benzoin solution in acetonitile-d3) measured with the parameters given above. The polarization of the aldehyde proton of benzaldehyde increases with time while the negative polarization of the benzyl proton of benzoin is created within the time resolution of the experiment. The aldehyde proton cannot derive from the solvent[43] which is perdeuterated. The precursor of benzaldehyde must be a freely diffusing radical abstracting a hydrogen. This is supported by the observation that the time constant of increase of polarization is strongly concentration dependent.

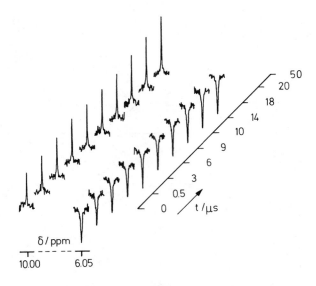

FIGURE 6.   Time-resolved CIDNP spectrum of $7 \cdot 10^{-3}\ M$ (deaerated) benzoin in acetonitrile-d3 ($\delta = 1.93$ ppm) solution. Aldehyde proton of benzaldehyde at 10.00 ppm, benzylproton of benzoin at 6.05 ppm. t is the delay between laser pulse and NMR pulse.

IV. CONCLUSIONS

The combination of a pulsed light source with a pulsed FT-NMR detection of CIDNP signals represents an alternative method to the laser flash spectroscopy with light absorption or emission to monitor transients. Application of lasers in time-resolved CIDNP spectroscopy of photochemical reactions in general and of those of photoinitiators in particular offers some distinct advantages but is subject to some limitations compared to conventional flash spectroscopy. Since the detection uses transitions between nuclear Zeeman levels in an external field, both sensitivity and time resolution are low compared to optical detection methods. In contrast to other spectroscopic methods, however, the possibility exists in CIDNP spectroscopy to eliminate all background signals by presaturation. Since the polarization is detected in the NMR spectrum of diamagnetic reaction products, the analytical information on the reaction is extremely high. The sign and intensity of polarization gives important information on the reaction mechanism, e.g., singlet vs. triplet precursor, geminate vs. escape products, or properties of intermediate free radicals. Time-resolved CIDNP spectra give information on the kinetics of fast radical reactions, but due to low sensitivity of the NMR experiment the precision of kinetic data may be low. In the CIDNP experiment molecular species are labeled in a unique way by spin polarization of their nuclei. Thus it is possible to study reactions leading to no permanent products such as, for example, the recombination to starting material, the degenerate electron exchange, or the formation of unstable primary products which rearrange to the ultimate products.

Support of this work by the Deutsche Forschungsgemeinschaft and Fond der Chemie is gratefully acknowledged.

REFERENCES

1. **Bargon, J., Fischer, H., and Johnsen, U.,** Kernresonanz-Emissionslinien während rascher Radikalreaktionen, *Z. Naturforsch.,* 22a, 1551, 1967.
2. **Ward, H. R. and Lawler, R. G.,** Nuclear magnetic resonance emission and enhanced absorption in rapid organometallic reactions, *J. Am. Chem. Soc.,* 89, 5518, 1967.
3. **Closs, G. L.,** A mechanism explaining nuclear spin polarizations in radical combination reactions, *J. Am. Chem. Soc.,* 91, 4552, 1969.
4. **Kaptein, R. and Oosterhoff, J. L.,** Chemically induced dynamic nuclear polarization. II, *Chem. Phys. Lett.,* 4, 195, 1969.
5. **McLauchlan, K. A.,** The effect of magnetic fields on chemical reactions, *Sci. Prog., Oxf.,* 67, 509, 1981.
6. **Closs, G. L. and Czeropski, M. S.,** Amendment of the CIDNP phase rules. Radical pairs leading to triplet states, *J. Am. Chem. Soc.,* 99, 6127, 1977.
7. **Kaptein, R.,** Simple rules for chemical induced dynamic nuclear polarization, *Chem. Commun.,* 732, 1971.
8. **Cocivera, M.,** Optically induced Overhauser effect in solution. Nuclear magnetic resonance emission, *J. Am. Chem. Soc.,* 90, 3261, 1968.
9. **Closs, G. L. and Closs, L. E.,** Induced dynamic nuclear spin polarization in reactions of photochemically and thermally generated triplet diphenylmethylene, *J. Am. Chem. Soc.,* 91, 4549, 1969.
10. **Lehnig, M. and Fischer, H.,** Chemisch induzierte Kernpolarisation von Phenylhalogeniden bei der Photolyse von Aroylperoxiden, *Z. Naturforsch.,* 24a, 1771, 1969.
11. **Freed, J. H. and Pedersen, J. B.,** The theory of chemically induced dynamic spin polarization, *Adv. Magn. Reson.,* 8, 1, 1976.
12. **Muus, L. T., Atkins, P. W., McLauchlan, K. A., and Pedersen, J. B., Eds.,** *Chemically Induced Magnetic Polarization,* D. Reidel, Dordrecht, Netherlands, 1977.
13. **Molin, Yu. N., Ed.,** *Spin Polarization and Magnetic Effects in Radical Reactions,* Elsevier, Amsterdam, 1984.
14. **Hageman, H. J.,** Photoinitiators for free radical polymerisation, *Prog. Org. Coat.,* 13, 123, 1985.
15. **Ernst, R. R., Bodenhausen, G., and Wokaun, A.,** *Principles of Nuclear Magnetic Resonance in One and Two Dimensions,* (Int. Ser. Monogr. Chem., Vol. 14), Clarendon Press, Oxford, 1987, chap. 1.
16. **Schäublin, S., Wokaun, A., and Ernst, R. R.,** Pulse techniques applied to chemically induced dynamic nuclear polarization, *J. Magn. Reson.,* 27, 273, 1977.
17. **Closs, G. L. and Miller, R. J.,** Laser flash photolysis with NMR detection. Microsecond time-resolved CIDNP: separation of geminate and random-phase processes, *J. Am. Chem. Soc.,* 101, 1639, 1979.
18. **Lerman, C. L. and Cohn, M.,** Laser photo-CIDNP NMR of pyridoxal phosphate and pyridoxyllysine residues: an extrinsic probe for nonaromatic amino acid residues in proteins, *Biochem. Biophys. Res. Commun.,* 97, 121, 1980.
19. **McCord, E. F., Bucks, R. R., and Boxer, S. G.,** Laser chemically induced dynamic nuclear polarization study of the reaction between photoexcited flavins and tryptophan derivatives at 360 MHz, *Biochemistry,* 20, 2880, 1981.
20. **Miller, R. J. and Closs, G. L.,** Application of Fourier transform-NMR spectroscopy to submicrosecond time-resolved detection in laser photolysis experiments, *Rev. Sci. Instrum.,* 52, 1876, 1981.
21. **Closs, G. L., Miller, R. J., and Redwine, O. D.,** Time resolved CIDNP: applications to radical and biradical chemistry, *Acc. Chem. Res.,* 18, 196, 1985.
22. **Läufer, M.,** Increasing the time resolution of flash CIDNP by numerical analysis: photoreduction of anthraquinone by *N,N*-dimethylaniline, *Chem. Phys. Lett.,* 127, 136, 1986.
23. **Closs, G. L. and Miller, R. J.,** Laser flash photolysis with NMR detection. Submicrosecond time-resolved CIDNP: kinetics of triplet states and biradicals, *J. Am. Chem. Soc.,* 103, 3586, 1981.
24. **Closs, G. L. and Redwine, O. D.,** Cyclization and disproportionation kinetics of triplet generated, medium chain length, localized biradicals measured by time resolved CIDNP, *J. Am. Chem. Soc.,* 107, 4543, 1985.
25. **Läufer, M. and Dreeskamp, H.,** The CIDNP-detected laser-flash photolysis of benzyl ketones, *J. Magn. Reson.,* 60, 357, 1984.
26. **Vollenweider, J.-K., Fischer, H., Henning, J., and Leuschner, R.,** Time resolved CIDNP in laser flash photolysis of aliphatic ketones. A quantitative analysis, *Chem. Phys.,* 97, 217, 1985.
27. **Turro, N. J., Zimmt, M. B., and Gould, I., R.,** Dynamics of micellized radical pairs. Measurement of micellar exit rates of benzylic radicals by time-resolved flash CIDNP and optical spectroscopy, *J. Am. Chem. Soc.,* 105, 6347, 1983.
28. **Closs, G. L. and Sitzmann, E. V.,** Measurement of degenerate radical ion-neutral molecule electron exchange by microsecond time-resolved CIDNP. Determination of relative coupling constant of radical cations of chlorophylls and derivatives, *J. Am. Chem. Soc.,* 103, 3217, 1981.
29. **Hore, P. J., Zuiderweg, E. R. P., Kaptein, R., and Dijkstra, K.,** Flash photolysis NMR CIDNP time dependence in cyclic photochemical reactions, *Chem. Phys. Lett.,* 83, 376, 1981.

30. **Kaptein, R.,** Structural information from photo-CIDNP in proteins, in *Nuclear Magnetic Resonance Spectroscopy in Molecular Biology*, Pullman, B., Ed., D. Reidel, Dordrecht, Netherlands, 1978, 211.

31. **Scheffler, J. E., Cottrell, C. E., and Berliner, L. J.,** An inexpensive, versatile sample illuminator for photo-CIDNP on any NMR spectrometer, *J. Magn. Reson.*, 63, 199, 1985.

32. **Cocivera, M. and Trozzolo, A. M.,** Photolysis of benzaldehyde in solution studied by nuclear magnetic resonance spectroscopy, *J. Am. Chem. Soc.*, 92, 1772, 1970.

33. **Closs, G. L. and Paulson, D. R.,** Application of the radical theory of chemically induced dynamic nuclear spin polarization (CIDNP) to photochemical reactions of aromatic aldehydes and ketones, *J. Am. Chem. Soc.*, 92, 7229, 1970.

34. **Hutton, R. S., Roth, H. D., and Bertz, S. H.,** Nuclear spin polarization effects in systems with large hyperfine couplings. Limitations of Kaptein's rules, *J. Am. Chem. Soc.*, 105, 6371, 1983.

35. **Baumann, H., Müller, U., Pfeifer, D., and Timpe, H.-J.,** Photoinduzierte Zersetzung von Aryldiazoniumsalzen durch Benzoinderivate, *J. Prakt. Chem.*, 324, 217, 1982.

36. **Lewis, F. D., Lauterbach, R. T., Heine, H.-G., Hartmann, W., and Rudolph, H.,** Photochemical α-cleavage of benzoin derivatives. Polar transition states for free-radical formation, *J. Am. Chem. Soc.*, 97, 1519, 1975.

37. **Kornis, G. and DeMayo, P.,** Photochemical syntheses. IX. The conversion of dibenzoylmethane to tribenzoylethane, *Can. J. Chem.*, 42, 2822, 1964.

38. **Adam, S., Güsten, H., Steenken, S., and Schulte-Frohlinde, D.,** Photolyse von Benzoinalkyläthern in sauerstofffreien Lösungen, *Liebigs. Ann. Chem.*, 1831, 1974.

39. **Borer, A., Kirchmayer, R., and Rist, G.,** CIDNP-investigation of photoinduced polymerisation, *Helv. Chim. Acta*, 61, 305, 1978.

40. **Hagemann, H. J. and Overeem, T.,** Trapping of primary radicals from photoinitiators by 2,2,6,6-tetramethylpiperidinoxyl, *Makrom. Chem., Rapid Commun.*, 2, 719, 1981.

41. **Closs, G. L.,** CIDNP in reactions of carbenes, azocompounds and photoexcited carbonyl compounds, in *Chemically Induced Magnetic Polarization*, Lepley, A. R. and Closs, G. L., Eds., John Wiley & Sons, New York, 1973, 120.

42. **Paul, H. and Fischer, H.,** Elektronenspinresonanz freier Radikale bei photochemischen Reaktionen in Lösung, *Helv. Chim. Acta*, 56, 1575, 1973.

43. **Bak, C., Praefcke, K., Muszkat, K. A., and Weinstein, M.,** CIDNP studies of hydrogen abstraction by aroyl radicals in Norrish Type I processes, *Z. Naturforsch.*, 32b, 674, 1977.

Chapter 3

STUDIES OF RADICALS AND BIRADICALS IN POLYMERS AND MODEL COMPOUNDS BY NANOSECOND LASER FLASH PHOTOLYSIS AND TRANSIENT ABSORPTION SPECTROSCOPY

Paritosh K. Das

TABLE OF CONTENTS

ABSTRACT

In recent years, nanosecond laser flash photolysis coupled with kinetic absorption spectrophotometry has been extensively applied in mechanistic studies of radical-related photoprocesses in polymers and model compounds. Representative results concerning transient spectra and reactivity of biradical and radical intermediates relevant to photochemistry of simple compounds and photodegradation/stabilization of polymers will be reviewed.

I. INTRODUCTION AND SCOPE

Thermally or photochemically generated radicals and radical adducts are integrally involved in various stages of free-radical polymerization processes. Less obvious, and often recognized only through detailed mechanistic photochemical and photophysical studies, is the fact that radicals and biradicals play important roles as intermediaries in photostabilization, photo-oxidation, and photodegration phenomena of polymeric systems.[1-5] Not surprisingly, during the past decade, time-resolved emission and absorption techniques covering pico- to millisecond time domains have been widely used to obtain intimate information on kinetics and mechanisms of photolytic bond scission processes in polymers. In particular, the transient absorption spectroscopy following photoexcitation by laser pulses has led to the identification of short-lived radical and biradical intermediates as well as their excited-state precursors, and to the measurement of absolute rates of elementary reactions involving these species. These results, combined with steady-state quantum yield and quenching studies, have provided strong bases for understanding the stabilization and degradation behavior of polymers in relationship to light.

The photoinduced formation of radicals and biradicals in polymers is best exemplified by Norrish type I and type II processes leading to photodegradation in a carbonyl containing polymer. These are shown in Scheme I. In fact, much of the time-resolved work based on laser flash photolysis has been confined to carbonyl-containing polymers; this led to the observation of acyl-type radicals and 1,4-biradicals and made possible the measurement of absolute kinetics of reactions involving them.

The material presented in this chapter has been gleaned primarily from the literature of the past decade. It is limited to transient absorption phenomena in fluid solutions at nanosecond and longer time scales following laser pulse excitation and is meant to be representative rather than exhaustive. The behavior or radicals and biradicals derivable from polymers and a limited number of related model compounds has been covered. No attempt has been made to include such topics as application of polymeric environments to generate and stabilize radicals and radical ions (via photoinduced redox reactions), charge separation in biologically significant polymers/aggregates, radical-pair dynamics in polymers, and long-lived or delayed emission process related to triplet decay or triplet-triplet annihilation of polymer-bound chromophores and of molecules hosted in solid polymeric matrices. However, for several relevant systems, the involvement of triplet states in the processes of bond homolysis and intramacromolecular energy migration, as studied by nanosecond laser spectroscopy, has been briefly discussed.

II. NANOSECOND LASER FLASH PHOTOLYSIS TECHNIQUE

The instrumentation for laser flash photolytic generation of transient species and subsequent detection of the latter based on kinetic absorption spectrophotometry is now well developed. Descriptions of nanosecond laser flash photolysis systems that are in use in various laboratories are available in the literature.[6-13] The basic features of these systems are the following: pulsed laser source, analyzing light source (usually pulsed) and associated

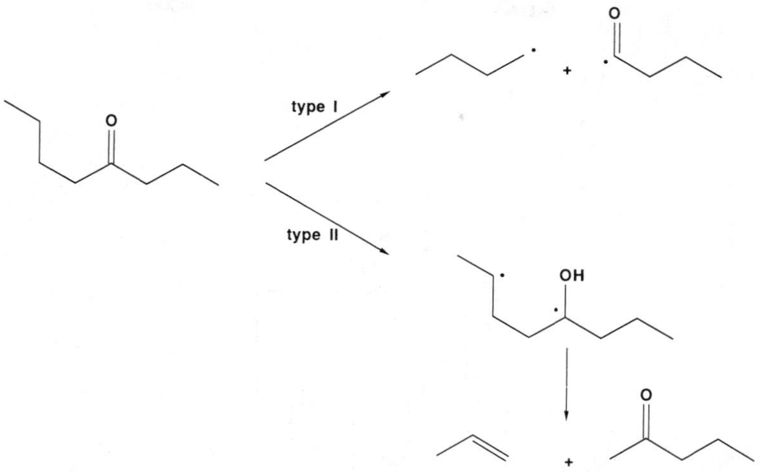

SCHEME I.   Photodegradation in polymers via Norrish type I and II photoprocesses.

optics and power supply, detector (monochromator, and photomultiplier tube with fast response), fast digitizer, computer and accessory equipments for data display, processing and storage, as well as for control of experiments and sequencing of events. Almost invariably, the same optical setup permits time-resolved measurements of both emission and absorption.

The schematic of one of the multiuser laser flash photolysis systems currently operational at the Radiation Laboratory is shown in Figure 1. The pulsed excitation is effected by two gas laser sources, namely, Molectron UV-400 (nitrogen, 337.1 nm, 2 to 3 mJ per pulse, ~8 ns) and Lambda-Physik EMG 101 MSC (inert gas/halogen mixtures, 10 to 120 mJ per pulse, ~20 ns); the two sources are often used in tandem for experiments in which a transient species produced by the first laser pulse is photoexcited by a second laser pulse (temporally separated from the first by nanosecond-to-millisecond intervals). The kinetic spectrophotometer consists of a 1000-W Xe lamp (pulsed over ~3 ms during experiment), the output of which is passed through an IR filter (H_2O/D_2O) and then focused onto the pinhole on one side of the cell holder. After passage through the solution under examination contained in a cell (usually rectangular) in the holder, the analyzing light is focused onto the entrance slit of a double monochromator (Instruments, SA). The spectrally resolved light emerging from the exit slit is directed toward a RCA 666S photomultiplier tube (PMT) subjected to variable high voltage (450 to 650 V) through six or seven stages of dynodes. The signal from the PMT is fed into a back-off circuit unit[15] to separate the transient component, $I_s(t)$, from the steady-state one, I_o. The I_o and $I_s(t)$ signals are read and processed by a sample-and-hold unit and a Tektronix 7912AD digitizer, respectively, which in turn are interfaced with an LSI-11 microprocessor. In addition to processing data and allowing routine kinetic fits, the microprocessor controls the experiment by timing various events, namely, pulsing the xenon lamp, opening the electromechanical shutters for the analyzing light beam and the laser pulses, and triggering the lasers, the sample-and-hold unit, and the digitizer. For storage and future use for involved kinetic fits, the data are transferred to a time-shared, multiuser PDP 11/55 or VAX computer system.

In a typical experiment, solutions of absorbances 0.01 to 1.0 at the wavelength and along the direction of exciting laser pulse are used in rectangular quartz cells. The geometry of configuration between the excitation laser pulse and the analyzing light is either right angle or front face; in the latter, the angle between the directions of the monitoring light and the laser pulse is usually 15 to 20°. The front-face geometry is particularly suited for

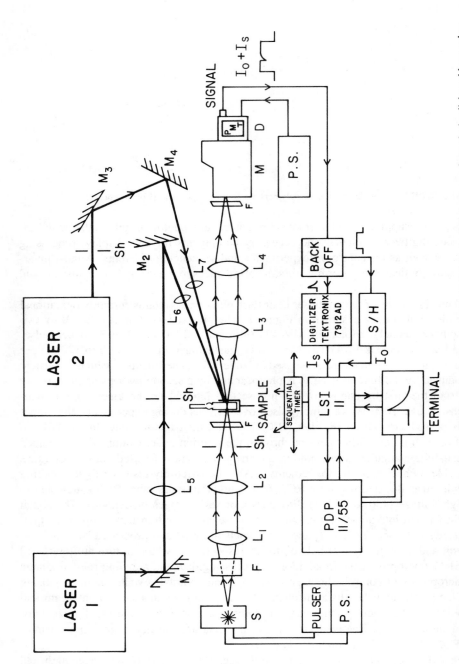

FIGURE 1. Schematic diagram for a laser flash photolysis setup. F: filters in analyzing light path, L_1 to L_7: lenses for analyzing light and laser pulses, LASER 1,2: pulsed laser sources, M: monochromator, M_1 to M_4: mirrors for laser pulses, P.S.: Power supplies for lamp and PMT, PMT: photomultiplier tube, PULSER: pulser for monitoring lamp, S: analyzing light source, S/H: sample and hold for reading I_o, and Sh: electromechanical shutters for analyzing light and laser sources.

emission measurements and for experiments with solutions of high ground-state absorbances, although the problem of laser-induced scattered light is more serious in this configuration. The absorbance changes [$\Delta A(t)$] due to a laser-flash-photolytic transient species are calculated as follows:

$$\Delta A(t) = \ell(\epsilon^* - \epsilon^G) \, C^*(t) \qquad (1a)$$

$$\Delta A(t) = -\log\left[1 - \frac{I_s(t)}{I_o} \right] \qquad (1b)$$

In Equation 1a, ℓ is the cell path length, C^* is the concentration of transient, and ϵ^* and ϵ^G are the extinction coefficient of transient and substrate (ground state), respectively, at the monitoring wavelength. For point-by-point transient spectra, the recommended practice is to use a flow cell with arrangements for continuously draining and replenishing the solutions. For kinetics measurements that do not require too many laser shots, static cells are occasionally used, although photodegradation of substrates and accumulation of photoproducts can easily lead to spurious results, especially in experiments with polymers.

III. PHOTOGENERATED BIRADICALS

With the advent of laser flash photolysis in the 1970s, there has been an upsurge of interest in biradicals derived from intramolecular hydrogen abstraction in carbonyl compounds (Norrish type II photoreaction).[16-21] A large volume of data is now available concerning absolute rates of decay and of bimolecular reactions of photogenerated biradicals of various types, e.g., Norrish type I and II classes, Paterno-Buchi biradicals, and trimethylenemethane biradicals. Several studies have also addressed the behavior of Norrish type II macrobiradicals produced in carbonyl-containing polymers. In fluid solutions, the biradicals are either observed directly in terms of transient absorption or implicated indirectly form anticipated reactions with a trapping agent. A common example of the latter is the electron transfer from the ketyl radical site of a type II biradical to methyl viologen (paraquat, PQ^{+2}); the spectrophotometric observation of resultant reduced viologen radical ($PQ^{+\cdot}$) has proved to be very useful in measuring kinetics associated with this class of biradical.[18-26]

A. BIRADICALS IN MODEL COMPOUNDS

Kinetics data on biradicals from relatively simple compounds, obtained by the laser flash photolysis technique, are now fairly extensive and have formed the subject matter of several recent reviews.[18-21] Because of this, as well as of the fact that nearly all of the laser flash photolysis studies concerning radicals and biradicals in polymers have focused on systems containing the carbonyl chromophore, we have restricted this section to only representative biradicals derived from Norrish type I and II photoreactions. The details of these processes are shown in Schemes II to IV.

Transient absorption spectra assignable to biradicals have been observed in the course of many of the laser photolysis studies on appropriate carbonyl substrates. By following the kinetics of decay of the absorptions in the presence as well as absence of quenchers, it has been possible to measure the rates of intra- and intermolecular reactions of biradicals. Quite often, use has been made of a given trapping reaction of a biradical (e.g., electron transfer to paraquat, see Equation 2) and the product of this reaction (e.g., $PQ^{+\cdot}$) has been monitored by transient absorption spectroscopy.

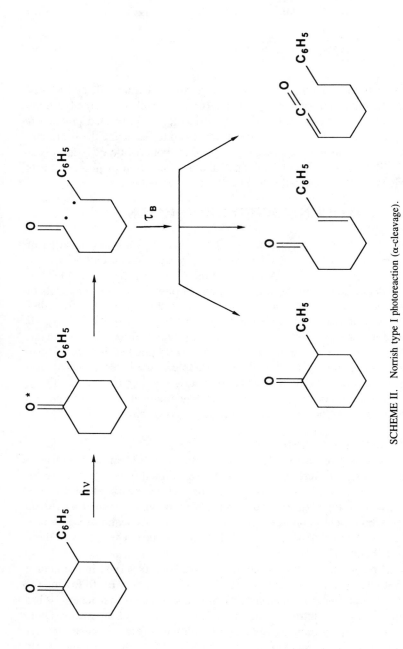

SCHEME II.   Norrish type I photoreaction (α-cleavage).

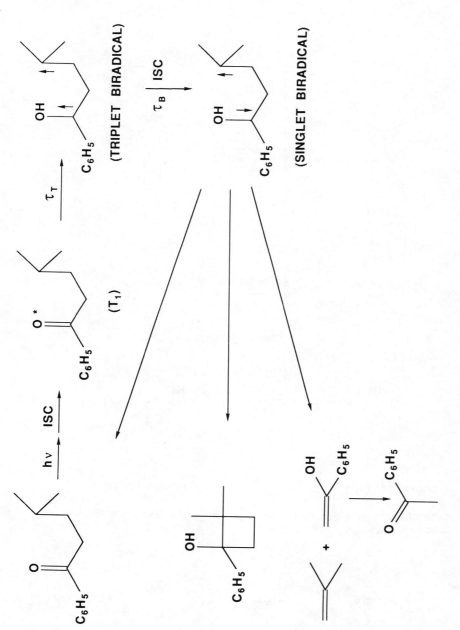

SCHEME III. Norrish type II photoreaction (γ-hydrogen abstraction) from a carbonyl triplet.

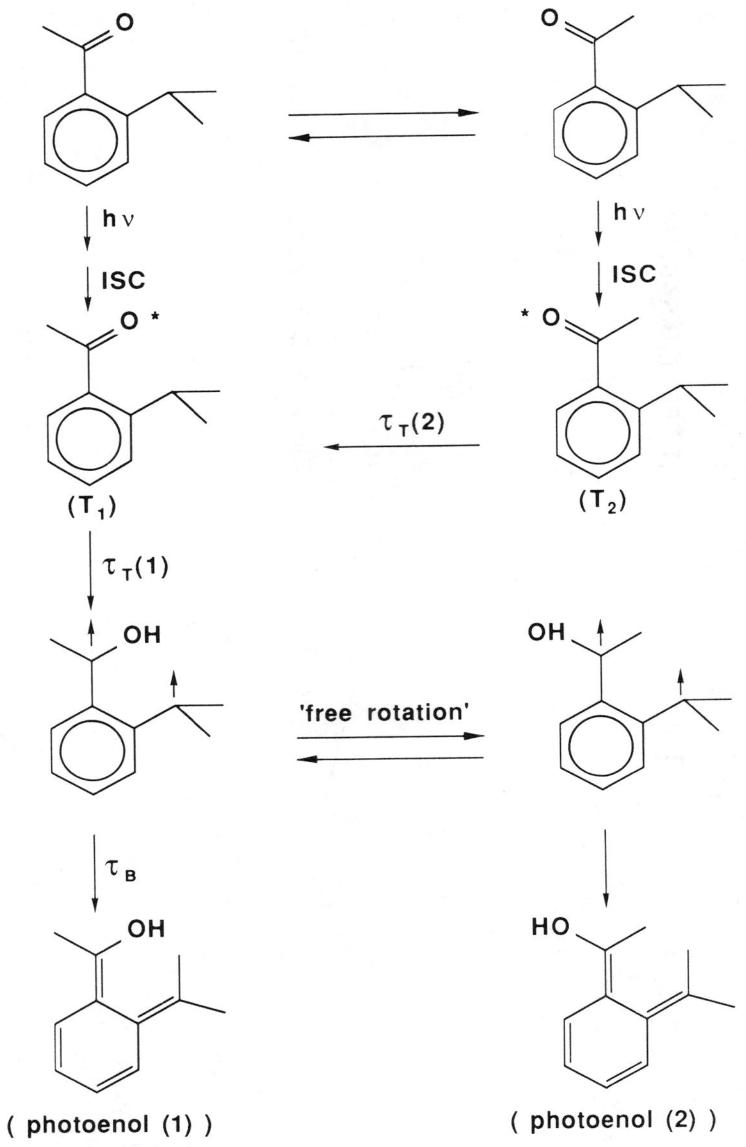

SCHEME IV.   Norrish type II photoreaction (photoenolization) from an *o*-alkyl-substituted aromatic ketone triplet.

$$(2)$$

The common practice has been to measure the pseudo-first-order rate constants (k_{obsd}) for the formation of the product, monitored at or near its absorption maxima (~390 and ~615 nm for PQ$^{+\cdot}$)), as a function of trapping agent concentration ([Q]). The intercept and slope of the linear plot of k_{obsd} against [Q] give the unimolecular decay rate constant ($1/\tau_B$) of the biradical and the rate constant (k_r) of its reaction with the trapping agent (see Equation

3). The deviation of biradical decay or product formation kinetics from a single exponential behavior is usually indicative of the multiplicity of reacting species in a given photoreaction.

$$k_{obsd} = \tau_B^{-1} + k_r[Q] \tag{3}$$

In Table 1 we have compiled a representative cross-section of kinetic information on biradicals photogenerated from simple carbonyl systems.

Based on the literature surveyed in order to prepare Table 1, the following general observations can be made regarding biradicals derived from carbonyl compounds. (1) For most systems, these transients display broad, featureless absorptions at 300 to 450 nm. For example, the absorption spectrum of the Norrish type II biradical from γ-methylvalerophenone exhibits a maximum at 415 nm in methanol (τ_B = 98 ns, ϵ_{max} = 800 M^{-1} cm^{-1}).[45] The spectra of Paterno-Buchi 1,4-biradical from benzhydryl phenacyl ether,[41] conformationally locked Norrish II biradical from *cis*-1-benzoyl-2-benzhydrylcyclohexane,[32] and Norrish I 1,6-biradical from α-benzhydrylcyclohexanone[40] are characterized by maxima at 325 to 335 mn in methanol and heptane; these are primarily attributable to the radical center at the benzhydryl terminus. (2) Thanks to efficient intersystem crossing in carbonyl compounds, biradicals are produced via short-lived triplet states and, hence, may be considered to be in the triplet configuration following its formation. The involvement of triplet precursors is established by their direct observation as well as by indirect quenching studies using dienes and naphthalenes. (3) The general mechanism of decay of triplet biradicals (BR) appears to be as follows: ^{3}BR → ^{1}BR → product(s). Electronic, conformational, and spatial factors affect the spin-orbit coupling mechanism of intersystem crossing in triplet biradicals. (4) The lifetimes of biradicals vary over a wide range (nanoseconds to microseconds, see Table 1) and are usually subject to solvent effects. Polar solvents that can engage in hydrogen bonding with the OH group stabilize Norrish type II biradicals. (5) Biradicals containing the ketyl radical site are excellent electron donors to viologens. The bimolecular rate constants for reactions with paraquat in methanol and aqueous acetonitrile are in the limit of diffusion control (Table 1). (6) Paramagnetic species, namely, oxygen, nitroxy radicals, and copper acetylacetonate complex, are efficient quenchers of biradicals (Table 1) and may significantly affect their relative partitioning into various products (Schemes II to IV). (7) The biradicals from photoenolizable *o*-alkyl-substituted aromatic carbonyl compounds may be viewed as the triplet photoexcited states of the photoenols; evidence for this is obtained from triplet energy transfer (Equation 4) to long-chain polyenes (carotenoids).[34] (8) Hydrogen abstraction (Equation 5) by type II biradicals from good hydrogen donors, e.g., 1-octane-thiol and tri-*n*-butylstannane, can occur efficiently (e.g., k_r = 10^6 to 10^7 M^{-1} s^{-1} for biradicals from valerophenone and γ-methylvalerophenone in benzene with 1.2 M pyridine).[46]

$$\tag{4}$$

$$\tag{5}$$

TABLE 1

Kinetic Behavior of Representative Biradicals Photogenerated from Carbonyl Compounds in Solution

Source compound	Laser wavelength (nm)	Solvent	Temp (K)	τ_B (μs)	τ_T (ns)	k_r (10^9 M^{-1} s^{-1}) PQ^{+2}	k_r (10^9 M^{-1} s^{-1}) O$_2$	Ref.
Valeraldehyde	337	MeCN/H$_2$O (9:1)	295	2.0	35	3.7		27
2-Propylcyclohexanone	337	MeCN/H$_2$O (9:1)	295	2.2	1	3.6		28
2-Isobutylcyclohexane	337	MeCN/H$_2$O (9:1)	295	3.0	0.7	4.2		28
Valerophenone	347	Benzene	293	0.060	4			13,29
	337	MeCN/H$_2$O (9:1)	295	0.10	8.7	2.4		22
γ-Methylvalerophenone	266	Methanol	295	0.093	16	3.6	6.2	22,41
	337	Methanol	295	0.097	4.7	4.6	6.5	23,44
		Toluene-d$_6$	302.6	0.025				19
		MeCN/H$_2$O (4:1)	295	0.081	2.5		2.4	30
1,4-Diphenyl-1-hydroxy-4-butanone	337	SDS micelle	300	0.092				31
γ-Phenylbutyrophenone	266	Methanol	295	0.083	1.5			22
		n-Heptane	295	0.055				41
		Methanol	295	0.147				41
γ,γ-Diphenylbutyrophenone	266	n-Heptane	295	0.088				32,41
		Methanol	295	0.166				32,41
cis-1-Benzoyl-2-benzhydrylcyclohex-ane	266	n-Heptane	295	0.069				32
		Methanol	295	0.242				32
Benzhydrylphenacylether	266	n-Heptane	295	0.0016	~0.12			32
		Methanol	295	0.0013	~0.12			32
o-Phthalaldehyde	337	Methanol	295	1.61	6, 36	0.76		33
o-Tolualdehyde	337	MeCN/H$_2$O (4:1)	295	1.50	1.1, 95	6.2		26
o-Methylacetophenone	337	Toluene	295	0.145	0.14, 9.3			34
		Methanol	295	0.300		4.8	4.6	25

Compound	λ	Solvent	T				Ref.
Mesitylaldehyde	265/353	MeCN/H₂O (4:1)	295	0.581	0.5, 34	6.5	26
	337	Cyclohexane	295	0.130			35
		MeCN/H₂O (4:1)	295	1.09	1.6		26
5,8-Dimethyltetralone	337	Toluene	295	0.310	0.6	5.6	34
		MeCN/H₂O (4:1)	295	0.926	4.5		26
2-Methylbenzophenone	337	MeCN/H₂O (4:1)	295	0.024			26
2,4,6-Triisopropylbenzophenone	347	Ethanol	295	0.026			36
	266	MeCN	295	16.9			37—39
		Methanol	295	31.3			37—39
		Heptane	295	31.3			37—39
2-Phenylcyclohexanone	265	Heptane	295	0.049			40
		Methanol	295	0.050			40
2,2-Diphenylcyclohexanone	265	Heptane	295	0.056			40
		Methanol	295	0.067			40
Mesitylacetophenone	337	Methanol	300	0.015	<1		42,43
2,4,6-Triisopropyl phenylacetophenone	337	Methanol	300	0.050	<1		42,43

B. MACROMOLECULES

Type II biradicals are known as efficient initiators of methyl methacrylate polymerization.[47] More importantly, type II photofragmentation (Scheme I) is a commonly recognized process in polymer photochemistry[1,17,18,48] and has traditionally been subjected to steady-state and time-resolved studies in the context of photodegradation of poly(phenyl vinyl ketone), PPVK.[11,12,48-63] As noted by Bays et al.,[63] the findings on photochemistry of carbonyl-containing polymers may have bearings on related noncarbonyl polymers because the carbonyl chromophore may be easily introduced into a polymer through oxidative reactions in the course of polymerization, purification, and processing.

The macrobiradicals derived from PPVK poly(*o*-tolyl vinyl ketone) (PTVK), and related copolymers have been studied in detail, primarily by monitoring their transient absorptions at 350 to 600 nm following laser flash photolysis of the polymers.[13,19,55,59-66] At least in one case, namely, a copolymer of methyl vinyl ketone and *o*-tolyl vinyl ketone, electron transfer from the resultant macrobiradical to methyl viologen has been observed ($k_r = 1.7 \times 10^9$ M^{-1} s^{-1} in 4:3:3 methanol/acetone/acetonitrile).[64,65] The kinetic data on polymer-derived biradicals and their triplet precursors are summarized in Table 2. It is noted that the biradicals in PTVK polymers, derived from intrachromophore *o*-methyl hydrogen abstraction, are significantly longer lived. These provide pathways for regeneration of the polymer through reketonization of photoenols and hence improve polymer photostability considerably. On the other hand, in the case of PPVK, the carbonyl triplet abstracts hydrogen from the polymer backbone leading to shorter-lived biradicals which, because of subsequent bond scission, cause degradation in the polymer. The relative order in τ_B's in the case of PPVK and PTVK Table 2) is similar to that observed for model compounds (e.g., between valerophenone and *o*-methylacetophenone, τ_B in methanol is longer for the latter, see Table 1).

Naito and Schnabel[67] have been able to demonstrate the formation of an 1,4-biradical upon 265-nm laser flash photolysis of poly(3-buten-2-one) (PMVK) in dichloromethane. The transient absorption spectrum attributable to this species shows a weak maximum at ~300 nm. It decays with a lifetime of about 40 ns (in dichloromethane). On the basis of results from steady-state quenching experiments using 1-*cis*,3-*cis*-cyclooctadiene, the lifetime of triplet from PMVK is estimated to be much shorter (~6 ns in ethyl acetate at 298 K).[50]

IV. RADICAL IN POLYMERS

The kinetics of chain propagation and termination reactions of polymer radicals are of primary interest in polymer chemistry and have been studied extensively under steady-state photolysis of initiators.[68,69] The spectrophotometric observation of "saturated" carbon-centered transient macroradicals in the course of flash photolysis is difficult because of the weak character of the absorption as well as the location of the latter in the UV region. Naito and Schnabel[67] carried out 265-nm laser flash photolysis of poly(4,4-dimethyl-1-penten-3-one) (PBVK), poly(3-methyl-3-buten,2-one) (PMIK), and 3,3-dimethyl-2-butanone (a model compound) in dichloromethane solutions and detected the absorption spectra of acyl and acetyl radicals originating from α-cleavage. At λ > 300 nm, these species show absorption maxima at 320 and ~540 nm. The acyl macroradial from PBVK decays with a rate constant of 4×10^6 s^{-1}. In comparison, the decay of the acetyl radicals from PMIK and 3,3-dimethyl-2-butanone is slower by an order of magnitude (rates = $(1.5$ to $2.0) \times 10^5$ s^{-1}).

Optical measurements[70] following 265-nm laser flash excitation of poly(styrene sulfone) (PSS) in dioxane have revealed the formation of a short-lived species ($\tau < 25$ ns, $\lambda_{max} = 500$ nm) assigned as a singlet excimer. The latter acts as a precursor in the polymer chain scission process leading to a macroradical characterized by an absorption band below 400 nm ($\lambda_{max} \sim 320$ nm) and a half-life of 40 ms. Similar observations regarding excimer-

TABLE 2
Kinetic Behavior of Macrobiradicals Photogenerated from Carbonyl Containing Polymers

Source polymer	Laser wavelength (nm)	Solvent	Temp (K)	τ_B (μs)	τ_T (ns)	Ref.
Poly(phenyl vinyl ketone)	347	Benzene	295	0.059	7	67
	337	Benzene	295	0.067	55	63,64
	347	Benzene	293	0.060		13,55
	337	Benzene	295	0.076		59,60
Poly(o-tolyl vinyl ketone)	337	Toluene-d$_8$	300	0.20		19,66
	337	Benzene	293	0.20	2	63,64
Copolymer of methylvinyl ketone and o-tolyl vinyl ketone	337	Methanol/acetone/MeCN 4:3:3	295	0.37	1.8	64,65
Copolymer of phenyl vinyl ketone and o-tolyl vinyl ketone		Benzene	295	0.32		64,65
	337	Benzene	293	0.075—0.25	16—46	63,64
Poly (3-buten-2-one)	265	Dichloromethane	293	0.04		68

TABLE 3
Reactivity of Benzoyl and Diphenylhydroxymethyl Radicals with
Monomers at Room Temperature

Radical	Solvent	Monomer	Rate (M^{-1} s^{-1})	Ref.
$\overset{\cdot}{C_6H_5CO}$	Benzene	Styrene	4.7×10^5	82
		Methyl methacrylate	6.3×10^5	82
		Acrylonitrile	1.6×10^6	82
$(C_6H_5)_2\overset{\cdot}{C}OH$	Tetrahydrofuran	Vinyl acetate	5.5×10^3	83
		Methyl methacrylate	9.0×10^3	83
		Acrylonitrile	3.8×10^3	83

mediated bond cleavage have been made with polysterene in dioxane.[71] The homolytic main chain scission is a common phenomenon for silicon polymers, as shown by a recent laser flash photolysis study[72] of poly[dimethylsilylene-co-methyl (1-naphthyl) silylene] (PSN). In addition to a naphthalene-like triplet (τ = 3 μs, λ_{max} = 460 nm), the spectrum of which is broadened by the polymeric environment, a long-lived transient (τ = 300 μs, λ_{max} = 380 nm) is observed upon 337-nm laser excitation of PSN in tetrahydrofuran. The latter species has been identified as a polymer-derived silyl radical bearing a naphthyl group.

Significant results on dynamics of laser flash photolytic macroradicals have been obtained by Schnabel and co-workers employing time-resolved light scattering (LS) intensity measurements.[54,56,73-79] The LS method allows interesting investigations of disentanglement diffusion of fragments generated by main-chain scission and of elemetary processes leading to the main-chain scission. Another application of the LS technique is the measurement of rates of conformational changes due to *cis* → *trans* photoisomerization of appropriate chromophores in main chains, or in pendant groups, of polymers (e.g., polyamides with azobenzene groups in the backbone).[74]

For obvious reasons, considerable interest has been shown in laser-photolytic transient behaviors of radicals or radical ions from photoinitiators.[80-90] Investigations of benzoin and various benzoin derivatives have led to the spectral identification of benzoyl, and, accompanying, radicals produced as a result of α-cleavage (Equation 6).

$$C_6H_5 - \overset{\overset{\displaystyle O}{\|}}{C} - \overset{\overset{\displaystyle OH}{|}}{\underset{\underset{\displaystyle R_1}{|}}{C}} - R_2 \quad \xrightarrow{\ h\nu\ } \quad C_6H_5 - \overset{\overset{\displaystyle O}{\|}}{C}\cdot \quad + \quad \cdot\overset{\overset{\displaystyle OH}{|}}{\underset{\underset{\displaystyle R_1}{|}}{C}} - R_2 \qquad (6)$$

The absorption spectrum of benzoyl radical consists of a weak band system ($\lambda_{max} \sim 450$ nm) in the visible (in addition to stronger absorptions at < 340 nm).[81,85] Similarly, benzophenone and benzophenone derivatives including several water-soluble ionic ones have been subjected to laser photolysis in order to obtain spectral and kinetic information on radicals and radical anions produced as a result of electron and/or hydrogen-atom transfer from suitable donors.[86-90] Of particular interest is the reactivity of the primary initiator radicals with monomers. A sample of related kinetic data is presented in Table 3. The laser spectroscopy study of photoinitiators is treated in detail elsewhere in this monograph.

V. TRIPLET PRECURSORS

The involvement of triplets in the photochemistry of carbonyl compounds (including polymers) is established either by observing transient absorptions associated with them or via energy transfer quenching by suitable acceptors (e.g., styrenes, dienes, and naphthalenes). For systems in which extensive photochemistry renders triplet lifetimes very short, picosecond laser absorption spectroscopy is called for to monitor the triplet directly. An early example is provided by the picosecond measurement of triplet lifetimes for PPVK systems by Faure et al.[13,57,58] Short-lived carbonyl-type triplets are, however, conveniently probed by laser flash photolysis of carbonyl systems in the presence of dienes and naphthalenes. With naphthalenes as energy acceptors, their characteristic triplet-triplet (T-T) absorptions (λ_{max} = 415 to 430 nm) can be easily monitored. From the dependencies of naphthalene triplet absorbance changes (ΔA_T) on ground-state naphthalene concentration, it is possible to obtain information on the lifetime of the donor triplet. The intercept-to-slope ratios of double-reciprocal plots based on Equation 10 give the products ($k_q^T \tau_T$) of the triplet lifetime (τ_T) and the rate constant (k_q^T) of energy transfer to Np.

$$>\text{C=O} \xrightarrow{h\nu} \xrightarrow{\text{ISC}} >\text{C=O* (T}_1\text{)} \tag{7}$$

$$>\text{C=O*} \xrightarrow{1/\tau_T} >\text{C=O, \quad or products} \tag{8}$$

$$>\text{C=O*} + \text{Np} \xrightarrow{k_q^T} >\text{C=O} + {}^3\text{Np*} \tag{9}$$

$$\frac{1}{\Delta A_T} = \text{const} \left[1 + \frac{1}{k_q^T \tau_T [\text{Np}]} \right] \tag{10}$$

Since the triplet energies (E_T) of naphthalene derivatives are lower than those of carbonyl compounds by $\geq$10 kcal mol^{-1}, the energy transfer process (Equation 9) may be considered to be diffusion controlled. Thus, assuming an appropriate value for k_q^T, one can approximately calculate τ_T from the $k_q^T \tau_T$ data. Furthermore, the comparison of the intercept of the plot based on Equation 10 with that of a similar one in the case of a standard compound (e.g., benzophenone) allows estimation of triplet yields (ϕ_T). Note that in ϕ_T measurements the solutions of the standard and the compounds in question are to be optically matched at the laser excitation wavelength.

With styrenes or dienes as quenchers, the acceptor triplets cannot be easily observed because of their short lifetimes and/or inaccessible absorption regions. However, for these quenchers, the progressive decrease in absorbance changes (ΔA_p) due to a triplet-mediated photoproduct (e.g., radical, biradical, or PQ$^{+\cdot}$ via e$^-$-transfer from biradical) can be measured as a function of the quencher concentration and k_q^T values can be obtained from the slopes of Stern-Volmer plots (based on Equation 11).

$$\frac{\Delta A_p^\circ}{\Delta A_p} = 1 + k_q^T \tau_T [Q] \tag{11}$$

The direct spectrophotometric observation and/or the indirect methods discussed above have allowed τ_T's to be measured for most of the carbonyl systems that have been subjected to laser flash photolysis investigation for energy migration and radical/biradical processes. The τ_T data are included in Tables 2 and 3. Generally speaking, the magnitude of τ_T reflects the importance of intramolecular photochemical pathway(s) (bond scission and hydrogen abstraction) available to a triplet in a given conformation and in a given medium, and sheds light on electronic, structural, and environmental factors that come into play in triplet-mediated photoreactions.

For several polymers, in which intramolecular photochemistry is not important, relatively long-lived triplets have been studied by laser flash photolysis to understand the mechanism of intra- and intermolecular energy migration.[91-96] For example, the triplets of homopolymers of *p*-methoxyacrylophenone and its copolymers with phenyl vinyl ketone, styrene, and methyl methacrylate are characterized by long lifetimes and relatively high extinction coefficients (λ_{max}^T = 390 nm and ~630 nm, τ_T = 0.2 to 9 μs in benzene, chloroform, or chlorobenzene at 300 K). These have been examined[91] to show the importance of intramacromolecular energy migration (e.g., with a hopping frequency of 4.3 × 10^{11} s^{-1} in poly(*p*-methoxyacrylophenone) in chloroform at −60°). In a detailed study[92] of the quenching of triplet macromolecules by small molecules, it has been shown that energy migration in macromolecules containing carbonyl chromophores is primarily responsible for decreased bimolecular interaction of this chromophore with small-molecule quenchers (namely, oxygen, nitroxy radicals, dienes, and 1-methylnaphthalene). The β-phenylpropiophenone moiety, anticipated to act as an energy sink through its unusually short-lived, unreactive triplet state,[95] has been incorporated into copolymers of phenyl vinyl ketone with methacrylate esters of *p*-hydroxy-substituted derivatives of β-phenylpropiophenone. This has been shown to result in only modest decrease in photodegradation quantum yields (relative to PPVK).[94] The lack of stability against photodegradation is explained in terms of structural constraints on the intramacromolecular energy transfer process; that is, the distances between the β-phenylpropiophenone moieties and their nearest-neighbor benzoyl groups are too large for the triplet excitation transfer to be efficient (via exchange interaction).[94]

In the course of laser flash photoexcitation of copolymers of phenyl vinyl ketone, 2-vinyl naphthalene, and methyl methacrylate, the intramolecular triplet energy transfer from the initially excited benzoyl chromophore to the naphthalene moiety is shown[97,98] to take place by two distinct mechanisms. The static component ("jump", >10^8 s^{-1}) is interpreted to occur between close neighbors along the chain as well as when loops in the polymer bring the donor and acceptor moieties together. The dynamic component ("growth", 10^6 to 10^7 s^{-1}, dependent on naphthalene concentration in copolymers)) is ascribed to the diffusional motion of segments within the polymer. The investigation of chain flexibility and dynamics of intramacromolecular end-to-end collisions by laser spectroscopic measurement of second-order decay (T-T annihilation) of triplet absorption is exemplified by poly(*p*-vinylbenzophenone)[99] and α,ω-dianthrylpolystyrenes.[74,100] For the latter polymers, the rate constants for intramolecular end-to-end collisions have been measured to be in the range 7 × 10^3 to 1.5 × 10^5 s^{-1} in benzene at 295 K and 1.5 × 10^3 to 8.1 × 10^4 s^{-1} in cyclohexane at 307 K.

VI. SUMMARIZING REMARKS

In conjunction with transient absorption spectroscopy, laser flash photolysis has proved to be a valuable tool in mechanistic studies of photoinduced bond scission and intramolecular hydrogen abstraction processes in polymers and model compounds. Besides absorption-spectral characterization of photogenerated radicals and biradicals, data have been obtained concerning absolute rates of their intra- and intermolecular elementary processes. The results are relevant to the understanding of mechanisms of photoinitiation of polymerization, as well as photodegradation and photostabilization of polymers. Laser spectroscopic studies of macromolecular triplets, short-lived as well as relatively persistent, have thrown light on the existence of photochemical pathways for degradation in polymers and on the process of energy migration and transfer. Time-resolved light scattering intensity measurements following laser flash photolysis have been successfully utilized in elucidating the dynamics of disentanglement diffusion of macroradical fragments produced by main-chain photocleavage as well as in obtaining information on primary photoprocesses causing the main-chain scission.

ACKNOWLEDGMENTS

This work was supported by the Office of Basic Energy Sciences, U.S. Department of Energy. This is Document No. SR-113 from the Notre Dame Radiation Laboratory.

REFERENCES

1. **Ranby, B. and Rabek, J. F., Eds.,** *Photodegradation, Photooxidation and Photostabilization of Polymers,* Wiley-Interscience, New York, 1974.
2. **Grassie, N. and Scott, G.,** Photodegradation, in *Polymer Degradation and Stabilization,* Cambridge University Press, London, 1985, 69.
3. **Scott, G.,** Photodegradation and photosensitization of polymers, *J. Photochem.,* 25, 83, 1984.
4. **Geuskens, G. and Lu-Yinh, Q.,** Photooxidation of polymers. VII. A reinvestigation of the photo-oxidation of polystyrene based on a model compound study, *Eur. Pol. J.,* 18, 307, 1982.
5. Papers presented at the 179th National Meeting of the American Chemical Society, Houston, Texas, 1980, *Org. Coat, Plast. Chem.,* 42, 1980.
6. **Scaiano, J. C.,** Solvent effects in the photochemistry of xanthone, *J. Am. Chem. Soc.,* 102, 7747, 1980.
7. **Scaiano, J. C., Tanner, M., and Weir, D.,** Exploratory study of the intermolecular reactivity of excited diphenylmethyl radicals, *J. Am. Chem. Soc.,* 107, 4396, 1985.
8. **Nagarajan, V. and Fessenden, R. W.,** Flash photolysis of transient radicals. I. X_2^- with X = Cl, Br, I and SCN, *J. Phys. Chem.,* 89, 2330, 1985.
9. **Bensasson, R. V., Land, E. J., and Truscott, T. G.,** *Flash Photolysis and Pulse Radiolysis,* Pergamon Press, Elmsford, NY, 1983.
10. **Rodgers, M. A. J.,** Instrumentation for the generation and detection of transient species, *Primary Photoprocesses Biol. Med.,* Bensasson, R. V., Jori, J., Land, E. J., and Truscott, T. G., Eds., Plenum Press, New York, 1985, 1.
11. **Beck, G., Dobrowolski, G., Kiwi, J., and Schnabel, W.,** On the kinetics of polymer degradation in solution. II. Laser flash photolysis studies on poly(phenyl vinyl ketone) and butyrophenone, *Macromolecules,* 8, 9, 1975.
12. **Kiwi, J. and Schnabel, W.,** Laser flash photolysis studies on homo- and copolymers of phenylvinylketone in solution, *Macromolecules,* 8, 430, 1975.
13. **Faure, J., Fouassier, J. P., Lougnot, D. H., and Salvin, R.,** Picosecond and nanosecond kinetics of the Norrish type II photoscission in polymers, *Nouv. J. Chim.,* 1, 15, 1977.
14. **Masuhara, H.,** An introduction to transient absorption spectroscopy and nonlinear photochemical behavior of polymer systems, in *Photophysical and Photochemical Tools in Polymer Science,* D. Reidel, Dordrecht, Netherlands, 1986, 43.
15. **Janata, E.,** Baseline compensation circuit for measurement of transient signals, *Rev. Sci. Instrum.,* 57, 273, 1986.
16. **Wagner, P. J. and Zepp, R. G.,** Trapping by mercaptons of the biradical intermediates in type II photoenolization, *J. Am. Chem. Soc.,* 94, 287, 1972.
17. **Wagner, P. J., Kelso, P. A., and Zepp, R. G.,** Type II photoprocesses of phenyl ketones. Evidence for a biradical intermediate, *J. Am. Chem. Soc.,* 94, 7480, 1972.
18. **Scaiano, J. C., Lissi, E. A., and Encina, M. V.,** Chemistry of the biradicals produced in the Norrish type II reaction, *Rev. Chem. Intermediates,* 139, 1978.
19. **Scaiano, J. C.,** Does intersystem crossing in triplet biradical generate singlets with conformational memory?, *Tetrahedron,* 38, 819, 1982.
20. **Scaiano, J. C.,** Laser flash photolysis studies of the reactions of some 1,4-biradicals, *Acct. Chem. Res.,* 15, 252, 1982.
21. **Scaiano, J. C.,** Biradicals, in *Radical Reaction Rates in Liquids,* Fischer, H., Ed., (Landolt-Börnstein, New Series, II, Vol. 13E), Springer-Verlag, Berlin, Chap. 11.
22. **Small, R. D., Jr. and Scaiano, J. V.,** Photochemistry of phenyl alkyl ketones. The lifetime of the intermediates biradicals, *J. Phys. Chem.,* 81, 2126, 1977.
23. **Small, R. D. and Scaiano, J. C.,** Reaction of type II biradicals with paraquat ions. Measurement of biradical lifetimes, *J. Phys. Chem.,* 81, 2126, 1977.
24. **Small, R. D. and Scaiano, J. C.,** One electron reduction of paraquat dication by photogenerated biradicals, *J. Photochem.,* 6, 453, 1977.

25. **Small, R. D., Jr. and Scaiano, J. C.,** Role of biradical intermediates in the photochemistry of *o*-methylacetophenone, *J. Am. Chem. Soc.*, 99, 7713, 1977.

26. **Das, P. K., Encinas, M. V., Small, R. D., Jr., and Scaiano, J. C.,** Photoenolization of *o*-alkylsubstituted carbonyl compounds. Use of electron transfer processes to characterize transient intermediates, *J. Am. Chem. Soc.*, 101, 6965, 1979.

27. **Encinas, M. V. and Scaiano, J. C.,** Formation, trapping, and lifetime of the biradials generated in the photochemistry of valeraldehyde, *J. Am. Chem. Soc.*, 100, 7108, 1978.

28. **Encinas, M. V. and Scaiano, J. C.,** Photochemistry of α-substituted cyclohexanones. Chemistry of the intermediate type II biradicals, *J. Chem. Soc. Perkin Trans. 2*, 56, 1980.

29. **Faure, J., Fouassier, J.-P., Lougnot, D.-J., and Salvin, R.,** Photophysical aspects of molecules and macromolecules in solution, *Eur. Polym. J.*, 13, 891, 1977.

30. **Small, R. D., Jr. and Scaiano, J. C.,** Electron transfer reactions of the biradicals produced in the Norrish type II processes, *J. Phys. Chem.*, 82, 2662, 1978.

31. **Scaiano, J. C. and Selwyn, J. C.,** Triplet energy migration between carbonyl chromophores in micellar solution, *Photochem. Photobiol.*, 34, 29, 1981.

32. **Caldwell, R. A., Dhawan, S. N., and Majima, T.,** Lifetime of a conformationally constrained Norrish II biradical. Photochemistry of *cis*-1-benzoyl-2-benzyhydrylcyclohexane, *J. Am. Chem. Soc.*, 106, 6454, 1984.

33. **Scaiano, J. C., Encinas, M. V., and George, M. V.,** Photochemistry of *o*-phthalaldehyde, *J. Chem. Soc. Perkin Trans. 2*, 724, 1980.

34. **Kumar, C. V., Chattopadhyay, S. K., and Das, P. K.,** Triplet excitation transfer to caretenoids from biradical intermediates in Norrish type II photoreactions of *o*-alkylsubstituted aromatic carbonyl compounds, *J. Am. Chem. Soc.*, 105, 5142, 1983.

35. **Haag, R., Wirz, J., and Wagner, P. J.,** The photoenolization of 2-methylacetophenone and related compounds, *Helv. Chim. Acta*, 60, 2595, 1977.

36. **Nakayama, T., Hamanoue, K., Hidaka, T., Okamoto, M., and Teranishi, H.,** Photoenolization of 2-methylbenzophenone studied by picosecond and nanosecond laser spectroscopy, *J. Photochem.*, 24, 71, 1984.

37. **Hayashi, H., Nagakura, S., Ito, Y., Umehara, T., and Matsuura, T.,** Lase-photolysis study of biradical formation form the triplet state of 2,4,6-trisopropylbenzophenone, *Chem. Lett.*, 939, 1980.

38. **Ito, Y., Nishimura, H., Umahara, Y., Yamada, Y., Tone, M., and Matsuura, T.,** Intramolecular hydrogen abstraction from triplet states of 2,4,6-trisopropylbenzophenones: importance of hindered rotation in excited states, *J. Am. Chem. Soc.*, 105, 1590, 1483.

39. **Ito, Y., Nishimura, H., and Matsuura, T.,** Effect of ring substituents on the lifetimes of biradicals photogenerated from 2,4,6-triisopropylbenzophenones, *J. Chem. Soc. Chem. Commun.*, 1187, 1981.

40. **Caldwell, R. A., Sakuragi, H., and Majima, T.,** Direct observation and chemistry of triplet 1,6-biradicals in the Norrish I Reaction, *J. Am. Chem. Soc.*, 106, 2471, 1984.

41. **Caldwell, R. A., Majima, T., and Pac, C.,** Some structural effects on triplet biradical lifetimes. Norrish II and Paterno-Buchi biradicals, *J. Am. Chem. Soc.*, 104, 629, 1982.

42. **Scaiano, J. C. and Wagner, P. J.,** Excited-state chemistry of a 1,5-biradical: laser-induced ejection of a 1,3-biradical, *J. Am. Chem. Soc.*, 106, 4626, 1984.

43. **Meador, M. A. and Wagner, P. J.,** 2-Indanol formation from photocyclization of α-arylacetophenones, *J. Am. Chem. Soc.*, 105, 4484, 1983.

44. **Small, R. D., Jr. and Scaiano, J. C.,** Interaction of oxygen with transient biradicals photogenerated from γ-methylvalerophenone, *Chem. Phys. Lett.*, 48, 354, 1977.

45. **Small, R. D., Jr. and Scaiano, J. C.,** Direct detection of the biradicals generated in the Norrish type II reaction, *Chem. Phys. Lett.*, 50, 431, 1977.

46. **Encinas, M. V., Wagner, P. J., and Scaiano, J. C.,** Hydrogen abstraction by biradicals. Reactions with tri-*n*-butylstannane and octanethiol, *J. Am. Chem. Soc.*, 102, 1357, 1980.

47. **Hamity, M. and Scaiano, J. C.,** Trapping by methyl methacrylate of the biradicals derived from the photolysis of some alkyl aryl ketones. A novel initiation polymerization, *J. Photochem.*, 4, 229, 1975.

48. **DeSchryver, F. C. and Smets, G.,** in *Reactivity Mechanism and Structure in Polymer Chemistry*, Jenkins, A. D. and Ledwith, A., Eds., John Wiley & Sons, New York, 1974, chap. 14.

49. **Golemba, F. J. and Guillet, J. E.,** Photochemistry of ketone polymers. VII. Polymers and copolymers of phenyl vinyl ketone, *Macromolecules*, 5, 212, 1972.

50. **Dan, E., Somersall, A. C., and Guillet, J. E.,** Photochemistry of ketone polymers. IX. Triplet energy transfer in poly(vinyl ketones), *Macromolecules*, 6, 228, 1973.

51. **Dan, E. and Guillet, J. E.,** Photochemistry of ketone polymers. X. Chain scission reactions in the solid state, *Macromolecules*, 6, 230, 1973.

52. **Somersall, A. C., Dan, E., and Guillet, J. E.,** Photochemistry of ketone polymers. XI. Phosphorescence as a probe of subgroup motion in polymers at low temperatures, *Macromolecules*, 7, 233, 1974.

53. **Lukac, I., Hrdlovic, P., Manasek, Z., and Bellus, D.,** Influence of free and copolymerized triplet quenchers on the photolysis of poly(vinyl phenyl ketone) in solutions, *J. Polym. Sci. Part A,* 9, 69, 1971.
54. **Beck, G., Kiwi, J., Lindenau, D., and Schnabel, W.,** On the kinetics of polymer degradation in solution. I. Laser flash photolysis and pulse radiolysis studies using the light scattering detection method, *Eur. Polym. J.,* 10, 1069, 1974.
55. **Kiwi, J. and Schnabel, W.,** Laser flash photolysis studies on homo- and copolymers of phenyl vinyl ketone in solution, *Macromolecules,* 9, 468, 1976.
56. **Dobrowlski, G., Kiwi, J., and Schnabel, W.,** On the kinetics of polymer degradation in solution. IV. Laser flash photolysis studies of copolymers containing phenyl vinyl ketone using the light scattering detection method, *Eur. Polym. J.,* 12, 657, 1976.
57. **Faure, J., Fouassier, J.-P., and Lougnot, D.-J.,** Direct measurements of carbonyl triplet lifetime in the nanosecond range: investigation in a polymer solution, *J. Photochem.,* 5, 13, 1976.
58. **Faure, J.,** Laser spectroscopical methods for the study of primary processes during the photodegradation, *Pure Appl. Chem.,* 49, 487, 1977.
59. **Small, R. D., Jr. and Scaiano, J. C.,** Importance of intermolecular biradical reactions in polymer photochemistry. Poly(phenyl vinyl ketone), *Macromolecules,* 11, 840, 1978.
60. **Encinas, M. V., Funabashi, K., and Scaiano, J. C.,** Triplet energy migration in polymer photochemistry. A model for the photodegradation of poly(phenyl vinyl ketone) in solution, *Macromolecules,* 12, 1167, 1979.
61. **David, C., Demarteau, W., and Geuskens, G.,** Energy transfer in polymers. III. Polyphenylvinylketone-naphthalene system, *Eur. Polym. J.,* 6, 1405, 1970.
62. **Scaiano, J.-C. and Stewart, L. C.,** The triplet lifetime of poly(phenyl vinyl ketone). A laser flash photolysis study, *Polymer,* 23, 913, 1982.
63. **Bays, J. P., Encinas, M. V., and Scaiano, J. C.,** Photochemistry of polymers and copolymers of phenyl vinyl ketone and o-tolyl vinyl ketone, *Macromolecules,* 13, 815, 1980.
64. **Scaiano, J. C., Bays, J. P., and Encinas, M. V.,** Photoenolization in polymers, in *Photodegradation and Photostabilization of Coatings,* (ACS Symp. Ser., Vol. 151) Pappas, S. P. and Winslow, F. H., Eds., American Chemical Society, Washington, D. C., 1981, chap. 2
65. **Bays, J. P., Encinas, M. V., and Scaiano, J. C.,** Laser flash photolysis study of o-tolyl vinyl ketone-methyl vinyl ketone polymers. Electron donor properties of the intermediate microbiradicals, *Polymer,* 21, 283, 1980.
66. **Bays, J. P., Encinas, M. V., and Scaiano, J. C.,** Photoenolization in polymers. A simple way to reduce photodegradation, *Macromolecules,* 12, 348, 1979.
67. **Naito, I. and Schnabel, W.,** Laser flash photolysis studies of some aliphatic ketone polymers and the model compound, 3,3-dimethyl-2-butanone, *Polym. J.,* 16, 81, 1984.
68. **Griller, D.,** Carbon-centered radicals: radical-radical reactions, in *Radical Reaction Rates in Liquids,* Fischer, H., Ed., Landolt-Börnstein, New Series II, Vol. 13A, Springer-Verlag, Berlin, 1984, ch. 1.
69. **Lorand, J. P.,** Carbon-centered radicals: radical-molecule addition reactions, in *Radical Reaction Rates in Liquids,* Fischer, H., Ed., Landolt-Börnstein, New Series, II, Vol. 13A, Springer, Berlin, 1984, ch. 2.
70. **Horie, K. and Schnabel, W.,** Photolysis degradation of poly(olefin sulphones): a laser flash photolysis study, *Polym. Photochem.,* 2, 419, 1982.
71. **Tagawa, S. and Schnabel, W.,** On the mechanism of the photolysis of polystyrene in chloroform solution. Laser flash photolysis studies, *Makromol. Chem. Rapid Commun.,* 1, 345, 1980.
72. **Todesco, R. V. and Kamat, P. V.,** Excited-state behavior of poly[dimethylsilylene-co-methyl(1-naphthyl)silylene], *Macromolecules,* 19, 196, 1986.
73. **Schnabel, W.,** Photochemical transformations of polymers — investigations of rapid reactions, *Pure Appl. Chem.,* 51, 2373, 1979.
74. **Schnabel, W.,** Laser flash photolysis studies of the dynamics of polymers in solution, in *Photophysics of Synthetic Polymers,* Phillips, D. and Roberts, A. J., Eds., Science Reviews, Northwood, England, 1982, 55.
75. **Schnabel, W.,** Photodegradation of polymers in solution as studied by laser flash photolysis, *Polym. Eng. Sci.,* 20, 688, 1980.
76. **Beavan, S. W., Beck, G., and Schnabel, W.,** On the kinetics of polymer degradation in solution. VII. Laser flash photolysis of poly-α-methyl styrene in solution, *Eur. Polym. J.,* 14, 385, 1978.
77. **Beavan, S. W. and Schnabel, W.,** On the kinetics of polymer degradation in solution. VIII. Laser flash photolysis of polysterene, *Macromolecules,* 11, 782, 1978.
78. **Schnabel, W.,** High energy radiation — and laser induced reactions in polymers as studied by light scattering measurements, *Polym. Mater. Sci. Eng.,* 55, 208, 1986.
79. **Lindenau, D., Beavan, S. W., Beck, G., and Schnabel, W.,** On the kinetics of polymer degradation in solution. VI. Laser flash photolysis and pulse radiolysis studies of polymethyl-vinylketone in solution using the light scattering detection method, *Eur. Polym. J.,* 13, 819, 1977.
80. **Kuhlmann, R. and Schnabel, W.,** Flash photolysis investigation on primary processes of the sensitized polymerization of vinylmonomers, *Angew. Makromol. Chem.,* 70, 145, 1978.

81. **Kuhlmann, R. and Schnabel, W.,** Flash photolysis investigation on primary processes of the sensitized polymerization of vinyl monomers. II. Experiments with benzoin and benzoin derivatives, *Polymer,* 18, 1163, 1977.

82. **Salmassi, A., Eichler, J., Herz, C. P., and Schnabel, W.,** On the photolysis of 1-phenyl-2-hydroxy-2-methyl-propanone-1 in the presence of methylmethacrylate, styrene and acrylonitrile: laser flash photolysis studies, *Polymer Photochem.,* 2, 209, 1982.

83. **Kuhlmann, R. and Schnabel, W.,** Laser flash photolysis investigations on primary processes of the sensitized polymerization of vinyl monomers. I. Experiments with benzophenone, *Polymer,* 17, 419, 1976.

84. **Pappas, S. P., Pappas, B. C., Gatechair, L. R., and Schnabel, W.,** Photoinitiation of cationic polymerization. II. Laser flash photolysis of diphenyliodonium salts, *J. Polym. Sci. Polym. Chem., Ed.,* 22, 69, 1984.

85. **Fouassier, J.-P. and Merlin A.,** Laser investigation of Norrish type I photoscission in the photoinitiator irgacure (2,2-dimethoxy-2-phenylacetophenone), *J. Photochem.,* 12, 17, 1980.

86. **Fouassier, J.-P., Lougnot, D.-J., Zuchowicz, I., Green, P.-N., Timpe, H.-J., Kronfeld, K.-P., and Müller, V.,** Photoinitiation mechanism of acrylamide polymerization in the presence of water-soluble benzophenones, *J. Photochem.,* 36, 347, 1987.

87. **Fouassier, J.-P. and Lougnot, D.-J.,** Ionic photoinitiators for radical polymerization in direct micelles: the role of the excited states, *J. Appl. Polym. Sci.,* 32, 6209, 1986.

88. **Fouassier, J.-P., Lougnot, D.-J., and Zuchowicz, I.,** Reverse micelle radical photopolymerization of acrylamide: excited states of ionic initiators, *Eur. Polym. J.,* 22, 933, 1986.

89. **Lougnot, D. J., Jacques, P., Fouassier, J. P., Casal, H. L., Nguyen, K.-T., and Scaiano, J. C.,** New functionalized water-soluble benzophenones: a laser flash photolysis study, *Can. J. Chem.,* 63, 3001, 1985.

90. **Merlin, A., Lougnot, D.-J., and Fouassier, J. P.,** Laser spectroscopy of substituted benzophenone used as photoinitiators of vinyl polymerization, *Polym. Bull.,* 2, 847, 1980.

91. **Scaiano, J. C. and Selwyn, J. C.,** Energy transfer and migration processes in the photochemistry of polymers and copolymers of *p*-methoxyacrylophenone, *Macromolecules,* 14, 1723, 1981.

92. **Scaiano, J. C., Lissi, E. A., and Stewart, L. C.,** Quenching of triplet macromolecules by small molecules. The role of energy migration, *J. Am. Chem. Soc.,* 106, 1539, 1984.

93. **Hrdlovic, P., Scaiano, J. C., Lukac, I., and Guillet, J. E.,** Transient spectroscopy and kinetics of poly-(1-(4-substituted-phenyl)-2-propen-1-ones), *Macromolecules,* 19, 1637, 1986.

94. **Leigh, W. J., Scaiano, J. C., Paraskevopoulos, C. I., Charette, G. M., and Sugamori, S. E.,** Photodegradation of poly(phenyl vinyl ketones) containing β-phenyl propiophenone moieties, *Macromolecules,* 18, 2148, 1985.

95. **Netto-Ferreira, J. C., Leigh, W. J., and Scaiano, J. C.,** Laser flash photolysis study of the photochemistry of ring-substituted β-phenyl propiophenones, *J. Am. Chem. Soc.,* 107, 2617, 1985.

96. **Wismontski-Knittel, T. and Kilp, T.,** Intramolecular quenching of carbonyl triplets by β-phenyl rings, *J. Phys. Chem.,* 88, 110, 1984.

97. **Das, P. K., Encinas, M. V., and Scaiano, J. C.,** Intramolecular energy transfer in polymers containing benzoyl and naphthalene moieties, *J. Photochem.,* 12, 357, 1980.

98. **Das, P. K. and Scaiano, J. C.,** Laser flash photolysis study of polymers containing benzoyl and naphthalene groups, *Macromolecules,* 14, 693, 1981.

99. **Schnabel, W.,** Laser flash photolysis of poly(*p*-vinylbenzophenone) in solution. Intramolecular triplet deactivation processes, *Makromol. Chem.,* 180, 1487, 1979.

100. **Horie, K., Schnabel, W., Mita, I., and Ushiki, H.,** Rates of intramolecular collision between terminal groups of α,ω-dianthrylpolystyrene in benzene and cyclohexane solutions as studied by triplet-triplet absorption measurements, *Macromolecules,* 14, 1422, 1981.

Chapter 4

STEPWISE MULTIPHOTON PROCESSES AND THEIR APPLICATIONS IN POLYMER CHEMISTRY

W. Grant McGimpsey

TABLE OF CONTENTS

ABSTRACT

This chapter describes the detection of stepwise multiphoton processes by a "controlled two-photon" (two-laser) flash photolysis technique. This technique yields direct time-resolved evidence for the occurrence of multiphoton absorption processes and can be used to investigate the multiphoton specificity of chemical systems. Multiphoton specific (MPS) systems exhibit chemical reactivity only following stepwise absorption of two or more photons; that is, these sytems are, ideally, chemically inert under one-photon irradiation conditions. Specific areas of application for MPS systems in polymer chemistry are discussed and examples of compounds exhibiting a high degree of multiphoton specificity are described.

I. INTRODUCTION: STEPWISE MULTIPHOTON ABSORPTION PROCESSES

The technique of laser flash photolysis has been used for over 2 decades in the investigation of the kinetic and spectroscopic properties of short-lived transient intermediates produced in photochemical reactions in solution.[1] The monochromatic quality of laser radiation as well as the short duration, high-intensity pulses supplied by commercially available lasers have made these instruments popular as excitation sources. Organic chemists interested in the mechanistic details of photochemical reactions routinely turn to the use of lasers in order to produce large concentrations of transient species. It is not widely recognized, however, that the high photon flux responsible for large transient concentrations can often lead, inadvertently, to the occurrence of stepwise multiphoton absorption processes. These processes take place during pulsed laser irradiation when transient species produced by the absorption of a laser photon, e.g., excited singlet (S_1) and triplet (T_1) states, free radicals, etc. ("one-photon" species), reach sufficiently high local concentrations *during* the laser pulse to compete efficiently with the ground state precursor molecules for incident laser photons. This situation is partially described mathematically in Equation 1:

$$\epsilon_T^\lambda \cdot C_T \approx \epsilon_P^\lambda \cdot C_P \qquad (1)$$

where the subscripts T and P refer to the transient and precursor molecule respectively; ϵ is the extinction coefficient at the laser wavelength λ, and C is the concentration of the species indicated. For the equality in Equation 1 to be met, not only are high transient concentrations required, a condition which is routinely encountered under laser photolysis conditions, but also the transient species must possess an extinction coefficient, at the laser wavelength, which is significantly greater than that of the ground state precursors. For transients normally produced during photolysis of organic substrates, this is not an unusual characteristic, particularly when the wavelength of the laser radiation falls in the near UV region of the spectrum.

Not explicit in Equation 1 is the additional requirement that transient formation must take place on a time scale comparable with the duration of the laser pulse. For most laser flash photolysis systems used in the investigation of organic photochemical problems this corresponds to 10^{-9} to 10^{-8} s. If significant concentrations of transients are not produced within this short time interval, no multiphoton absorption can take place. This condition limits the types of transients which can participate in multiphoton absorption to excited singlet states, most triplet states, and chemically distinct species (radicals, radical ions, carbenes, etc.) which are produced "instantaneously" from excited singlet and triplet state precursors.

The absorption of a laser photon by a one-photon transient results in the production of an excited transient species (two-photon transient). For example, absorption of a photon by

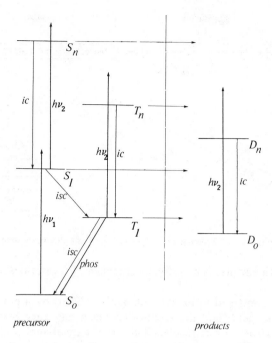

precursor   *products*

FIGURE 1. Stepwise multiphoton absorption processes in four-level, two-photon chemical systems: $h\nu_1 \equiv$ synthesis laser photon; $h\nu_2 \equiv$ transient excitation laser photon; S_1, T_1, and $D_1 \equiv$ one-photon transients; S_n, T_n, and $D_n \equiv$ two-photon transients; IC $\equiv$ internal conversion; ISC $\equiv$ intersystem crossing; and Phos $\equiv$ phosphorescence.

an excited singlet (S_1) or triplet (T_1) state produces the corresponding upper excited state, S_n or T_n. Figure 1 shows a four-level, two-photon energy diagram which illustrates these absorption processes. As noted above, species which are produced as a result of a chemical reaction following one-photon absorption can also compete for the available incident laser photons. Thus, absorption by ground state (D_0) radicals produced from excited singlet and triplet state precursors will result in the production of excited radicals (D_n).

The decay of excited (two-photon) transient species can follow a variety of efficient pathways including radiative processes as well as inter- and intramolecular radiationless processes. For example, $S_2 \rightarrow S_0$ fluorescence has been reported for a variety of thioketones,[2-9] azulenes,[10-12] and aromatic compounds.[13-15] Triplet-triplet ($T_2 \rightarrow T_1$)[16] and doublet-doublet[17-35] ($D_1 \rightarrow D_0$, $D_2 \rightarrow D_0$) luminescence has also been reported. Upper excited states and other excited transients can also be deactivated by radiationless processes such as internal conversion (where Franck-Condon factors are large) intersystem crossing ($T_n \rightarrow S_1$), energy or charge transfer to the solvent, and irreversible chemical reactions.

II. TIME-RESOLVED DETECTION OF MULTIPHOTON ABSORPTION

The occurrence of stepwise multiphoton processes can be established both directly and indirectly using the technique of laser flash photolysis.

A. INDIRECT EVIDENCE OF MULTIPHOTON ABSORPTION

A schematic of a typical nanosecond laser flash photolysis system is shown in Figure 2. While a complete, detailed description of the components of this system, including a

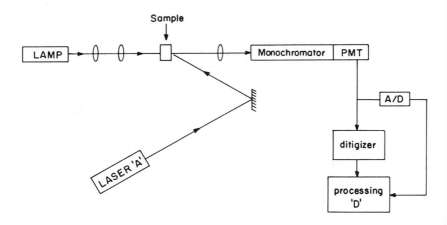

FIGURE 2.   Schematic of a typical laser flash photolysis setup.

number of elegant refinements, can be found elsewhere,[36] a brief description will be given here.

Sample solutions contained in quartz cells are irradiated by a pulsed laser. The angle between the excitation pulse and the monitoring beam can vary between 90° (side face excitation) and ~10° (front or rear face excitation). For small angles, the irradiated volume is small and hence this geometry is preferred for low energy lasers. Broad band continuous wave monitoring light enters the cell and the transmitted light is wavelength sorted by a monochromator and detected by a photomultiplier. Temporal changes in the transmitted light intensity which occur as a result of transient production (and ground state depletion) and decay (as well as ground state recovery or product formation) are recorded by a transient digitizer and kinetic traces are then transferred to a computer for storage and analysis.

With this system it is possible to obtain indirect time-resolved evidence for multiphoton absorption by measuring the dependence of the transient yield on the laser dose. A linear dependence indicates that the efficiency of transient production is independent of laser dose, i.e., all absorption processes taking place involve the ground state precursor molecule, while a negative deviation from linearity indicates a decrease in efficiency with increasing laser dose and is indicative of multiphoton absorption processes.

For example, Figure 3 shows the dependence of the maximum triplet optical density for 4-acetylbiphenyl[37] as a function of a percentage of the maximum laser dose. (The maximum optical density is directly proportional to the triplet yield.) Attenuation of the laser is achieved with a calibrated set of neutral density filters. At low laser dose (low triplet optical density) there is only a small negative deviation from linearity. However, as the laser energy and therefore the triplet concentration increases, the deviation also increases. In fact this deviation is so pronounced that fewer triplets are detected at 100% laser dose than at 66% laser dose. Thus, at high doses, the efficiency of triplet production decreases, indicating that proportionately fewer photons are absorbed by ground state molecules. This observation implies that another species competes with the ground state for the available laser photons.

A disadvantage of this indirect approach is that the competing species cannot be identified from the evidence obtained. While in this particular case it is probable that the competing species is the triplet state itself (since it is known to possess a large extinction coefficient at the laser wavelength), in other systems it is possible that the singlet state (the precursor of the triplet state) or another chemical species produced in a side reaction from the singlet or triplet may be the transient responsible for multiphoton absorption.

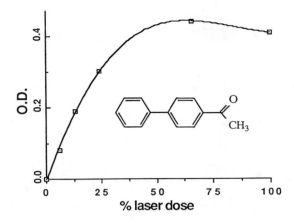

FIGURE 3.   Dependence of 4-acetylbiphenyl triplet yield on laser dose. Yield measured as maximum triplet-triplet absorption. Dose measured as a percentage of maximum energy.

B. DIRECT EVIDENCE FOR MULTIPHOTON ABSORPTION: THE CONTROLLED TWO-PHOTON (TWO-LASER) TECHNIQUE

The identity of a transient species which is responsible for multiphoton absorption processes can be determined directly by performing a ''controlled two-photon'' or two-laser experiment. In this experiment, the pulses from one laser are used to produce one-photon transients. Before these species decay completely they are excited by the pulses from a second laser. Figure 4a shows a schematic of a two-laser flash photolysis system. Figure 4b (top) shows the timing sequence of a two-laser experiment. At the bottom of Figure 4b (trace A) the kinetic profile of a transient species is shown. The growth of the transient absorption (which in this example is instantaneous) is represented by a vertical line coinciding with the first laser pulse (synthesis pulse). Following the synthesis pulse, the transient decays spontaneously. However, before complete decay, the second laser (transient excitation) is fired resulting in bleaching of the transient absorption. This permanent bleaching process occurs when the transient which absorbs a photon is promoted to an excited transient state and then participates in a decay process which does not regenerate the original transient. Kinetic trace B in Figure 4b shows growth of the absorption of a new transient species which occurs as a consequence of the bleaching process. This absorption may be due to the excited transient itself or another species formed as a result of the decay of the excited transient, e.g., chemical reaction leading to radical formation.

The two-laser technique has several advantages over a one-laser experiment in which both transient production and excitation take place during one-laser pulse: (1) the wavelength of the maximum transient absorption may not coincide with the synthesis laser wavelength. In fact the transient absorption at this wavelength may be minimal and hence little multiphoton absorption will take place during a one-laser experiment. However, in a two-laser experiment, if the wavelength of the transient excitation laser is tunable (e.g., if a dye laser is used), it is possible to maximize the extent to which multiphoton processes take place by tuning the dye laser to the absorption maximum of the transient. A hypothetical absorption spectrum illustrating this situation is shown in Figure 5. The synthesis laser wavelength falls within the absorption band of the ground state precursor molecule. Absorption of the synthesis pulse yields a transient species exhibiting an absorption spectrum different from that of the ground state. The transient excitation wavelength is then tuned to the maximum of this band. The spectral characteristics of the excited transient or of a chemical species produced as a result of transient excitation may also be quite different from those of the transient itself.

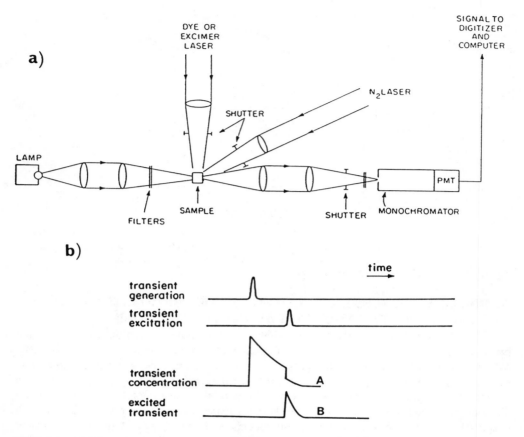

FIGURE 4.   (a) Schematic of a two-laser flash photolysis setup. (b) Timing sequence and transient kinetic behavior in a two-laser experiment. A ≡ one-photon transient absorption showing bleaching and B ≡ excited transient absorption or absorption due to new chemical species.

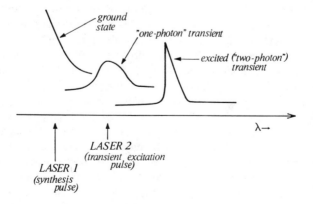

FIGURE 5.   Hypothetical absorption spectrum showing absorption bands of ground state precursor molecule, one-photon transient, and excited transient species.

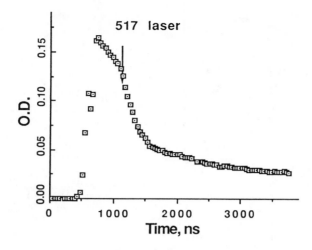

FIGURE 6.   Decay of benzophenone triplet-triplet absorption monitored at 535 nm obtained under two-laser (308 nm + 517 nm) irradiation conditions (in benzene).

Absorption or emission due to these species is also shown in Figure 5; (2) the time delay between the transient production and excitation is variable in a two-laser experiment. This removes the requirement that the transient must be produced during the synthesis laser pulse in order that multiphoton absorption can take place and therefore allows excitation of species produced at relatively slow (less than diffusion controlled) rates; (3) the time-resolved effects of transient excitation can be observed directly (bleaching of transient absorption, growth of new absorptions).

III. EXAMPLES OF MULTIPHOTON ABSORPTION IN ORGANIC SYSTEMS: USE OF THE TWO-LASER TECHNIQUE

A. BENZOPHENONE

Pulsed UV laser photolysis of benzene solutions containing millimolar concentrations of benzophenone results in the efficient ($\Phi_{ISC} = 1.0$) production of the benzophenone triplet state, T_1, which is characterized by its triplet-triplet absorption spectrum (λ_{max} 525 nm).[38] The triplet decays by a combination of first and second order processes over a period of a few microseconds. When the triplet yield (maximum optical density) is plotted against the laser dose, a small negative deviation from linearity is observed, indicating the occurrence of multiphoton absorption. This multiphoton absorption was confirmed by performing a two-laser experiment in which transient excitation was facilitated by a flashlamp-pumped dye laser tuned to 517 nm.[39] The kinetic profile obtained under two-laser irradiation conditions is shown in Figure 6. Transient excitation results in the efficient irreversible bleaching of the benzophenone triplet absorption. While this observation indicates that upper excited triplet states decay by processes other than relaxation back to the lowest triplet state, the lack of any new transient absorptions associated with triplet bleaching prevents identification of these processes on the basis of time-resolved evidence alone. However, when these observations are combined with the results of studies indicating that (1) no product formation takes place as a result of triplet bleaching and (2) no bleaching is observed in nonaromatic solvents, it is possible to attribute the observed bleaching to solvent interactions which lead to deactivation of upper excited triplet states resulting in ground state repopulation. While the nature of these solvent interactions have yet to be verified, two-laser experiments performed with a variety of chemical systems[40] indicate that triplet-triplet energy transfer from

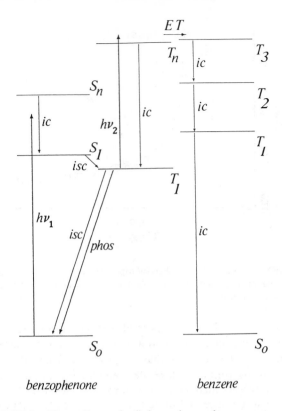

FIGURE 7.   Energy diagram for the benzophenone-benzene system showing laser excitation and energy transfer from an upper excited triplet state
of benzophenone to the triplet manifold of benzene; $h\nu_1 \equiv 308$ nm, $h\nu_2$
$\equiv 517$ nm, IC $\equiv$ internal conversion; ISC $\equiv$ intersystem crossing; ET $\equiv$
energy transfer; and Phos $\equiv$ phosphorescence.

upper excited benzophenone triplet states to the triplet manifold of benzene may be involved.
Figure 7 shows an energy diagram illustrating this energy transfer process.

B. 9-HALOFLUORENES

While the time-resolved evidence obtained for benzophenone indicated that multiphoton
absorption involving the triplet state takes place (triplet bleaching), no direct information
could be obtained concerning the fate of the excited transient (upper triplet state) produced
by the transient excitation pulse. For the 9-chloro- and 9-bromofluorene systems, however,
the time-resolved evidence obtained not only indicates the occurrence of multiphoton absorption but also the chemistry resulting from transient excitation.[37]

UV laser photolysis of 9-chloro- or 9-bromofluorene in benzene or cyclohexane results
in the production of a transient species which has been tentatively identified as a rearranged
photoproduct resulting from a [1,3] sigmatropic shift of the halogen[40a] (λ_{max} ~360 nm,
shoulder at 430 nm). Dye laser photolysis ($\lambda \sim 430$ nm) leads to the observation of irreversible
bleaching (concurrent with the dye laser pulse) of this absorption when it is monitored at
410 nm. When the experiment is performed in benzene (Figure 8), the bleaching at 410 nm
is accompanied by a growth with a maximum absorption at 550 nm. This wavelength
corresponds to the absorption maximum of the bromine atom-benzene π-complex.[41] If the
same experiment is performed in cyclohexane solvent (Figure 9), no 550-nm growth is
observed. However, a growth is observed with a maximum absorption at 498 nm. This
wavelength corresponds to the maximum for the 9-fluorenyl radical.[42] These observations

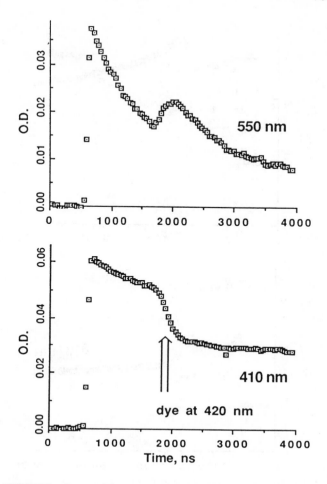

FIGURE 8.   Decay of absorption transient produced following excitation of 9-chloro- or 9-bromofluorene in benzene under 2-laser (308 nm + 420 nm) irradiation conditions monitored at 550 nm (top) and 410 nm (bottom). Top trace shows growth of bromine atom-benzene π-complex concurrent with dye laser pulse. Bottom trace shows dye laser induced transient bleaching.

confirm that transient excitation, forming an excited transient, results in the homolytic cleavage of the carbon-bromine bond (Scheme I). Bond cleavage yields Br˙ and the 9-fluorenyl radical. When the experiment is performed in benzene, Br˙ is scavenged by the solvent resulting in π-complex formation.

IV. MULTIPHOTON SPECIFIC (MPS) MATERIALS

The two-laser technique described above can be used to investigate the behavior of MPS materials. An MPS chemical system is one which undergoes a chemical reaction only following the stepwise absorption of two or more photons; that is, one-photon absorption results in the production of excited states (S_1, T_1) which, due to the energy requirements of the reaction, are chemically inert. However, absorption of a second photon by these excited states yields upper states possessing sufficient energy to facilitate the reaction.

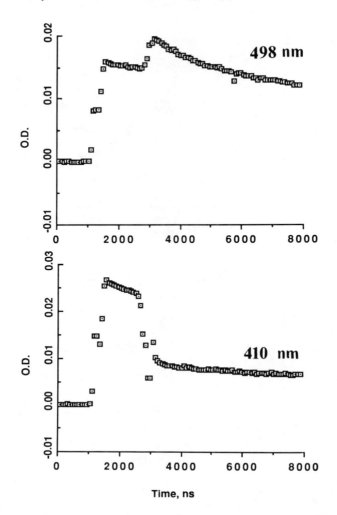

FIGURE 9.   Decay of absorption transient produced following excitation
of 9-chloro- or 9-bromofluorene in cyclohexane under two-laser (308 nm
+ 420 nm) irradiation conditions monitored at 498 nm (top) and 410 nm
(bottom). Top trace shows growth of 9-fluorenyl radical absorption. Bottom
trace shows transient bleaching.

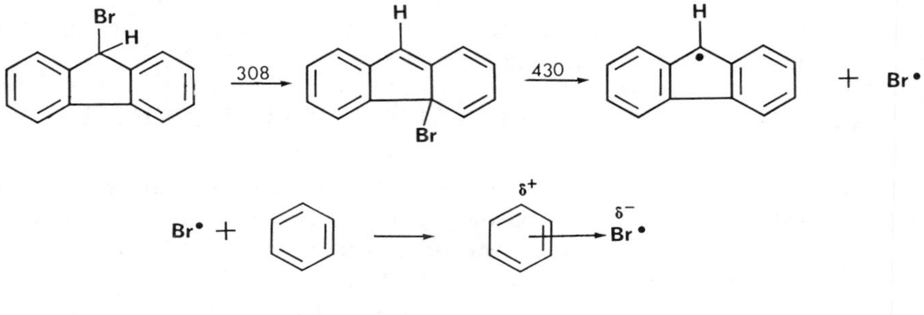

SCHEME I

A. APPLICATIONS OF MPS SYSTEMS IN POLYMER CHEMISTRY
1. Three-Dimensional Photochemical Machining (PCM)

PCM is a technique currently being developed by a number of industrially based groups, including that of Schwerzel et al.[43,44]

The technique involves the exposure of a polymer material which has been doped with an MPS compound to two-laser beams which intersect at a point within the volume of the material. The chemical reaction which takes place as a result of the multiphoton excitation of the MPS dopant (ideally this only takes place in those regions irradiated by both lasers) causes changes in the physical properties of the polymer. These changes may involve either degradation or cross-linking reactions. If degradation occurs, the material which is degraded becomes soft and easily dissolved and can be removed leaving behind those regions of the polymer substrate which were not irradiated by both lasers. If, on the other hand, cross-linking takes place, the material becomes hard and insoluble in those regions irradiated by both lasers. Thus, the remaining material (that which was not subjected to multiphoton absorption) can be removed.

If the movement of both lasers relative to each other and to the substrate are controlled by computer, it is possible to produce three-dimensional objects of high complexity. This technique is expected to have a wide range of applications in the casting industry. To date, the major problem associated with this technique involves the identification of potential dopants with sufficiently high multiphoton specificity, that is, it is difficult to prevent the occurrence of one-photon chemical reactions from taking place in volumes of the polymer material irradiated by only one laser.

2. Holography

Bräuchle and co-workers[45-50] have also investigated the applicability of MPS systems to the field of holography. In this technique, the interference patterns formed by temporally and spatially coherent object and reference waves are impressed on a photosensitive recording medium. (Where the bright regions of the interference pattern fall on the recording medium, photochemical reactions take place, causing a modulation of the absorption or index of refraction.) The information which is recorded can then be "read" by illuminating the holographic material with a "reconstruction" wave identical to the original reference wave. This results in the reproduction of the original object wave, but unfortunately can also gradually lead to photochemical erasure of the hologram. However, it is possible, by using an MPS recording medium, to limit the extent of degradation. Bräuchle et al.[46] describe an experiment using this approach. α-Diketones such as biacetyl (which is known to have a degree of multiphoton specificity) were dissolved in a variety of alkyl-α-cyanoacrylates. These compounds undergo rapid anionic initiated polymerization during sample preparation. The polymerized material was irradiated with a CW mercury lamp (UV-vis wavelengths) which facilitates excitation of the α-diketones to their lowest excited triplet states. During this irradiation one-photon reaction from the triplet states may take place (to varying extents), but no hologram is recorded. The triplet states of α-diketones possess absorption bands in the 700- to 1000-nm wavelength range, and thus the wavelength of the reference and object waves were chosen to fall within these absorption bands. Hence, triplet state excitation is facilitated by the reference and object waves and results in the occurrence of MPS chemistry which, in turn, leads to hologram formation. However, this is a two-photon hologram which cannot be erased by the reconstruction wave alone. (The wavelength of the reconstruction wave possesses sufficient energy to cause triplet state excitation; however, without the accompanying UV lamp irradiation, no triplet states are produced and hence no chemistry can take place.)

The major problem encountered in the investigation of this two-photon holographic technique is, again, the identification of materials possessing high multiphoton specificity. The occurrence of one-photon side reactions decreases the efficiency of hologram production.

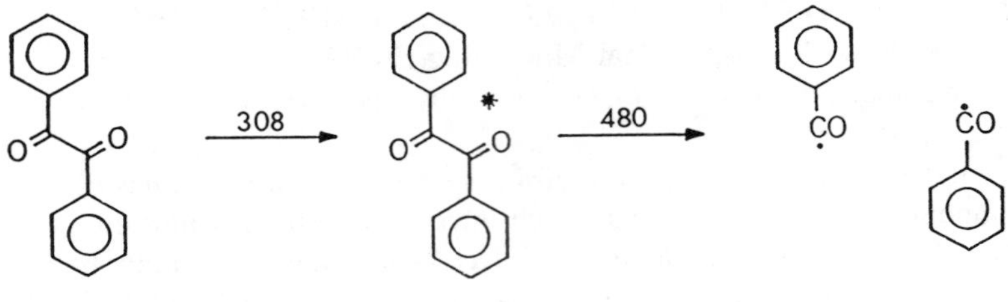

SCHEME II

B. MPS CHEMICAL SYSTEMS

Recently, a number of MPS materials have been identified (by the two-laser technique) which under the correct conditions exhibit a high degree of specificity. Two examples of such materials are given below.

1. Benzil

The α-cleavage of benzil is known to be very inefficient following one-photon excitation.[51] The reason for this lack of reactivity can be understood on the basis of the energy requirements for the cleavage reaction. One-photon UV excitation results in the production of the lowest excited triplet state, T_1 ($\Phi_{ISC} = 0.92$).[52] The energy stored in this state is ~54 kcal/mol,[53] while the C(O)-C(O) bond energy is ~66 kcal/mol. Hence α-cleavage taking place from the T_1 state is endothermic by ~12 kcal/mol. However, reexcitation of the T_1 state by a second photon results in the production of an upper excited state possessing more than sufficient energy to facilitate α-cleavage (see Scheme II). This MPS cleavage reaction was confirmed using the two-laser technique already described.[54] Figure 10 shows the time-resolved results obtained for one- and two-laser experiments. In the one-laser experiment (trace A) the UV pulses from an excimer laser (308 nm, 4 ns) produce the benzil T_1 state (monitored near its absorption maximum at 480 nm) which decays spontaneously over a period of several microseconds. In the two-laser experiment (trace B) the excimer laser pulse was followed, after a delay of ~1 μs, by the pulse from a dye laser tuned to the benzil triplet absorption maximum. Irreversible bleaching of the triplet absorption, concurrent with the dye laser pulse, was observed. While this time-resolved evidence does not directly confirm the two-photon-induced α-cleavage of benzil, it does indicate that a photochemical process occurs following triplet state excitation which does not result in the regeneration of the T_1 state. Confirmation for the bond cleavage process was obtained by performing product analyses on samples which had been irradiated by several hundred pairs of excimer and dye laser pulses (dye laser pulse followed excimer pulse after 1-μs delay). These samples contained equimolar concentrations of benzil and 4,4'-dimethylbenzil (the latter was shown to have triplet kinetics and spectral characteristics as well as triplet bleaching behavior nearly identical to that of benzil). Analysis indicated the presence of the cross-product, 4-methylbenzil, which could only be produced by α-cleavage of both benzil and dimethylbenzil followed by recombination of the radical products (see Scheme III). One-photon irradiation of the same mixture of benzil and dimethylbenzil by UV lamps did not yield the cross-product. This gives further confirmation for the high multiphoton specificity of benzil.

2. 1,3-Di-1-Naphthylpropan-2-One

A second example of a chemical system possessing a relatively high degree of multiphoton specificity is 1,3-di-1-naphthylpropan-2-one.[55] This system is similar to benzil, in that α-cleavage is inefficient following excitation to the lowest triplet state. Again this inefficiency is due to the fact that the energy stored in the T_1 state is less than that of the C-C(O) bond energy. Triplet excitation, however, yields an upper state with sufficient energy

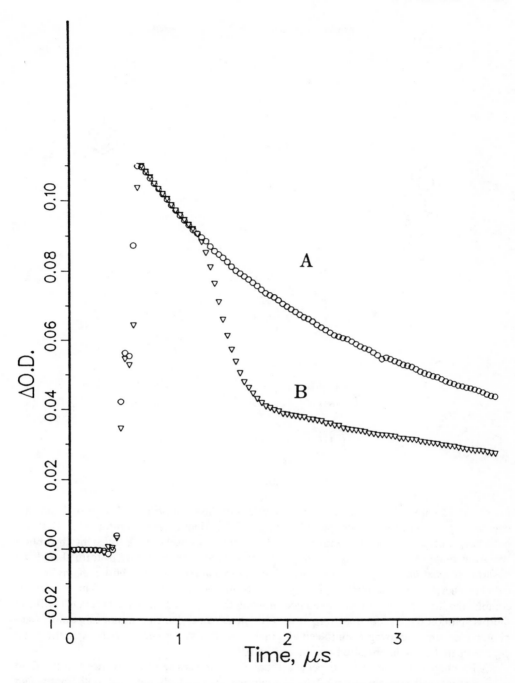

FIGURE 10.   Decay of benzil triplet-triplet absorption monitored at 470 nm under one- (A) (308 nm) and two-(B) (308 + 480 nm) laser irradiation conditions. (From McGimpsey, W. G. and Scaiano, J. C., *J. Am. Chem. Soc.*, 109, 2179, 1987. With permission.)

to facilitate the cleavage reaction (see Scheme IV). Unlike the time-resolved results obtained in the benzil system which did not reveal the upper excited state decay process responsible for permanent triplet bleaching, results for this system give direct evidence for α-cleavage. Figure 11 shows the decay of the triplet state monitored at (a) 430 and (b) 370 nm under one-(A) and two-(B) laser irradiation conditions (308-nm excimer laser + 420-nm dye

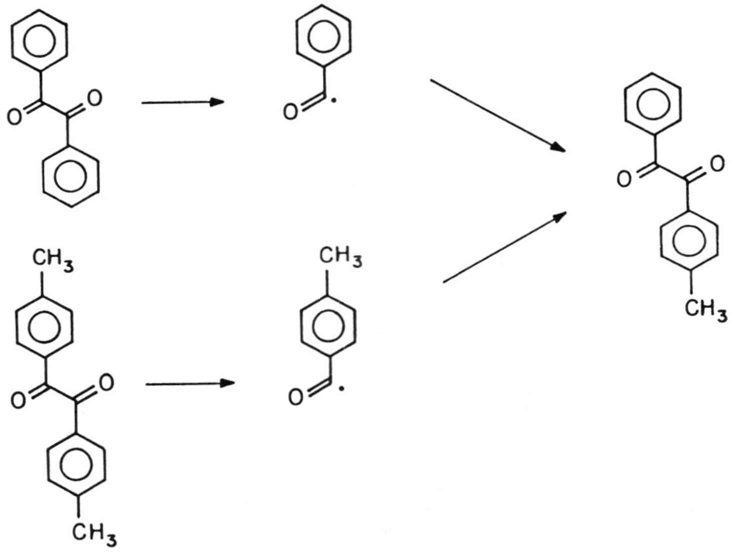

SCHEME III

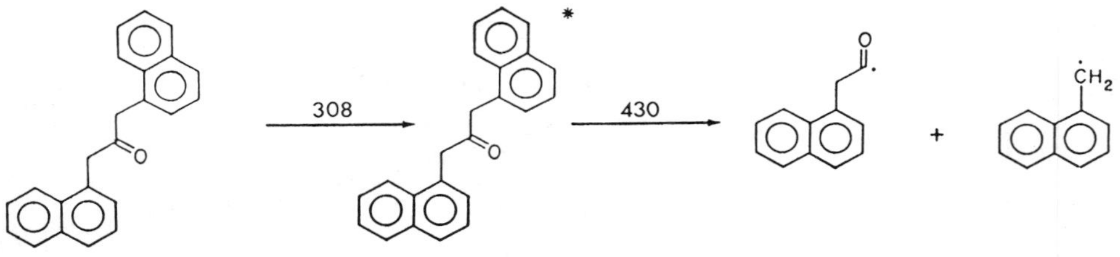

SCHEME IV

laser). At 430 nm, which is close to the maximum of the triplet-triplet absorption band, the dye laser pulse causes irreversible bleaching. If the kinetics are monitored at 370 nm, however, a growth associated with the dye laser pulse is observed. A transient absorption spectrum obtained after the dye laser pulse shows that a new species is formed with an absorption spectrum (λ_{max} 370 nm) nearly identical to that recorded for the 1-naphthylmethyl radical generated by one-photon photolysis of 1-chloromethylnaphthalene. This evidence, combined with two-laser product evidence showing the expected cleavage products confirms the two-photon α-cleavage reaction. It must be noted that, due to a minor amount of one-photon cleavage occurring from the singlet manifold, this system does not possess the same degree of multiphoton specificity as that of benzil.

MPS chemical systems are expected to have a wide variety of applications in the field of industrial polymer chemistry. Indeed their use in the techniques of holography and photochemical machining are presently being investigated. However, much wider applications are anticipated, particularly for use as photoinitiators in techniques requiring a high degree of spatial control and resolution (e.g., photoresists). The two-laser technique described here represents an important tool in the investigation of potential MPS systems.

ACKNOWLEDGMENT

The author wishes to acknowledge Mrs. H. Letaif for the typing of this manuscript.

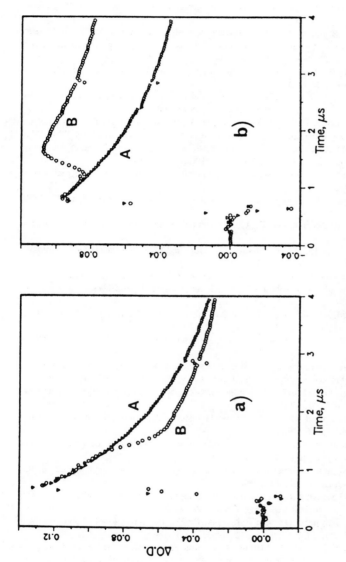

FIGURE 11.   Decay of di-1-naphthylpropan-2-one triplet-triplet absorption monitored at 430 nm (a) and 370 nm (b) obtained under one- (A) laser (308 nm) and two- (B) laser (308 + 420 nm) irradiation conditions. (From McGimpsey, W. G. and Scaiano, J. C., *J. Am. Chem. Soc.*, 109, 2179, 1987. With permission.)

REFERENCES

1. For a description of flash photolysis techniques see **Rabek, J. F.**, *Experimental Methods in Photochemistry and Photophysics: Part 2*, Wiley-Interscience, New York, 1982, 822.
2. **Hui, M. H., De Mayo, P., Suau, R., and Ware, W. R.**, Thione photochemistry: fluorescence from higher excited states, *Chem. Phys. Lett.*, 31, 257, 1975.
3. **Maciejewski, A. and Steer, R. P.**, Photophysics of the second excited singlet states of xanthione and related thiones in perfluoroalkane solvents, *J. Am. Chem. Soc.*, 105, 6738, 1983.
4. **Maciejewski, A., Safarzadeh-Amiri, A., Verral, R. E., and Steer, R. P.**, Radiationless decay of the second excited singlet states of aromatic thiones: experimental verification of the energy gap law, *Chem. Phys.*, 87, 295, 1984.
5. **Maciejewski, A., Demmer, D. R., James, D. R., Safarzadeh-Amiri, A., Verral, R. E., and Steer, R. P.**, Relaxation of the second excited singlet states of aromatic thiones: the role of specific solute-solvent interactions, *J. Am. Chem. Soc.*, 107, 2831, 1985.
6. **Maciejewski, A. and Steer, R. P.**, Bimolecular quenching of the second excited singlet state of tetramethylindanethione, *J. Photochem.*, 24, 303, 1984.
7. **Maciejewski, A. and Steer, R. P.**, Effect of solvent on the subnanosecond decay of the second excited singlet state of tetramethylindanethione, *Chem. Phys. Lett.*, 100, 540, 1983.
8. **Anderson, R. W., Jr., Hochstrasser, R. M., and Pownall, H. J.**, Picosecond absorption spectra of the second excited state of a molecule in the condensed phase: xanthione, *Chem. Phys. Lett.*, 43, 224, 1976.
9. **Falk, K. J., Knight, A. R., Maciejewski, A., and Steer, R. P.**, Concerning the lifetime of the second excited singlet state of adamantanethione, *J. Am. Chem. Soc.*, 106, 8292, 1984.
10. **Murata, S., Iwanaga, C., Toda, T., and Kokubun, H.**, Fluorescence yields of azulene derivatives, *Chem. Phys. Lett.*, 13, 101, 1972.
11. **Beer, M. and Longuet-Higgins, H. C.**, Anomalous light emission of azulene, *J. Chem. Phys.*, 23, 1390, 1955.
12. **Eaton, D. F., Evans, T. R., and Leermakers, P. A.**, The anomalous electronic properties of azulenes, *Mol. Photochem.*, 1, 347, 1969.
13. **Hirayama, F., Gregory, T. A., and Lipsky, S.**, Fluorescence from highly excited states of some aromatic molecules in solution, *J. Chem. Phys.*, 58, 4696, 1973.
14. **Easterly, C. E. and Christophorou, L. G.**, Fluorescence emission from the first- and second-excited π-singlet states of aromatic hydrocarbons in solution, and their temperature dependences, *J. Chem. Soc. Faraday Trans. 2*, 70, 267, 1974.
15. **Dawson, W. R. and Kropp, J. L.**, Radiationless deactivation and anomalous fluorescence of singlet 1,12-benzperylene, *J. Phys. Chem.*, 73, 1752, 1969.
16. **Gillispie, G. D. and Lim, E. C.**, Further results on the triplet-triplet fluorescence of anthracenes, *Chem. Phys. Lett.*, 63, 355, 1979.
17. **Meisel, D., Das, P. K., Hug, G. L., Bhattacharyya, K., and Fessenden, R. W.**, Temperature dependence of the lifetime of benzyl and other arylmethyl radicals, *J. Am. Chem. Soc.*, 108, 4706, 1986.
18. **Bromberg, A., Schmidt, K. H., and Meisel, D.**, Photophysics and photochemistry of arylmethyl radicals in liquids, *J. Am. Chem. Soc.*, 107, 83, 1985.
19. **Bromberg, A., Schmidt, K. H., and Meisel, D.**, Photochemistry and photophysics of phenylmethyl radicals, *J. Am. Chem. Soc.*, 106, 3056, 1984.
20. **Scaiano, J. C., Tanner, M., and Weir, D.**, Exploratory study of the intermolecular reactivity of excited diphenylmethyl radicals, *J. Am. Chem. Soc.*, 107, 4396, 1985.
21. **Weir, D. and Scaiano, J. C.**, Substituent effects on the lifetime and fluorescence of excited diphenylmethyl radicals in solution, *Chem. Phys. Lett.*, 128, 156, 1986.
22. **Weir, D., Johnston, L. J., and Scaiano, J. C.**, Excited states properties of arylmethyl radicals containing naphthyl, phenanthryl and biphenyl moieties, *J. Phys. Chem.*, 92, 1742, 1988.
23. **Kelley, D. F., Milton, S. V., Huppert, D., and Rentzepis, P. M.**, Formation kinetics and spectra of aromatic radicals in solution, *J. Phys. Chem.*, 87, 1842, 1983.
24. **Hilinski, E. F., Huppert, D., Kelley, D. F., Milton, S. V., and Rentzepis, P. M.**, Photodissociation of haloaromatics: detection, kinetics, and mechanism of arylmethyl radical formation, *J. Am. Chem. Soc.*, 106, 1951, 1984.
25. **Tokamura, K., Udagawa, M., and Itoh, M.**, Two-step laser excitation fluorescence study of the reaction of the 1-naphthylmethyl radical produced in the 248 nm KrF laser photolysis of 1-(chloromethyl)-naphthalene in hexane at room temperature, *J. Phys. Chem.*, 89, 5147, 1985.
26. **Johnston, L. J. and Scaiano, J. C.**, Generation, spectroscopy, and reactivity of excited 1-naphthylmethyl radicals, *J. Am. Chem. Soc.*, 107, 6368, 1985.
27. **Tokamura, K., Mizukami, N., Udagawa, M., and Itoh, M.**, Doublet-doublet fluorescence and coupling reactions of 9-anthrylmethyl radical in fluid hexane solution studied by two-step laser excitation fluorescence spectroscopy, *J. Phys. Chem.*, 90, 3873, 1986.

28. **Razi Naqvi, K. and Wild, U. P.**, Room temperature fluorescence of the diphenyl ketyl radical, *Chem. Phys. Lett.*, 41, 570, 1976.

29. **Johnston, L. J., Lougnot, D. J., and Scaiano, J. C.**, Isotope and temperature effects on the lifetime of excited diphenyl ketyl radicals, *Chem. Phys. Lett.*, 129, 205, 1906.

30. **Johnston, L. J., Lougnot, D. J., and Scaiano, J. C.**, Photochemistry of diphenyl ketyl radicals: spectroscopy, kinetics and mechanisms, *J. Am. Chem. Soc.*, 110, 518, 1988.

31. **Okamura, T. and Yip, R. W.**, Fluorescence of the hindered phenoxyl radical, *Bull. Chem. Soc. Jpn.*, 51, 937, 1978.

32. **Bhattacharyya, K., Das, P. K., Fessenden, R. W., George, M. V., Gopidas, K. R., and Hug, G. L.**, Photophysics of furanoxy radicals. Fluorescence and triplet-doublet energy transfer, *J. Phys. Chem.*, 89, 4164, 1985.

33. **Jinguji, M., Imamura, T., Obi, K., and Tanaka, I.**, Emission lifetimes of thiophenoxy radicals at 77 K, *Chem. Phys. Lett.*, 109, 31, 1984.

34. **Morine, G. A. and Kuntz, R. R.**, Spectral shifts of the *p*-aminophenylthiyl radical absorption and emission in solution, *Chem. Phys. Lett.*, 67, 552, 1979.

35. **Smirnov, V. A. and Plotnikov, V. G.**, The luminescence-spectroscopic properties of aromatic radicals and biradicals, *Russ. Chem. Rev.*, 55, 929, 1986.

36. **Scaiano, J. C.**, Solvent effects in the photochemistry of xanthone, *J. Am. Chem. Soc.*, 102, 7747, 1980.

37. **McGimpsey, W. G. and Scaiano, J. C.**, unpublished results.

38. **Carmichael, I. and Hug, G. L.**, Triplet-triplet absorption spectra of organic molecules in condensed phases, *J. Phys. Chem. Ref. Data*, 15, 1, 1986.

39. **McGimpsey, W. G. and Scaiano, J. C.**, Photoexcitation of benzophenone triplets: a two-photon pathway for ground state repopulation, *Chem. Phys. Lett.*, 138, 13, 1987.

40. **McGimpsey, W. G. and Scaiano, J. C.**, Study of energy transfer from upper states in solution, *J. Am. Chem. Soc.*, 110, 2299, 1988.

40a. A similar transient has been observed following photolysis of 9-hydroxyfluorene in methanol. **Gaillard, E., Fox, M. A., and Wan, P.**, A kinetic study of the photosolvolysis of 9-fluorenol, *J. Am. Chem. Soc.*, 111, 2180, 1989.

41. **Bossy, J. M., Bühler, R. E., and Ebert, M.**, Transient bromine atom charge-transfer complexes observed by pulse radiolysis, *J. Am. Chem. Soc.*, 92, 1099, 1970.

42. **Zupancic, J. J. and Schuster, G. B.**, Chemistry of fluorenylidene: direct observation of, and kinetic measurements on, a singlet and a triplet carbene at room temperature, *J. Am. Chem. Soc.*, 102, 5958, 1980.

43. **Schwerzel, R. E., Wood, V. E., McGinniss, V. D., and Verber, C. M.**, Three-dimensional photochemical machining with lasers, *SPIE Appl. Lasers Ind. Chem.*, 458, 90, 1984.

44. This technique was originally described in U.S. Patents 4,041,476 (**Swainson, W. K.**, inventor); 3,609,706 (**Adamson, A. W.**, inventor); 3,829,838 (**Lewis, J. D., Verber, C. M., and McGhee, R. B.**, inventors).

45. **Bräuchle, C., Burland, D. M., and Bjorklund, G. C.**, Hydrogen abstraction by benzophenone studied by holographic photochemistry, *J. Phys. Chem.*, 85, 123, 1981.

46. **Bräuchle, C., Wild, U. P., Burland, D. M., Bjorklund, G. C., and Alvarez, D. C.**, A new class of materials for holography in the infrared, *J. Res. Dev.*, 26, 217, 1982.

47. **Bräuchle, C., Wild, U. P., Burland, D. M., Bjorklund, G. C., and Alvarez, D. C.**, Two-photon holographic recording with continuous-wave lasers in the 750—1100 nm range, *Optics Lett.*, 7, 177, 1982.

48. **Burland, D. M. and Bräuchle, C.**, The use of holography to investigate complex photochemical reactions, *J. Chem. Phys.*, 76, 4502, 1982.

49. **Bräuchle, C.**, Holography as a new tool for investigating photochemical reactions in the solid state, *Mol. Cryst. Liq. Cryst.*, 96, 83, 1983.

50. **Pinsle, J., Deeg, F. W., and Bräuchle, C.**, Two-photon four-level hologram recording in poly(alkyl-α-cyanoacrylates), *Appl. Phys. B*, 40, 77, 1986.

51. **Ledwith, A., Russell, P. J., and Sutcliffe, L. H.**, Radical intermediates in the photochemical decomposition of benzoin and related compounds, *J. Chem. Soc. Perkin Trans. 2*, 1925, 1972.

52. **Lamola, A. A. and Hammond, G. S.**, Mechanisms of photochemical reactions in solution. XXXIII. Intersystem crossing efficiencies, *J. Chem. Phys.*, 43, 2129, 1965.

53. **Evans, T. R. and Leermakers, P. A.**, Emission spectra and excited-state geometry of α-diketones, *J. Am. Chem. Soc.*, 89, 4380, 1967.

54. **McGimpsey, W. G. and Scaiano, J. C.**, A two-photon study of the "reluctant" Norrish type I reaction of benzil, *J. Am. Chem. Soc.*, 109, 2179, 1987.

55. **Johnston, L. J. and Scaiano, J. C.**, One- and two-photon processes in the photochemistry of 1,3-bis(1-naphthyl)-2-propanone: an example of a "reluctant" Norrish type I reaction, *J. Am. Chem. Soc.*, 109, 5487, 1987.

Chapter 5

APPLICATION OF LASER FLASH PHOTOLYSIS TO THE STUDY OF PHOTOPOLYMERIZATION REACTIONS IN NONAQUEOUS SYSTEMS

Wolfram Schnabel

TABLE OF CONTENTS

I. INTRODUCTION

A. GENERAL ASPECTS ON PHOTOPOLYMERIZATION
1. Importance of Light-Induced Polymerization for Application in Technical Processes

The striking advantage of light-induced over thermally initiated polymerizations is based on the fact that light-induced polymerizations can be carried out at rather low temperatures, i.e., at room or even lower temperatures. Therefore, various techniques were developed in modern times which utilize this advantage in technical processes such as in curing of coatings on wood, paper, metal, and plastics. There are other interesting applications concerning, for example, fabrication of laser vision video discs and curing of dental fillings.

Typical formulations used with photocurable systems contain mono-, di-, and trifunctional monomers, prepolymers having terminal functional groups, and an initiator. It is notable that photocurable formulations are free of solvents and need no thermal postcuring treatment which makes them safe from the environmental point of view and, in addition, helps in saving energy.

It should be pointed out that the role that light plays in the whole process of photopolymerization is restricted to the initiation step. Actually, it is the initiation step which has been investigated with the aid of the laser flash photolysis method and which is dealt with in this chapter. It was not intended to treat here other aspects of photopolymerization. For more comprehensive information the reader's attention is focused, therefore, to various books and reviews that were published in the past.[1-15]

2. Why Do We Need Photoinitiators?

As has been pointed out above, the mode of action of light in photopolymerization is restricted to the process of initiation, in which the so-called initiator is involved. The necessity of using initiators derives from the fact that light of wavelength 250 to 400 nm is used to generate reactive species capable of initiating the polymerization of monomers, i.e., free radicals or ions, and that most monomers are not efficiently decomposed into free radicals or ions upon irradiation with light of this wavelength range. Moreover, most monomers do not absorb light at $\lambda > 320$ nm, where commercial lamps predominantly emit. The requirements to an efficient initiator are, therefore, high quantum yield of radicals or ions, respectively, high reactivity of radicals or ions formed by the decomposition, or by the reaction of excited initiator molecules towards monomers and the incapability of excited initiator molecules to react with monomers forming unreactive products. A rather high absorptivity of the initiator in the near UV or visible wavelength range is a prerequisite for its utilization. In order to understand the mechanism of initiation, the photochemistry of initiators must be known. Actually, laser flash photolysis in conjunction with various detection methods (vide infra, see Section II.A) served as a very valuable tool in mechanistic and kinetic studies concerning the photolysis of photoinitiators and the reactions of initiator fragments or excited initiator molecules with polymerizable monomers. In this chapter such investigations are dealt with to some extent.

B. GENERAL TREATMENT OF LIGHT-INDUCED POLYMERIZATION
1. Radical Polymerization

The initiation of the polymerization of a monomer M is induced by free radicals R· which are produced by the photolysis of an initiator I.

$$R\cdot \; + \; M \xrightarrow{\;k_{R+M}\;} R\text{--}M\cdot \qquad (1)$$

There are two types of free radical initiators: (1) the cleavage type initiators, which spontaneously undergo cleavage into free radicals according to Reaction 2:

$$I \rightarrow I^* \xrightarrow{k_R} R_1^\cdot + R_2^\cdot \qquad (2)$$

and (2) the hydrogen-abstraction-type initiators which form radical pairs by reacting, in the electronically excited state, with hydrogen donors according to Reaction 3:

$$I^* + RH \rightarrow \cdot IH + R\cdot \qquad (3)$$

Reaction 3 frequently involves a transient state, that is, excited initiator molecules form complexes with the donor, so-called exciplexes. They can be of charge-transfer nature which may give rise to the formation of ion pairs, apart from the formation of free radicals. Ion pair formation is favored in polar media. The whole mechanism is schematically depicted by Reactions 4 and 5:

$$
\begin{array}{ccc}
 & & \longrightarrow R_1^\cdot + R_2^\cdot \qquad (4) \\
I^* + RH \rightarrow & (I^* \dots RH) & \\
 & \text{exciplex} & \\
 & \downarrow & \longrightarrow \cdot K^+ + \cdot A^- \qquad (5) \\
 & I + RH + \text{energy} &
\end{array}
$$

Here, $R_1^\cdot$ and $R_2^\cdot$ denote free radicals and $\cdot K^+$ and $\cdot A^-$ cation and anion radicals, respectively.

Once the free radicals, formed by the photoreactions of the initiator, have reacted with monomer molecules according to Reaction 1 the polymerization proceeds in the same manner as in the case of thermal or electrochemical initiation, that is, chain propagation and chain termination are generally not affected by light.

Monomers that polymerize by free radical mechanism are generally olefin derivatives of the structure:

$$
\begin{array}{c}
CH_2{=}CH \\
| \\
R
\end{array}
$$

Usually, these monomers are prone to react not only with free radicals but also with excited initiator molecules according to Reaction 6:

$$M + I^* \xrightarrow{k_q} \text{nonradical products} \qquad (6)$$

Reaction 6 can become important if I^* is relatively long-lived which usually applies to triplet excited initiators. One of the aims of flash photolysis studies was, therefore, to investigate the kinetics of the competing Reactions 1 and 6 in order to characterize the radical forming efficiency of certain initiators.

2. Cationic Polymerization

Apart from certain olefin derivatives various compounds of quite different chemical nature can be polymerized by an ionic mechanism. For example, oxiranes (epoxides) and other cyclic ethers such as tetrahydrofurane, can be polymerized by a cationic mechanism. The polymerization can be either induced by protons or by carbocations. Both species can be generated photochemically. Protons can be produced, for example, upon irradiation of phenyldiazonium hexafluorophosphate:[16-18]

$$Ph-N=N^+PF_6^- \xrightarrow{h\nu} PhF + N_2 + PF_5 \qquad (7)$$

PF$_5$ is a Lewis acid which reacts readily with water or alcohol forming a Brønstedt acid:

$$PF_5 + H_2O \rightarrow H^+PF_5OH^- \qquad (8)$$

$$PF_5 + ROH \rightarrow H^+PF_5OR^- \qquad (9)$$

Cationic polymerization of epoxides occurs via the following mechanism:

(10)

Apart from aryldiazonium salts also other onium salts such as diaryliodonium and triarylsulfonium salts have obtained prominence as photoinitiators in cationic polymerization. According to the mechanism postulated by McEwen[19] and Crivello[16] the primary step in the photolysis of diaryliodonium cation is homolytic bond cleavage to yield the aryliodonium radical cation, ArI‡, and an aryl radical. H-abstraction from the solvent or the monomer (RH) is postulated to result in the radical R· and the protonated aryl iodide which dissociates into H$^+$X$^-$ and ArI.

$$Ar_2I^+X^- \xrightarrow{h\nu} ArI^{\ddagger}X^- + \cdot Ar \qquad (11)$$

$$ArI^{\ddagger}X^- + RH \longrightarrow ArI^+-HX^- + R\cdot \qquad (12)$$

$$ArI^+-HX^- \longrightarrow ArI + H^+X^- \qquad (13)$$

Typical nonnucleophilic anions in these onium salts are BF$_4^-$, PF$_6^-$ AsF$_6^-$, and SbF$_6^-$

An elegant method to produce carbocations photochemically consists in the photolysis of certain free radical initiators in the presence of an appropriate onium salt.[12,13] Provided the redox potential of the onium cations are appropriate, they can oxidize carbon-centered radicals. For example, hydroxy alkyl radicals can be oxidized by diphenyliodonium ions. This way, the promotion of cationic polymerization by free radicals is possible. The photolysis of 2,2-dimethoxy-2-phenyl-acetophenone (DMPA) gives rise to the formation of carbocations according to the following mechanism:

$$\text{Ph-C-C-Ph} \xrightarrow{h\nu} \text{PhC·} + \text{·C-Ph}$$

(chemical scheme with OMe, O substituents)

$$\text{Ph}_2\text{I}^{\oplus} + \text{·C-Ph} \longrightarrow \text{PhI} + \text{·Ph} + \oplus\text{C-Ph}$$

(chemical scheme with OMe substituents)

Apart from diphenyliodonium ion also 4-(phenylthio)phenyldiphenylsulfonium ion (PTPDS) has been found to act as promotor of cationic polymerization.[13]

$$\text{Ph-S-} \langle \bigcirc \rangle \text{-} \overset{\oplus}{\text{S}}(\text{Ph})_2$$

(PTPDS)

II. EXPERIMENTAL TECHNIQUES

A. LASER FLASH PHOTOLYSIS AS A TOOL FOR KINETIC AND MECHANISTIC STUDIES

With the advent of high-power lasers producing single pico- or nanosecond light flashes in the UV and visible wavelength range, the possibility to detect short-lived intermediates generated by the photolysis of molecules was given. The application of laser flash photolysis for this purpose afforded detection methods of high sensitivity and high time resolution. Based on modern electronic techniques various detection methods were developed, which will be introduced briefly to the reader in the following sections.

B. OPTICAL ABSORPTION AND EMISSION DETECTION

In many cases, transients formed by the photolysis of a stable compound such as electronically excited molecules, free radicals, or instable molecules in the ground state possess absorption spectra which differ from the ground state spectrum of the irradiated compound. Changes in the optical absorption or the formation of new absorption bands then give evidence for the existence of transients. Time-resolved measurements of changes in the optical density result in kinetic data concerning formation and decay of the transients. A typical setup comprising a laser as the source of the photolyzing light, a cell containing the sample or the sample solution, and a monochromator/photomultiplier assembly is shown in Figure 1. In this case, the analyzing light beam, generated, for example, by a xenon lamp, passes the sample at an angle of 90° with respect to the laser light beam. The photomultiplier signal is directed to an oscilloscope or to a transient recorder which is connected to a computer.

When the analytical light source is not operated the same setup can be used to record luminescence (fluorescence from excited singlet states or phosphorescence from excited triplet states) emitted from the irradiated sample.

Quite frequently, ruby, neodymium YAG, or nitrogen gas lasers are used. With the aid of frequency multiplication photolyzing light of the following wavelengths can be generated by these lasers: 694 nm and 347 nm (ruby laser), 530 nm, 353 nm, and 265 nm (Nd-YAG laser), and 337 nm (nitrogen laser), In most laboratories, laser systems producing single flashes (duration: about 5 ps to about 20 ns) are used.

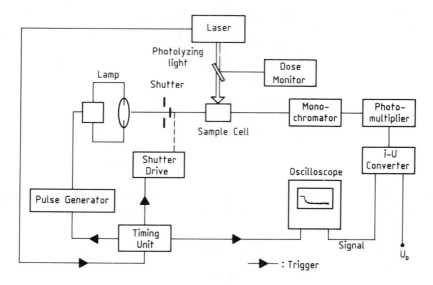

FIGURE 1.   Schematic illustration of a flash photolysis setup with optical absorption detection.

C. PHOTOCURRENT MEASUREMENTS

The formation of ionic species during or after the flash can be followed by electrical conductivity measurements. Both DC and AC methods have been developed.[20,21] Quartz cells containing a couple of platinum electrodes which are placed parallel to the laser light beam are used for these measurements.

D. DETECTION OF FREE RADICALS BY ESR MEASUREMENTS

Time-resolved electron spin resonance (ESR) spectroscopy has become a powerful tool for studying short-lived paramagnetic intermediates and has been applied successfully to evidence photochemically generated free radicals.[22-26] ESR spectrometers with time resolution down to 60 ns permitting the direct detection of rather short-lived free radicals generated by flash photolysis or pulse radiolysis were constructed recently. Notably, the application of this technique was facilitated by chemically induced dynamic electron polarization (CIDEP) of the radicals in cases where the photolysis proceeds via the excited triplet state. This effect frequently increases the intensity of the ESR signal of the free radicals by a factor of 10 to 100.

III. RADICAL POLYMERIZATION

As pointed out in the previous section, initiators are generally used to start light-induced radical polymerizations. In order to understand the mode of action of the initiators, their photochemical reactions were studied by the flash photolysis method. In these studies it was aimed at detecting precursors of free radicals and at measuring the reactivity of radical precursors and free radicals towards polymerizable monomers. In the following paragraphs such investigations concerning the two types of radical photoinitiators will be discussed.

A. CLEAVAGE-TYPE PHOTOINITIATORS

This class of initiators comprises compounds which decompose readily into free radicals upon electronic excitation. Regarding technical applications, most prominent representatives of this class are certain aromatic carbonyl compounds, especially benzoin and benzoin derivatives, benzil ketals, certain acetophenone derivatives, α-hydroxyalkylphenones, and

O-acyl-α-oximinoketones. Recently, also acylphosphine oxides were introduced as photoinitiators of appreciable technical importance.

1. Benzoin and Derivatives

The formation of radicals according to Reaction 14 was studied by irradiating B, BA, BME, and BIPE in benzene solution:[27,28]

$$
\begin{array}{ccc}
\text{O} & \text{OR} & \\
\| & | & \\
\text{Ph–C–C–Ph} & \xrightarrow{h\nu} & \text{Ph–C·} + ·\text{C–Ph} \\
| & & \\
\text{H} & & \\
\end{array}
\tag{14}
$$

		R
		H
Benzoin	(B)	H
Benzoinacetate	(BA)	COCH$_3$
Benzoinmethylether	(BME)	CH$_3$
Benzoinisopropylether	(BIPE)	CH(CH$_3$)$_2$

Evidence for the occurrence of Reaction 14 was obtained from product studies in the absence of vinyl compounds. Benzoyl radicals formed benzaldehyde via H-abstraction from surrounding molecules or scavengers, such as thiols, or formed benzil by dimerization. The α-substituted benzyl radical which is generated simultaneously with the benzoyl radical formed a pinacol derivative:

$$
\text{Ph–C·} + \text{RH} \rightarrow \text{Ph–C–H} + \text{R·}
\tag{15}
$$

$$
2\text{Ph–C·} \rightarrow \text{Ph–C–C–Ph}
\tag{16}
$$

$$
2\text{Ph–C·} \rightarrow \text{Ph–C–C–Ph}
\tag{17}
$$

Moreover, ^{1}H-NMR-CIDNP, ^{13}C-NMR-CIDNP, and ESR investigations including spin trapping experiments have confirmed the conclusions arrived at by photoproduct studies. This work has been reviewed in detail by Hageman.[4] In the flash photolysis studies in benzene solution it turned out that the lifetime of the excited state of the benzoin compounds depends strongly on the chemical nature of R. In the case of the benzoin ethers BE and BIPE absorption spectra of excited states could not be detected by irradiating solutions with 15-ns flashes of 347-nm light. According to Lewis et al.,[29] the lifetime of radical precursors should be shorter than 10^{-10} s, because α-cleavage of BE and BIPE could not be quenched by usual triplet quenchers. In the case of B and BA, the absorption spectra of triplets were observed.[27] Figure 2 shows the absorption spectrum of BA triplets and Figure 3 demonstrates that, in the case of B, the decrease in the T-T absorption formed during the 0.1-ns flash was correlated to the increase in the absorption of free radicals. The triplet decay rate

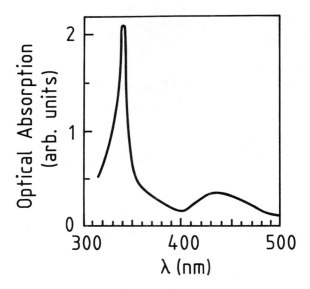

FIGURE 2.   Triplet-triplet optical absorption spectrum of benzoin acetate recorded in benzene solution at room temperature. $D_{abs} = 4.6 \times 10^{-5}$ einstein/l. [BA] $= 7.9 \times 10^{-4}$ mol/l. (From Kuhlmann, R. and Schnabel, W., *Polymer*, 18, 1163, 1977. With permission.)

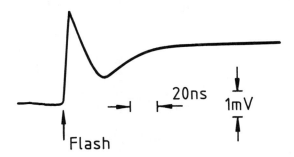

FIGURE 3.   Flash photolysis of benzoin in benzene solution. [B] $= 1.78 \times 10^{-3}$ mol/l. $\lambda_{inc} = 347$ nm. Kinetic trace, recorded at $\lambda_{obs} = 480$ nm, demonstrating the decay of the triplets, formed during the 0.1-ns flash, and the simultaneous formation of radicals. (From Kuhlmann, R. and Schnabel, W., *Polymer*, 18, 1163, 1977. With permission.)

constants were 1.1×10^8 s^{-1} (B) and 5.3×10^7 s^{-1} (BA). The triplet nature of these species was evidenced by the formation of the T-T absorption spectrum of naphthalene. Triplets of B and BA were quenched by naphthalene with the following rate constants: 7.7×10^9 l/mol s (B) and 5.9×10^9 l/mol s (BA).

With BA, BME, and BIPE a rapidly decaying (benzoyl) and a slowly decaying radical (α-substituted benzyl) were recognized. In the case of B only one decay mode of the transient absorption reflecting the formation of benzaldehyde by disproportionation of unlike radicals was observed. In the cases of BA, BME, and BIPE the combination of like radicals was predominant. The reaction of benzoyl radicals with styrene was evidenced ($k_{R+M} = 1.2 \times 10^5$ l/mol s), whereas the reaction of α-substituted benzyl radicals with this monomer could not be detected.[27] Rate constants of the reaction of benzoyl radicals with other monomers are listed in Table 3.[28] The values range from 2×10^4 to 1.2×10^5 l/mol s.

The fraction of initiator radicals f_P capable of initiating kinetic chains was derived from

TABLE 1
k_q/k_R Ratios (l/mol) for B, BA, and BME
Determined in Benzene Solution at Room
Temperature for Various Monomers with
the Aid of Equation 18

Monomer	Initiator		
	BA	B	BME
Styrene	90	6.8	0.17
Methylmethacrylate	15	0.4	0.02
Vinylacetate	1.1	$<10^{-2}$	$<10^{-2}$

polymerization rate determinations.[28] It turned out that only about 30% of the radicals formed by the photolysis of B, BA, or BME initiated polymerization, indicating again the low reactivity of α-substituted benzyl radicals towards olefinic monomers.

The efficiency of an initiator is largely determined by the extent to which its electronically excited states are quenched by monomer if the quenching reaction leads to nonradical products (see Reaction 6). Ratios of rate constants k_q and k_R were determined by measuring the radical yield at constant absorbed dose as a function of monomer concentration according to Equation 18:

$$O.D._0/O.D. = 1 + (k_q/k_R)[M] \tag{18}$$

Here, O.D. is the optical density of free radicals and the subscript 0 denotes the absence of monomer.

It can be seen from Table 1 that, apart from the system B/vinyl acetate, k_q/k_R is rather high in the case of B and BA. This means that these initiators are inappropriate to induce the radical polymerization. On the basis of k_q/k_R ratios only BME (and similarly also BIPE) is sufficiently efficient to be applied as initiator in technical processes. In cases where k_R could be measured (not with BEM and BIPE) k_q was calculated. Results are shown in Table 2, where it is seen that excited molecules of B and BA react with almost diffusion controlled rates with styrene and that they also possess a rather high reactivity towards methylmethacrylate (MMA) and acrylonitrile (AN). Quenching of excited BIPE and BME by acrylonitrile and vinylacetate was not detectable.

2. Deoxybenzoin and Derivatives

From product analyses carried out after continuous irradiation of deoxybenzoins in benzene solution, it was inferred that α-cleavage is the only chemical route in the deactivation of electronically excited deoxybenzoins:[30-32]

$$\text{Ph–C–CH–Ph} \xrightarrow{h\nu} \text{Ph–C·} + \text{·CH–Ph} \tag{19}$$

		R
deoxybenzoin	(DB)	H
methyldeoxybenzoin	(MDB)	CH_3
phenyldeoxybenzoin	(PDB)	Ph

A quantum yield of unity was reported for α-scission of MDB in benzene solution. In

TABLE 2
Rate constants k_q (l/mol s) of Quenching of Excited States of Benzoin Derivatives by Monomers in Benzene Solution at Room Temperature

Initiator		Monomer[a]			
		St	MMA	VA	AN
Ph—C—CH—Ph ‖ | O OH	B	8.1×10^9	4.8×10^8		

Ph—C—CH—Ph
 ‖ |
 O O—C—CH$_3$
          ‖
          O
	BA	4.8×10^9	8.0×10^8	6.0×10^7	1.3×10^9

Ph—C—CH$_2$—Ph ‖ O	DB	6.5×10^9	1.0×10^9	1.7×10^7	7.0×10^8
Ph—C—CH—Ph ‖ | O CH$_3$	MDB	1.5×10^9	2.0×10^8	$<10^7$	3.8×10^8
Ph—C—CH—Ph ‖ | O Ph	PDB	1.9×10^8	1.3×10^7	5.7×10^6	1.8×10^8

[a]   St: styrene, MMA: methylmethacrylate, VA: vinylacetate, and AN: acrylonitrile.

the presence of propanol-2 α-scission and reduction are competing chemical triplet deactivation routes.[32]

Flash photolysis of DB, MDB, and PDB with 25-ns flashes of 347-nm light yielded the absorption spectra of triplets.[33] Figure 4 shows transient absorption spectra recorded after irradiation of DB and PDB in benzene solution. Spectra obtained at t = 0 are ascribed to triplets and spectra recorded after the decay of the triplets to free radicals formed by α-scission. Actually, these spectra are mainly due to the substituted methyl radicals because benzoyl radicals absorb only very weakly at λ > 300 nm. The following k_R values (in s^{-1}) were determined by triplet lifetime measurements: 1.2×10^6 (DB), 3.7×10^6 (PDB), and 1.1×10^7 (MDB). The reactivity of the radicals PhĊH$_2$, PhĊH(CH$_3$), and PhĊH(Ph) towards olefinic monomers was not systematically investigated. It was found that diphenylmethyl radicals (Ph$_2$ĊH) react relatively slowly with MMA and AN: $k_{R+M} = 5.4 \times 10^4$ l/mol s and $k_{R+M} = 1.5 \times 10^3$ l/mol s. The relatively low reactivity of these radicals is probably a consequence of pronounced resonance stabilization.

In many cases, the utilization of deoxybenzoins as photoinitiators is strongly impeded by triplet quenching via monomer. Triplet quenching rate constants are compiled in the lower part of Table 2. In the case of the system DB/styrene, triplet quenching is a diffusion-controlled reaction as is expected on the basis of the triplet energies: $E_T = 260$ kJ/mol (styrene) and $E_T = 302$ kJ/mol (DB). The triplet energies of the other monomers examined in these studies are probably higher than those of the deoxybenzoins. In these cases the relatively high values of the quenching constants are explainable in terms of exciplexes that are formed as transient species in the reaction of excited deoxybenzoins with monomers.

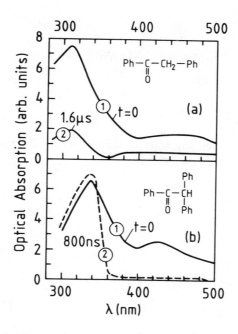

FIGURE 4. Transient optical absorption spectra recorded with deoxybenzoin (a) and α-phenyldeoxybenzoin (b) in benzene solution at room temperature. λ_{inc} = 347 nm. Flash duration: 25 ns. [DB] = 1 × 10^{-3} mol/l; [PDB] = 8 × 10^{-4} mol/l. D_{abs} = 5 × 10^{-5} einstein/l. (From Amirzadeh, G., Kuhlmann, R., and Schnabel, W., *J. Photochem.*, 10, 133, 1979. With permission.)

3. α-Hydroxyalkylphenones and Derivatives

Several hydroxyalkylphenones of the general structure

are efficient photoinitiators for curing formulations containing acrylic ester derivatives.[34] Most efficient are compounds with R_1 = H and R_2 = H or CH(CH$_3$)$_2$ and 1-benzoylcyclohexan-1-ol:[35]

α-Cleavage seems to be an important chemical deactivation route of electronically excited 1-phenyl-2-hydroxy-2-methylpropanone-1 (PHMP):

$$Ph-\underset{\underset{O}{\overset{|}{\underset{|}{CH_3}}}}{\overset{OH}{\overset{|}{C}}}-C-CH_3 \xrightarrow{h\nu} Ph-\underset{O}{\overset{OH}{C\cdot}} + \cdot\underset{CH_3}{\overset{|}{C}}-CH_3$$

(PHMP)   (20)

This was evidenced by scavenging benzoyl radicals with 2,2,6,6-tetramethylpiperidinoxyl (TMPO) and 1,1-diphenylethylene:[4,36,37]

$$Ph-\underset{O}{\overset{}{C\cdot}} + \cdot O-N \longrightarrow Ph-\underset{O}{\overset{}{C-O-N}} \tag{21}$$

$$Ph-\underset{O}{\overset{}{C\cdot}} + CH_2{=}\underset{Ph}{\overset{Ph}{\overset{|}{C}}} \longrightarrow Ph-\underset{O}{\overset{}{C}}-CH_2-\underset{Ph}{\overset{Ph}{\overset{|}{C\cdot}}} \tag{22}$$

Flash photolysis of PHMP in various solvents yielded evidence for the existence of the triplet state (lifetime: about 30 ns).[38] Figure 5a shows transient spectra recorded in benzene solution. The difference spectrum presented in Figure 5b was assigned to triplets. Upon irradiating PHMP in solution containing naphthalene, triplets of the latter were formed by energy transfer with a quantum yield $\phi(T) = 0.3$. Because of the relatively long lifetime of PHMP triplets, there is a good chance for the formation of ketyl radicals according to Reaction 23 in solvents of high hydrogen donor capability.

$$\prescript{3}{}{\left[Ph-\underset{\underset{O}{\overset{|}{\underset{|}{CH_3}}}}{\overset{CH_3}{\overset{|}{C}}}-C-OH \right]^*} + RH \longrightarrow Ph-\underset{HO}{\overset{CH_3}{\overset{|}{\underset{|}{\dot{C}}}}}-\underset{CH_3}{\overset{|}{C}}-OH \quad \cdot R \tag{23}$$

Transient absorption spectra consist in such cases of the overlapping spectra of various radicals and it was difficult, therefore, to detect the primary radicals by their optical absorption. Figure 6 shows a transient absorption spectrum recorded in aqueous solution, where triplets cannot abstract hydrogen from the solvent. This spectrum, with maxima at about 350 and 410 nm, should mainly be due to benzoyl radicals. The spectrum of benzoyl radicals was reported to possess absorption maxima at 368 and 460 nm in 3-methyl-3-pentanol solution.[39]

Certain aspects concerning the efficiency of PHMP as a photoinitiator were also investigated by flash photolysis.[40] In this work prominence was given to the reactivity of PHMP triplets towards typical olefinic monomers: styrene, MMA, and AN. Moreover, the reactivity of fragment radicals, generated by photolysis of the initiator, towards these monomers was studied. Rate constants of the reaction of triplets with monomers k_q were obtained by measuring fragment radical yields as a function of monomer concentration [M]. Straight lines were obtained by plotting $O.D._0/O.D.$ (at 320 nm) vs. [M] according to Equation 18. The slopes of the straight lines yielded k_q/k_R. With $k_R = 3.3 \times 10^9$ s^{-1} the following k_q values (in liters per mole per second) were calculated: 7.7×10^9 (St), 1.0×10^7 (AN), and 8.0×10^6 (MMA).

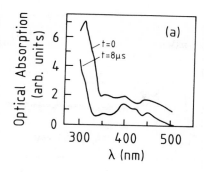

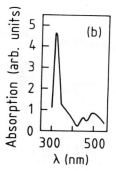

FIGURE 5. (a) Transient absorption spectra recorded with 1-phenyl-2-hydroxy-2-methylpropanone-1 (3×10^{-3} mol/l) in Ar-saturated benzene solution. (b) Difference of the two spectra shown in (a). λ_{inc} = 347 nm; flash duration: 25 ns. D_{abs} = 3.8×10^{-5} einstein/l. (From Eichler, J., Herz, C. P., Naito, I., and Schnabel, W., *J. Photochem.*, 12, 225, 1980. With permission.)

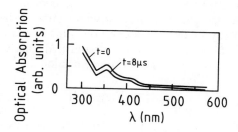

FIGURE 6. Transient absorption spectrum of 1-phenyl-2-hydroxy-2-methylpropanone-1 recorded in Ar-saturated aqueous solution. [Initiator] = 3×10^{-3} mol/l. λ_{inc} = 347 nm. Flash duration: 25 ns. D_{abs} = 3.8×10^{-5} einstein/l. (From Eichler, J., Herz, C. P., Naito, I., and Schnabel, W., *J. Photochem.*, 12, 225, 1980. With permission.)

With the three monomers examined new optical absorption spectra were formed at relatively long times after the flash. The new spectra were attributed to adducts of fragment radicals to monomers, for example, in the case of styrene to the radical:

$$
\begin{array}{c}
\mathrm{H} \\
| \\
\mathrm{R\text{-}CH_2\text{-}C\cdot} \\
| \\
\mathrm{Ph}
\end{array}
$$

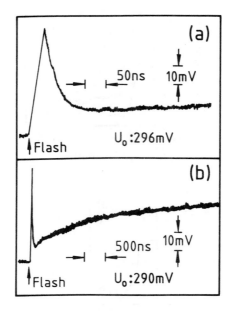

FIGURE 7. Irradiation of 1-phenyl-2-hydroxy-2-methylpropanone-1 (3 × 10^{-3} mol/l) in Ar-saturated benzene solution containing styrene (1 mol/l). Kinetic traces demonstrating the reaction of initiator fragment radicals with styrene at high (a) and at low (b) time-resolution. λ_{inc} = 347 nm; D_{abs} = 1.5 × 10^{-5} einstein/l. (From Salmassi, A., Eichler, J., Herz, C. P., and Schnabel, W., *Polym. Photochem.*, 2, 209, 1982. With permission.)

Typical results obtained with a benzene solution of PHMP containing 1 mol/l styrene are shown in Figure 7. During and after the flash the T-T absorption dropped rapidly, because styrene acts as a triplet quencher permitting only a small fraction of excited initiator molecules to decompose into fragments. The latter reacted with styrene as indicated by the increase in absorption after the decay of the T-T absorption. This increase followed pseudo-first-order kinetics. Upon measuring k_{pseudo} as a function of [M], the following values of k_{R+M} (in liters per mole per second) were obtained from Equation 24:

$$k_{R+M} = k_{pseudo}/[M] \tag{24}$$

4.7 × 10^5 (St), 6.3 × 10^5 (MMA), and 1.6 × 10^6 (AN). These values apply to integrated rate constants pertaining to both kinds of initiator fragment radicals. On the basis of the rate constants measured for benzoyl radicals (see Table 3) the k_{R+M} values for 2-hydroxy-propyl radicals presented in Table 4 were obtained. Notably, these rate constants are significantly higher than those of the reaction of the benzoyl radical. Actually, the high initiator efficiency of PHMP can be understood on the basis of k_{R+M} values. It is remarkable that in the case of PHMP not only both fragment radicals are reactive but that also one of the radicals, namely the 2-hydroxy-propyl radical, possesses a rather high reactivity towards olefinic monomers.

4. Acylphosphine Oxides and Derivatives

Acylphosphine oxides (APO) and acylphosphonates (APP) having the general structures

TABLE 3
Rate Constants k_{R+M} (l/mol s) of the Reaction of
Benzoyl Radicals with Various Monomers Obtained
in Benzene Solution at Room Temperature

Monomer	Initiator		
	BME	BA	B
Styrene	$(1.2 \pm 0.4)10^5$		
Methylmethacrylate	$(0.9 \pm 0.3)10^5$		
Vinylacetate	$(1.2 \pm 0.4)10^5$	$(1.2 \pm 0.4)10^5$	$(1.2 \pm 0.4)10^5$
Acrylonitrile	$(2.0 \pm 0.4)10^4$		

TABLE 4
Rate Constants k_{R+M} of the Reaction of
2-Hydroxy-Propyl and Benzoyl
Radicals with Olefinic Monomers
Measured in Benzene Solution at Room
Temperature

Monomer	Radical	
	Ph—C· $\parallel$ O	H_3C—Ċ—CH_3 $\mid$ OH
Styrene	1.2×10^5	3.5×10^5
Methylmethacrylate	0.9×10^5	5.4×10^5
Acrylonitrile	2.0×10^4	1.58×10^6

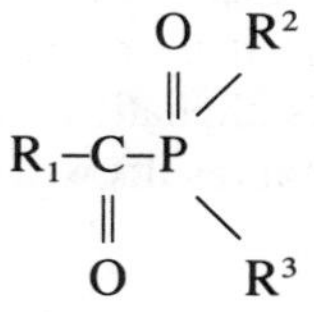

$$R_1\text{–}\overset{\overset{\displaystyle O}{\parallel}}{C}\text{–}\overset{\overset{\displaystyle R^2}{\diagup}}{\underset{\underset{\displaystyle R^3}{\diagdown}}{P}}$$

(APO)

$$R_1\text{–}\overset{\overset{\displaystyle O}{\parallel}}{C}\text{–}\overset{\overset{\displaystyle OR^2}{\diagup}}{\underset{\underset{\displaystyle OR^3}{\diagdown}}{P}}$$

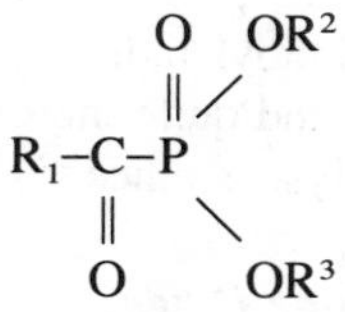

(APP)

were introduced as a new class of photoinitiators in the beginning of this decade.[41-43] Some of these compounds absorb light relatively strongly at wavelengths between 350 and 400 nm which makes them particularly suitable for the initiation of free radical polymerization of olefinic monomers in formulations containing TiO_2 pigments. Moreover, in formulations cured with the aid of an APO initiator yellowing occurs only to a negligible extent. Many of APO and APP compounds undergo α-scission after photoexcitation:

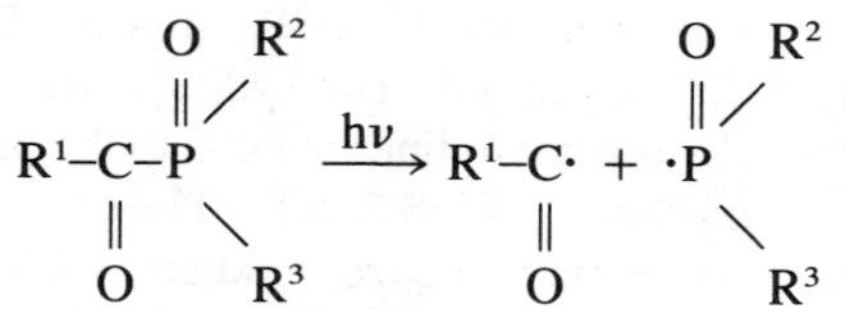

$$R^1\text{–}\overset{\overset{\displaystyle O}{\parallel}}{C}\text{–}\overset{\overset{\displaystyle R^2}{\diagup}}{\underset{\underset{\displaystyle R^3}{\diagdown}}{P}} \xrightarrow{h\nu} R^1\text{–}C\cdot \;+\; \cdot\overset{\overset{\displaystyle R^2}{\diagup}}{\underset{\underset{\displaystyle R^3}{\diagdown}}{P}} \tag{25}$$

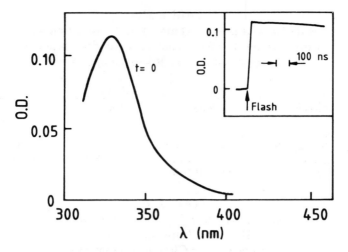

FIGURE 8.   Transient absorption spectrum of diphenylphosphinoyl radical, $Ph_2\dot{P}{=}O$. Radicals were formed during a 20-ns flash upon irradiating an Ar-saturated CH_2Cl_2 solution of PTDPO (7.6×10^{-4} mol/l). $D_{abs} = 3.0 \times 10^{-5}$ einstein/l. Inset: oscillogram depicting the increase in the O.D. during the flash. (From Sumiyoshi, T., Schnabel, W., and Henne, A., *J. Photochem.*, 32, 119, 1986. With permission.)

$$R^1{-}\underset{\underset{O}{\|}}{\overset{\overset{O}{\|}}{C}}{-}P\overset{OR^2}{\underset{OR^3}{}} \quad \xrightarrow{h\nu} \quad R^1{-}\underset{\underset{O}{\|}}{\overset{\overset{O}{\|}}{C}}\cdot \;+\; \cdot P\overset{OR^2}{\underset{OR^3}{}} \qquad (26)$$

Notably, Reactions 25 and 26 proceed, in many cases, with quantum yields between 0.3 and 1.0. Moreover, phosphinoyl radicals, generated by these reactions, are very reactive towards olefinic monomers and quite inert towards aliphatic amines and epoxides. This was concluded from flash photolysis studies.[44-53] Typical results will be presented in this section.

a. Benzoyldiphenylphosphine Oxides

Benzoyldiphenylphosphine oxide (BDPO), *p*-methylbenzoyldiphenylphosphine oxide (PTDPO), and 2,4,6-trimethylbenzoyldiphenylphosphine oxide (TMDPO) readily undergo α-scission according to Reaction 25 upon irradiation at $\lambda > 300$ nm.[45-47] This was evidenced by the formation of diphenylphosphinoyl radicals which possess a strong absorption band at about 335 nm ($\epsilon_{335nm} = 1.9 \times 10^4$ l/mol cm in benzene, at room temperature). Benzoyl radicals absorb comparatively weakly at this wavelength. Therefore, diphenylphosphine oxide radicals could be easily evidenced by optical absorption spectroscopy. Figure 8 shows a typical absorption spectrum obtained with PTDPO in CH_2Cl_2 solution at the end of a 20-ns flash ($\lambda_{inc} = 347$ nm). Extensive studies were performed with TMDPO. In this case phosphorus-centered radicals were detected by ESR upon irradiating TMDPO in benzene solution with a 20-ns flash of 337-nm light.[53] The ESR spectrum recorded at the end of the flash consisted of three lines. The outer two lines are due to the phosphorus-centered radical, $Ph_2\dot{P}{=}O$ (hyperfine coupling constant $a_P = 36.5$ mT). The central line is due to the carbon-centered radical, i.e., to the 2,4,6-trimethylbenzoyl radical, whose relatively weak hyperfine coupling was not resolved in this experiment. The formation of diphenylphosphinoyl radicals by the photolysis of BDPOs is a very rapid process (at least as far as BDPO, PTDPO, and TMDPO are concerned). The build-up lifetime is less than 1 ns as was revealed from flash photolysis experiments with 700-ps time resolution.[57] Therefore, it is difficult to decide whether the radicals have excited singlet and/or triplet states as precursors. Naphthalene quenching experiments[45] yielded naphthalene triplets which could have been formed, however, in this case, also by singlet quenching with subsequent intersystem crossing to the triplet state because relatively high naphthalene concentrations (>0.5 mol/l) had to be applied. In the photolysis of TMDPO in micellar solution of sodium dodecyl sulfate a significant magnetic field effect was observed concerning the decay of radical pairs;[56] in the absence of a magnetic field only one mode of radical pair recombination was observed and the lifetime of the radical pairs was about 140 ns. Under the influence of a magnetic field (1.2 T), on the other hand, an additional decay with a lifetime of 680 ns was detected. The appearance of the additional decay mode is explainable in terms of the relaxation mechanism, provided radical pairs are formed from triplet states. A triplet mechanism was also suggested recently[54] on the basis of ESR studies in conjunction with 20-ns flash photolysis of TMDPO in isopropanol solution at λ_{inc} = 308 nm. In this case also a three-line spectrum was recorded and a hyperfine coupling constant a_P = 37.3 mT, similar to that found in the other work,[53] was obtained. The short lifetime of the radical precursor excludes the possibility of excited state quenching by monomer. If the high quantum yield of α-cleavage, $\phi(\alpha)$, (see Table 5) and the high reactivity of BDPOs towards olefinic monomers (vide infra) are taken into account, it results that diphenylbenzoylphosphine oxides are very effective photoinitiators for the polymerization of these monomers. In practical applications TMDPO has obtained prominence among other phosphine oxides because of its pronounced stability, especially with respect to solvolysis.

It is interesting to note that in the case of o-methylbenzoyldiphenylphosphine oxide (OTDPO) there are two chemical routes of deactivation of excited states:[46] α-scission and enolization. This was revealed from flash photolysis. The mechanism is illustrated in Scheme I. From Figure 9 it is seen that a transient spectrum with two maxima, at 330 and 395 nm, was formed during the flash. The inset in Figure 9 shows that immediately after the flash the O.D. at 330 nm decreased rapidly to some extent with $\tau_{1/2} \approx$ 20 ns while the O.D. at 395 nm increased simultaneously. The remaining absorption spectrum exhibited wavelength-dependent decay kinetics: second-order decay at 330 nm and first-order decay at 395 nm. *trans*-Piperylene (*trans*-1,3-pentadiene) did not affect the decay at 395 nm but strongly accelerated the decay at 330 nm as can be seen from Figure 10. Whereas the rapidly disappearing band at 330 nm is assigned to diphenyl phosphinoyl radicals, the long-lived band at 395 nm is attributed to ground state enol molecules formed by interaction of excited carbonyl groups with hydrogens of the methyl group in ortho position. The very rapid absorption changes depicted in the inset of Figure 9 were thought to indicate the existence of a triplet state, namely the enol triplet, which is converted to the ground state enol with k = 3 × 10^7 s^{-1}. Piperylene reacts rather fast with diphenylphosphinoyl radicals (k = 1.6 × 10^9 l/mol s) and is unreactive towards the enol. It is notable that photoenolization was not observed with TMDPO which also possesses methyl groups in ortho position, although it was detected with trimethylbenzoylphosphonates (TMPDM, TMPDE, see Table 5). The different behavior might be correlated with the lifetime of the ketone triplets (0.3 ns in the case of TMDPO and >3 ns in the case of the phosphonates TMPDM and TMPDE). Probably, the interaction between methyl groups in ortho position and carbonyl groups involves triplet states. Enolization can, therefore, only become important if intersystem crossing is significant and if ketone triplets live longer than a few nanoseconds. Configuration may also play an important role. With TMDPO the position of the carbonyl is almost perpendicular to the trimethylphenyl moiety. Thus, interactions of excited carbonyl groups with o-methyl groups are strongly impeded. In the case of OTDPO, carbonyl and o-methyl groups are probably in positions much more favorable for interactions. In conclusion, substitution of hydrogens at the benzoyl group by methyl groups has a drastic effect in the case of para substitution:

TABLE 5
Photolysis of Acylphosphine Oxides and Acylphosphonates. Quantum Yields of Triplet Formation and α-Cleavage. Singlet and Triplet Lifetimes

	$\emptyset(\alpha)$	$\emptyset(T)$	τ_S (ns)	τ_T (ns)
BDPO	0.5			
PTDPO	1.0			
OTDPO	0.6			
TMDPO	0.5			
TMPDM	0.3	0.6	0.7	<20
TMPDE	0.3	0.6	0.7	<20
BPDE	0.0	0.9	≤1	24
DMPME	0.0			
DCPME	0.0			

TABLE 5 (continued)
Photolysis of Acylphosphine Oxides and Acylphosphonates.
Quantum Yields of Triplet Formation and α-Cleavage.
Singlet and Triplet Lifetimes

		Ø(α)	Ø(T)	τ_S (ns)	τ_T (ns)
TMDPS		0.3			
PDME		0.3	0.6	11	30
PDEE		0.3	0.6	11	30
PDPO		1.0	0.0		30

$\phi(\alpha) = 1.0$. The trimethyl substituted compound TMDPO undergoes α-scission with the same yield as the nonsubstituted compound. In the case of the *o*-substituted compound, enolization significantly competes with α-scission.

b. Benzoylphosphonic Acid Esters

Benzoylphosphonic acid diethyl ester (BDPE) did not undergo α-scission on irradiation at λ > 300 nm in dilute CH_2Cl_2 solution.[49] The singlet lifetime is short: $\tau < 1$ ns. Energy transfer to naphthalene occurs with $k_q = 4 \times 10^9$ l/mol s. From quenching studies $\phi(T) = 0.9$ was obtained. BPDE triplets undergo intramolecular hydrogen abstraction, as was revealed from flash photolysis studies.[49]

The dimethyl and diethyl esters of 2,4,6-trimethylbenzoylphosphonic acid, TMPDM and TMPDE, exhibited a behavior quite different from that of BPDE. Triplets were formed with $\phi(T) \approx 0.6$. Moreover, both compounds underwent α-scission with $\phi(\alpha) = 0.3$. Interestingly, α-scission occurs from the singlet state. The full mechanism of the photo-reactions is illustrated in Scheme II for the case of TMPDM. Ketone triplets, formed initially, are readily converted to biradicals by abstraction of hydrogens from methyl groups in ortho position. The biradicals have the character of enol triplets. On relaxation to the ground state both E- and Z-isomers are formed. In the case of the E-isomer, reketonization is a relatively slow process which can be kinetically discriminated from the much more rapid ketonization of the Z-isomer. Figure 11 shows typical results obtained in the flash photolysis of TMPDM in CCl_4 solution at room temperature. At the end of the 20-ns flash the spectrum of enol triplets was recorded. Depending on the solvent, enol triplets decayed with lifetimes of 45 to 60 ns. Simultaneously, the ground state spectrum of enols was formed. It was recorded 120 ns after the flash. This spectrum decayed in two modes with lifetimes depending on

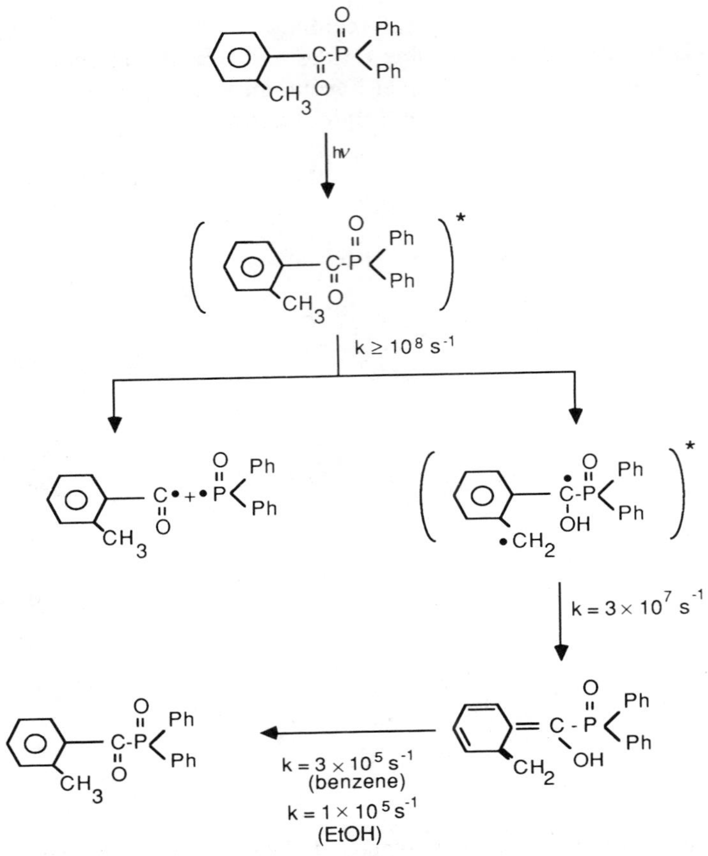

SCHEME   I

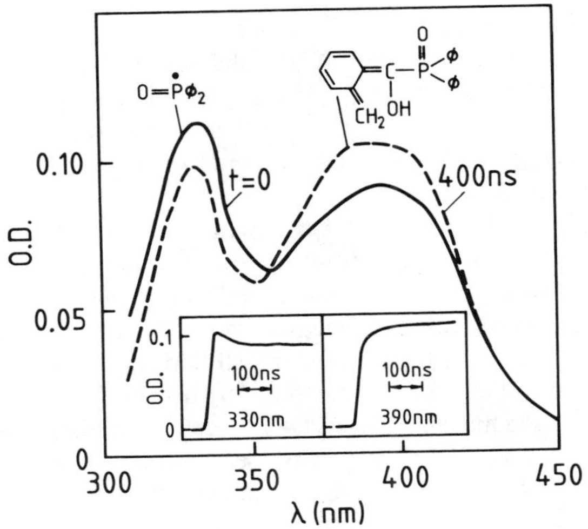

FIGURE 9.   Flash photolysis of OTDPO in Ar-saturated ethanolic solution at room temperature. [OTDPO] = 7.5×10^{-4} mol/l, λ_{inc} = 347 nm, flash duration: 20 ns. Transient optical absorption spectra recorded at the end of flash (t = 0) and 400 ns later. Insets: kinetic traces depicting changes in the O.D. at 330 and 390 nm. (From Sumiyoshi, T., Schnabel, W., and Henne, A., *J. Photochem.*, 32, 119, 1986. With permission.)

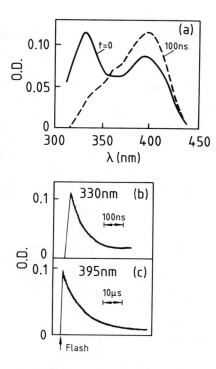

FIGURE 10. Flash photolysis of OTDPO in Ar-saturated ethanolic solution containing *trans*-piperylene (0.1 mol/l) at room temperature. [OTDPO] $= 7.5 \times 10^{-4}$ mol/l, $\lambda_{inc} = 347$ nm, flash duration: 20 ns. Transient optical absorption spectra recorded at the end of the flash (t = 0) and 100 ns later. (From Sumiyoshi, T., Schnabel, W., and Henne, A., *J. Photochem.*, 32, 119, 1986. With permission.)

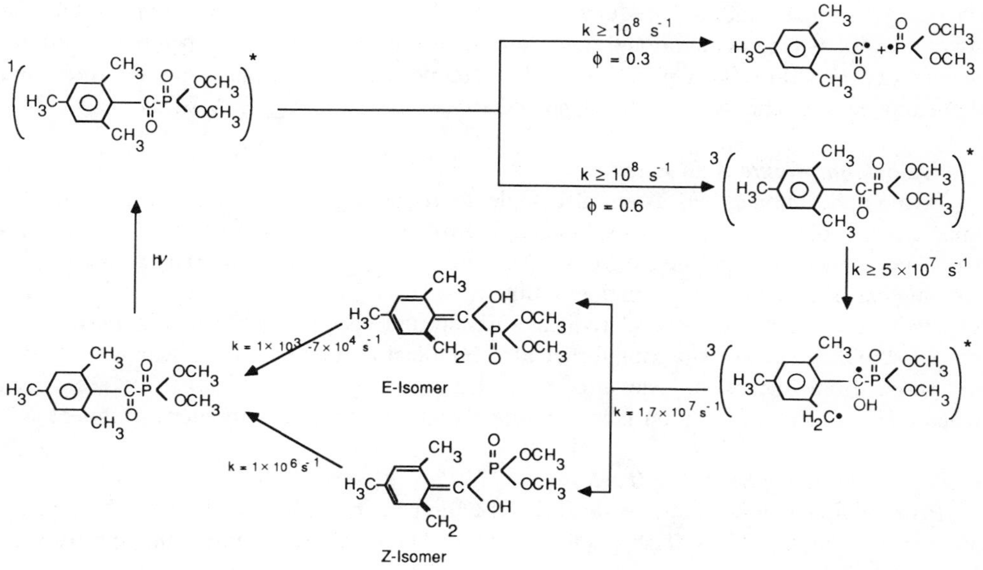

SCHEME   II

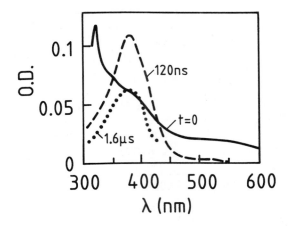

FIGURE 11.   Flash photolysis of TMPDM in Ar-saturated CCl_4 solution. [TMPDM] = 2.5 × 10^{-3} mol/l, λ_{inc} = 347 nm, D_{abs} = 5.2 × 10^{-5} einstein/l. Transient absorption spectra recorded at the end of flash (t = 0) and 0.12 and 1.6 μs after the flash. (From Sumiyoshi, T., Schnabel, W., and Henne, A., *J. Photochem.*, 30, 63, 1985. With permission.)

the chemical nature of the solvent: τ_{fast} = 0.29 to 1.0 μs and τ_{slow} = 15 to 870 μs. The spectrum recorded in CCl_4 solution 1.6 μs after the flash is the ground state spectrum of the E-isomer (vide: spectrum 3 in Figure 11). The enolization mechanism detected here is quite similar to that proposed by Haag et al.[58] for the photoreactions of 2-methylacetophenone and related compounds.

Evidence for the fact that α-scission is occurring via excited singlet states was obtained by experiments with styrene. It turned out that styrene quenched the formation of the absorption of enol triplets, that is, it reacted with ketone triplets. On the other hand, the extent of the formation of adducts of phosphinoyl radicals to styrene was not affected. Therefore, it was concluded that ketone triplets almost exclusively undergo enolization whereas phosphinoyl radicals are formed from singlets.[49] It should be noted that the lifetime of ketone triplets is rather short and, therefore, their transient absorption spectrum could not be recorded by 20-ns flash photolysis. The lifetime of ketone triplets was inferred from quenching experiments with naphthalene and *trans*-piperylene: τ_{ketone} < 20 ns.

c. Pivaloylphosphonic Acid Esters

Pivaloylphosphonic acid esters also undergo α-cleavage. $\phi(\alpha)$ ≈ 0.3 was found for dimethyl and diethyl pivaloylphosphonate, PDME and PDEE, respectively.[47] From naphthalene quenching experiments $\phi(T)$ ≈ 0.6 and τ_T ≈ 30 ns was obtained. Single photon counting experiments yielded a singlet lifetime τ_S ≈ 14 ns. Transient absorption spectroscopy revealed the existence of several radicals. Absorption spectra shown in Figure 12 were attributed to the dialkoxyphosphonyl radical, a biradical formed by intramolecular hydrogen abstraction from one of the ester groups, and the pivaloyl radical. As can be seen from Scheme III, biradical formation is an important route of triplet deactivation, in this case.

d. Pivaloyldiphenylphosphine Oxide

Pivaloyldiphenylphosphine oxide (PDPO) undergoes α-scission with a very high yield: $\phi(\alpha)$ ≈ 1.0. In the 20-ns flash photolysis of PDPO in dilute solution, the spectrum of diphenylphosphinoyl radical was recorded at the end of the flash. The singlet lifetime was determined by single photon counting: τ_S ≈ 30 ns. Singlet deactivation occurred almost exclusively via α-cleavage. This can be seen from Figure 13 where a kinetic trace demonstrating the formation of the absorption at λ = 330 nm due to diphenylphosphinoyl radicals

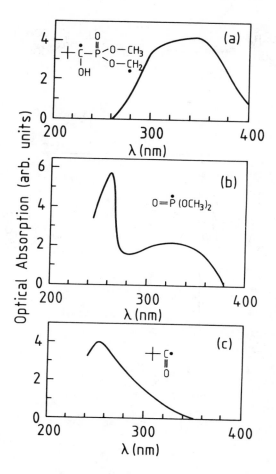

FIGURE 12. Flash photolysis of PDME in Ar-saturated dichloromethane solution at room temperature. [PDME] = 6.6×10^{-3} mol/l. λ_{inc} = 347 nm, D_{abs} = 1.3×10^{-4} einstein/l. Transient optical absorption spectra of radicals: (A): biradical, (B): dimethoxyphosphinoyl radical (O = $\dot{P}$(OCH$_3$)$_2$, and (C): pivaloyl radical (H$_3$C–C(CH$_3$)$_2$–$\dot{C}$=O). (From Sumiyoshi, T., Schnabel, W., and Henne, A., *J. Photochem.*, 32, 191, 1986. With permission.)

is shown. The lifetime of formation of about 30 ns is equal to that of the fluorescence decay. It was therefore concluded that α-cleavage of PDPO occurs from the excited singlet state.

e. The Reactivity of Phosphinoyl Radicals towards Olefinic Monomers

In the previous sections it was described how phosphinoyl radicals of different composition can be generated. In Scheme IV it is illustrated how phosphonyl radicals were produced to examine their reactivity towards various monomers.[48] In the cases of diphenylphosphonyl and phenylpropoxyphosphinoyl radicals, reaction rate constants were determined by measuring the rate of decay of the optical absorption of the radicals at λ = 330 nm as a function of monomer concentration. The other three radicals do not absorb characteristically light at λ > 300 nm. Therefore, the competition method illustrated in Scheme V was applied. This method is based on the fact that phosphinoyl radicals add to styrene forming a substituted styryl radical which absorbs strongly at λ = 322 nm, whereas adducts of phosphinoyl radicals to other monomers do not absorb or absorb only negligibly in this wavelength range. Figure 14 shows a typical example: the formation of the adduct radical in the reaction of styrene

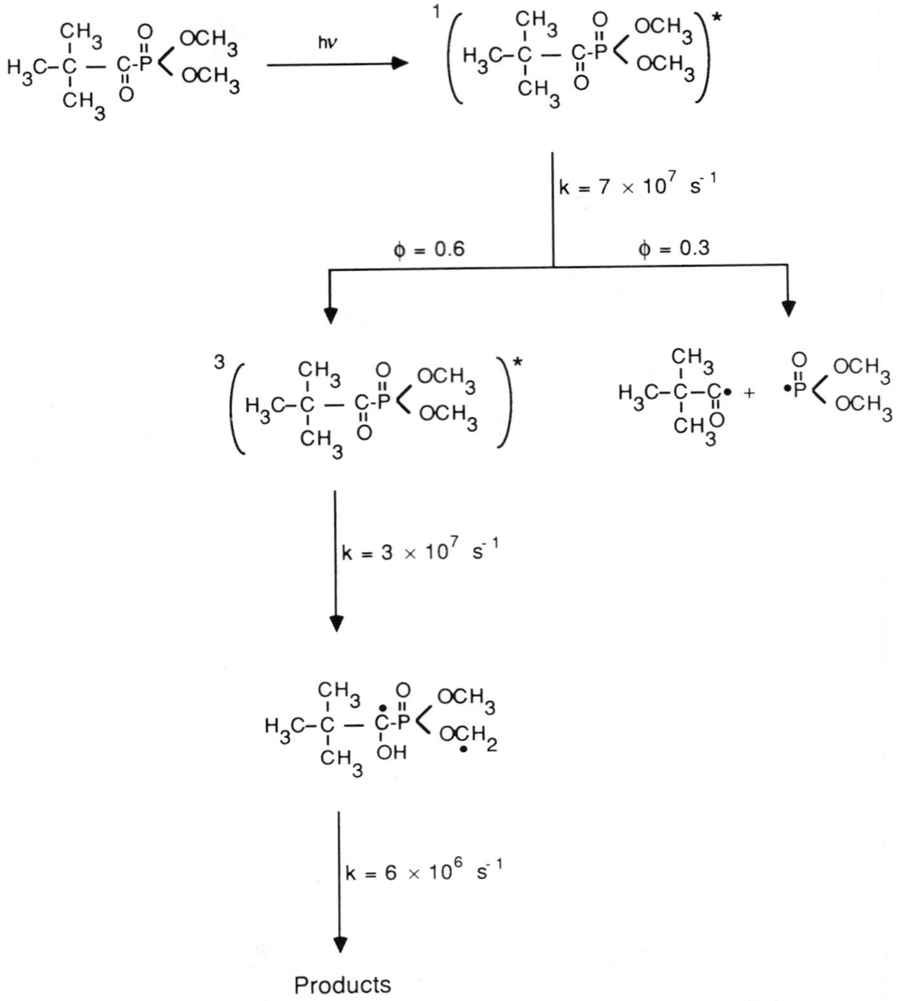

SCHEME   III

with diethoxyphosphinoyl radical. In Figure 15 pseudo-first-order rate constants of the formation of the optical density at $\lambda = 322$ nm were plotted vs. the styrene concentration. From the slope of the straight line the rate constant k_{R+St} was obtained. If, apart from styrene, an other monomer was added to the solution the optical density of styrene adduct radical was diminished as compared to the O.D. measured in the absence of the other monomer. Rate constants k_{R+M} were, then, determined according to Equation 27:

$$\frac{O.D._0}{O.D.} = 1 + \frac{k_{R+M}[M]}{k_{R+St}[St]} \tag{27}$$

O.D. and O.D.$_0$ denote the maximum optical density in the presence and absence of monomer, respectively. Rate constants obtained this way were compiled in Table 6. The high reactivity of phosphinoyl radicals towards methacrylonitrile, styrene, and MMA is obvious. Notably, dialkoxyphosphinoyl radicals exhibit the highest reactivity.

It should be pointed out that rate constants in the order of 10^8 l/mol s, as measured for the reaction of dimethoxyphosphinoyl radical with methacrylonitrile and styrene, are very

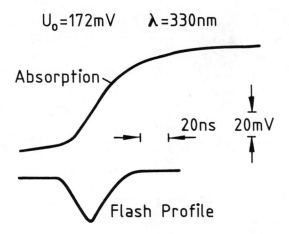

FIGURE 13. Flash photolysis of PDPO in Ar-saturated dichloromethane solution at room temperature. [PDPO] = 5 × 10⁻⁴ mol/l, λ_{inc} = 347 nm, D_{abs} = 7.4 × 10⁻⁵ einstein/l. Kinetic traces illustrating the formation of the absorption at 330 nm. The lower trace depicts the flash profile. (From Sumiyoshi, T., Schnabel, W., and Henne, A., *J. Photochem.*, 32, 191, 1986. With permission.)

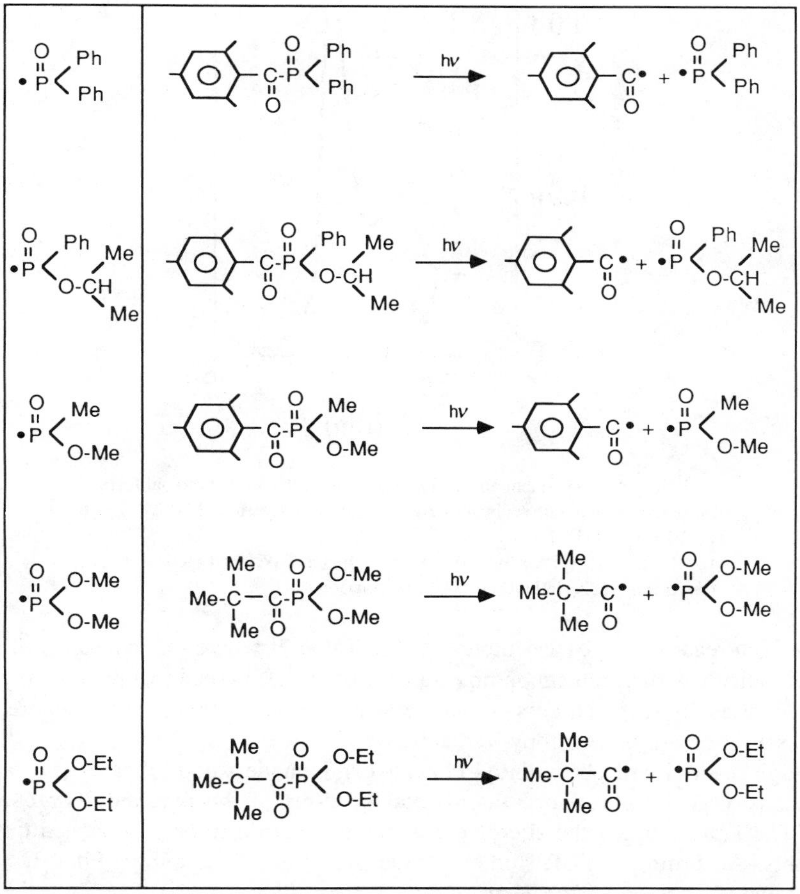

SCHEME IV

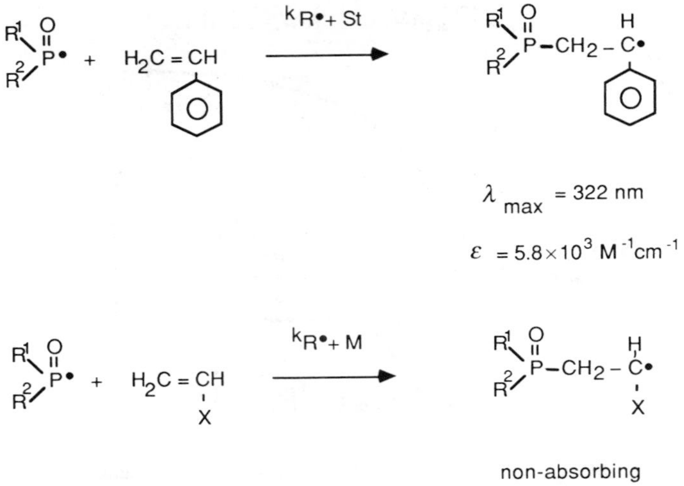

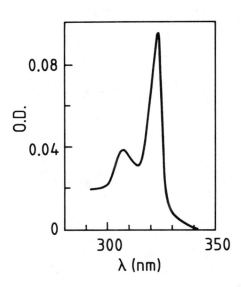

SCHEME V

FIGURE 14.   Transient absorption spectrum of the adduct radical formed by reaction of diethoxyphosphinoyl radical with styrene. [PDEE] = 3.6 × 10^{-3} mol/l, [St] = 1.4 × 10^{-2} mol/l, λ_{inc} = 347 nm, D_{abs} = 1.1 × 10^{-4} einstein/l. (From Sumiyoshi, T., Schnabel, W., and Henne, A., *J. Photochem.*, 32, 191, 1986. With permission.)

high. The high reactivity is best demonstrated in Table 7, where rate constants of reactions of various radicals with various monomers are compared. It is seen that phosphinoyl radicals are, in most cases, by several orders of magnitude more reactive than carbon-centered radicals such as benzoyl or 2-hydroxypropyl radicals.

The high reactivity of phosphinoyl radicals can be understood in terms of a rather high electron density at the phosphorus atoms and in terms of the pyramidal structure of the radicals. The latter permits the site of the unpaired electron to be approached quite freely by reactants. According to ESR studies[59,60] the pyramide of the radical $Ph_2\dot{P}=O$ should be rather flattened if compared to the pyramide of the radical $(H_3CO)_2\dot{P}=O$. On the basis of

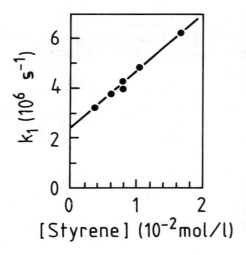

FIGURE 15. Plot of the pseudo-first-order rate constant of the formation of adduct radicals by reaction of dimethoxyphosphinoyl radicals with styrene as a function of the styrene concentration. Absorption measurements were carried out at $\lambda = 322$ nm. [PDME] $= 6.6 \times 10^{-3}$ mol/l, $D_{abs} = 1.3 \times 10^{-4}$ einstein/l. (From Sumiyoshi, T., Schnabel, W., and Henne, A., *J. Photochem.*, 32, 191, 1986. With permission.)

TABLE 6
Rate Constants k_{R+M} (l/mol) of the Reaction of Phosphinoyl Radicals with Various Monomers in Cyclohexane Solution at Room Temperature

Monomer	Radical $\overset{O}{\underset{\phi}{\overset{\|}{\cdot P}}}\diagup^{\phi}$	$\overset{O}{\underset{OCH(CH_3)_2}{\overset{\|}{\cdot P}}}\diagup^{\phi}$	$\overset{O}{\underset{OCH_3}{\overset{\|}{\cdot P}}}\diagup^{CH_3}$	$\overset{O}{\underset{OCH_3}{\overset{\|}{\cdot P}}}\diagup^{OCH_3}$	$\overset{O}{\underset{OC_2H_5}{\overset{\|}{\cdot P}}}\diagup^{OC_2H_5}$
MAN	$(5.0 \pm 0.2) \times 10^7$	$(4.6 \pm 0.2) \times 10^7$	$(4.5 \pm 0.2) \times 10^7$	$(9.2 \pm 1.3) \times 10^7$	$(1.1 \pm 0.1) \times 10^8$
ST	$(6.0 \pm 0.2) \times 10^7$	$(4.5 \pm 0.2) \times 10^7$	$(8.0 \pm 0.2) \times 10^7$	$(2.2 \pm 0.2) \times 10^8$	$(2.5 \pm 0.2) \times 10^8$
MMA	$(8.0 \pm 0.5) \times 10^7$	$(5.0 \pm 0.2) \times 10^7$	$(5.8 \pm 0.5) \times 10^7$	$(5.8 \pm 0.2) \times 10^7$	$(5.3 \pm 0.2) \times 10^7$
AN	$(2.0 \pm 0.1) \times 10^7$	$(2.0 \pm 0.1) \times 10^7$	$(1.8 \pm 0.3) \times 10^6$	$(5.8 \pm 0.2) \times 10^6$	$(2.6 \pm 0.2) \times 10^6$
MA	$(3.5 \pm 0.2) \times 10^7$	$(2.1 \pm 0.2) \times 10^7$	$(1.3 \pm 0.1) \times 10^7$	$(1.7 \pm 0.2) \times 10^7$	$(1.6 \pm 0.2) \times 10^7$
BVE	$(4.0 \pm 0.2) \times 10^6$	$(3.0 \pm 0.3) \times 10^6$	$(2.3 \pm 0.2) \times 10^6$	$(2.1 \pm 0.2) \times 10^7$	$(1.4 \pm 0.2) \times 10^7$
VA	$(1.6 \pm 0.1) \times 10^6$	$(1.3 \pm 0.1) \times 10^6$	$(8.2 \pm 0.4) \times 10^5$	$(2.9 \pm 0.2) \times 10^6$	$(1.8 \pm 0.2) \times 10^6$

MAN: methacrylonitrile, ST: styrene, MMA: methylmethacrylate, AN: acrylonitrile, MA: methyl acrylate, BVE: *n*-Butyl vinyl ether, and VA: vinyl acetate.

From Sumiyoshi, T. and Schnabel, W., *Makromol. Chem.*, 186, 1811, 1985. With permission.

these structural differences the higher reactivity of the dialkoxyphosphinoyl radical relative to that of the diphenylphosphinoyl radical can be explained. In this connection it is interesting to note that, upon substituting the oxygen in $Ph_2P=O$ by sulfur, the reactivity of the radical towards olefinic monomers was diminished.[50] This can be seen from Table 8, where rate constants of the reaction of monomers with diphenylphosphinoyl and diphenylthiophosphinoyl radicals, respectively, are listed. $Ph_2P=S$ radicals are 10 to 30 times less effective than

TABLE 7
Rate Constant k_{R+M} (l/mol s) of the Reaction of Various Radicals with Various Olefinic Monomers at Room Temperature

Monomer	Ph–Ċ–Ph | OH	H₃C–Ċ–CH₃ | OH	(S radical)	Ph–Ċ || O	Ph₂Ṗ=O
Styrene		3.5×10^5	7×10^3	1.2×10^5	6×10^7
Methylmethacrylate	9×10^3	5.4×10^5	4×10^2	0.9×10^5	6×10^7
Methylacrylate					2×10^7
Vinylacetate	6×10^3			2×10^5	2×10^6
Acrylonitrile	4×10^3	1.6×10^5		2×10^4	2×10^7
N-vinylpyrrolidone			4×10^2		
n-Butylvinylether			<10		5×10^6

TABLE 8
Rate Constants k_{R+M} of the Reaction of the Radicals Ph₂Ṗ=O and Ph₂Ṗ=S with Various Monomers (l/mol s)

Monomer	Ph₂Ṗ=O	Ph₂Ṗ=S
Styrene	4.6×10^7	4.0×10^6
Methylmethacrylate	4.1×10^7	1.9×10^6
Methacrylonitrile	1.9×10^7	9.1×10^5
Methylacrylate	1.7×10^7	6.2×10^5
Acrylonitrile	1.3×10^7	5.2×10^5
n-Butylvinylether	5.0×10^6	1.5×10^5
Vinylacetate	1.4×10^6	4.2×10^4

Note: Experiments were carried out in CH_2Cl_2 solution at room temperature.

From Sumiyoshi, T., Weber, W., and Schnabel, W., *Z. Naturforsch.*, 40a, 541, 1985. With permission.

Ph₂Ṗ=O radicals. The differences can be caused both by differences in the electron density distribution and by geometrical factors. With respect to steric effects it appears feasible that substitution of oxygen by the more voluminous sulfur brings about a structural alteration which probably reduces the accessibility of the site of the unpaired electron in the radical. In this work[50] diphenylthiophosphinoyl radicals were generated by irradiation of 2,4,6-trimethylbenzoyldiphenylphosphine sulfide ($\phi(\alpha) = 0.3$) with UV light according to Reaction 28:

$$R-\underset{\underset{O}{\|}}{\overset{\overset{S}{\|}}{C}}-P-Ph_2 \xrightarrow{h\nu} R-\underset{\underset{O}{\|}}{\overset{\overset{S}{\|}}{C}}\cdot \ + \ \cdot PPh_2$$

$$R \ = \ 2,4,6\text{-trimethylphenyl} \tag{28}$$

The optical absorption spectrum of thiophosphinoyl radical possesses a strong absorption band with a maximum at $\lambda = 340$ nm ($\epsilon_{340nm} = 1.2 \times 10^4$ l/mol cm) and a relatively weak band at about 500 nm.[50]

5. α-Keto-Oxime Esters

α-Keto-oxime esters of the structure shown below are quite efficient photoinitiators as far as curing of formulations containing acrylates or styrene is concerned.

$$\begin{array}{c} O \\ \| \\ C \qquad R^2 \\ {}^{\diagup} \ {}^{\diagdown} \ {}^{\diagup} \\ R^1 \qquad C \\ \| \\ N \qquad O \\ {}^{\diagdown} \quad \| \\ O-C-R^3 \end{array} \qquad \text{(E-isomer)}$$

Reportedly, the efficiency surpasses that of benzoin alkyl ethers.[61-63] However, the storage stability of α-keto-oxime esters in some formulations is insufficient. Product analysis and ESR spin trapping studies have led to the conclusion that α-keto-oxime esters are photo-fragmented upon irradiation with UV light of $\lambda > 300$ nm through primary N-O bond cleavage according to Reaction 29:[61,64]

$$R^1-\underset{\underset{O}{\|}}{C}-\underset{\underset{R^2}{|}}{C}=N-O-\underset{\underset{O}{\|}}{C}-R^3 \xrightarrow{h\nu} R^1-\underset{\underset{O}{\|}}{C}-\underset{\underset{R^2}{|}}{C}=N\cdot \ + \ \cdot O-\underset{\underset{O}{\|}}{C}-R^3 \tag{29}$$

Laser flash photolysis studies ($\lambda_{inc} = 347$ nm, flash duration: 25 ns) revealed that the photofragmentation of α-keto-oxime ester type initiators occurs very rapidly.[65] In experiments involving various compounds (see Table 9) it was not possible to evidence triplet excited states with the aid of quenchers or by luminescence measurements. Apart from ADO, which was the only compound not containing a phenyl group, all oxime esters formed, in benzene solution, a short-lived broad transient absorption band located between 400 and 650 nm. It decayed with a lifetime of about 10 ns and is tentatively attributed to excimers formed by interaction of excited oxime esters with benzene. This band was not observed when other solvents were used. Transient absorption spectra observed at the end of the flash at $\lambda < 400$ nm were due to free radicals. In the case of O-acetyl-diacetyl-mono-oxime ester (ADO, $R^1 = R^2 = R^3$ = methyl), the transient absorption spectrum shown in Figure 16 was recorded in cyclohexane solution at the end of the flash. It was ascribed to the acetyl radical, $CH_3-\overset{\cdot}{C}=O$ and its appearance was taken as evidence for α-cleavage being the primary step in the photofragmentation process:[65]

TABLE 9
α-Keto-Oxime Esters of the Structure

$$R^1-C-C-N=O-C-R^3$$
$$\underset{O}{\|}\ \underset{R^2}{|}\qquad \underset{O}{\|}$$

Examined in Flash Photolysis Studies

Denotation	R¹	R²	R³	ε347nm[a]
ADO	CH₃	CH₃	CH₃	13
BDO	CH₃	CH₃	C₆H₅	16
A1PPO	C₆H₅	CH₃	CH₃	113
B1PPO	C₆H₅	CH₃	C₆H₅	120
D1PPO	C₆H₅	CH₃	(CH₃)₂C–CN	100
C1PPO	C₆H₅	CH₃	(cyclobutane ring: CH₂–CH₂–C(CN)–CH₂)	142
E1PPO	C₆H₅	CH₃	C₂H₅O	97
A2PPO	CH₃	C₆H₅	CH₃	16
B2PPO	CH₃	C₆H₅	C₆H₅	37
ABO	C₆H₅	C₆H₅	CH₃	94
BBO	C₆H₅	C₆H₅	C₆H₅	116

[a]   Extinction coefficients in benzene at room temperature.

From Amirzadeh, G., Dissertation, Technische Universität, Berlin, 1981.

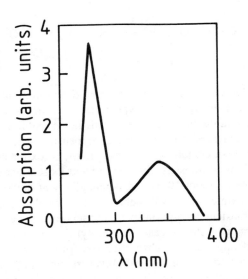

FIGURE 16.   Transient absorption spectrum recorded with ADO in cyclohexane solution at the end of the flash. [ADO] = 1.3×10^{-2} mol/l. λ_{inc} = 347 nm. D_{abs} = 2×10^{-5} einstein/l. Flash duration: 25 ns. (From Amirzadeh, G., Dissertation, Technische Universität, Berlin, 1981. With permission.)

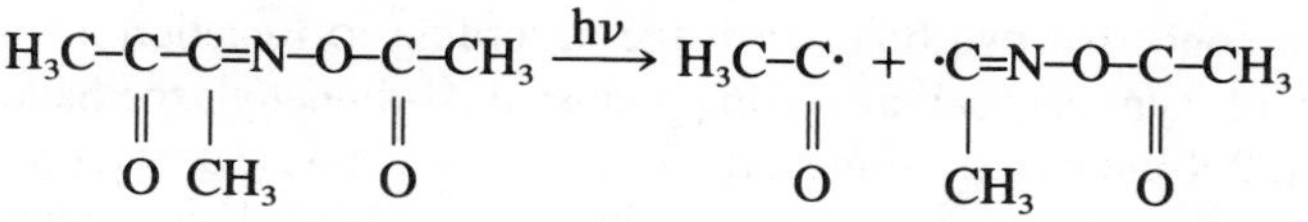

$$H_3C-C-C=N-O-C-CH_3 \xrightarrow{h\nu} H_3C-C\cdot + \cdot C=N-O-C-CH_3 \qquad (30)$$

(with the subscript groups: O, CH_3, O on the left; O, CH_3, O on the right)

Upon irradiating *O*-benzoyl-diacetyl-mono-oxime ester (BDO, $R^1 = R^2 =$ methyl, $R^3 =$ phenyl) a transient spectrum differing from that obtained with ADO was recorded at the end of the flash. It was concluded that, in this case, photofragmentation is initiated by cleavage of the N-O bond according to Reaction 29.[65] This implies that the chemical nature of R^3 strongly influences the fragmentation mechanism.

It should be noted that the assignment of transient absorption spectra obtained in that work was hampered in many cases by the instability of the primary radicals, which are prone to decompose into smaller fragments, e.g.:

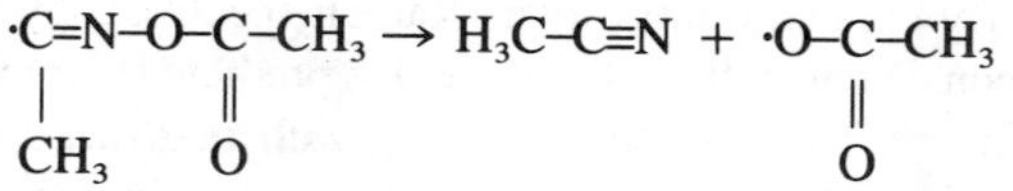

$$\cdot C=N-O-C-CH_3 \rightarrow H_3C-C\equiv N + \cdot O-C-CH_3 \qquad (31)$$

(with subscript groups: CH_3, O on the left; O on the right)

or:

$$CH_3-C-C=N\cdot \rightarrow CH_3-C\cdot + C\equiv N \qquad (32)$$

(with subscript groups: O CH_3 on the left; O CH_3 on the right)

Moreover, many experiments were carried out in benzene solution where transient radicals added quite rapidly to solvent molecules forming cyclohexadienyl-type adduct radicals that absorb rather strongly around 315 nm. Although these studies did not contribute greatly to the elucidation of the mechanism of the photolysis of α-keto-oxime esters, they yielded important information on the reactivity of fragment radicals towards two olefinic monomers: styrene (St) and MMA. Upon measuring the rate of decay of the transient absorption as a function of monomer concentration, bimolecular rate constants of the reaction of fragment radicals with these monomers were determined. Apart from the results obtained with ADO, k_{R+M} values ranging from 2.5×10^5 to 7.3×10^5 l/mols were found. These rate constants compare well with those of the reactions of 2-hydroxy-propyl-2 radicals with St and MMA (see Table 7). In the case of ADO, the decay of the spectrum of acetyl radicals was measured in the presence of St and MMA. The rate constants obtained here are $k_{R+St} = 1.1 \times 10^6$ and $k_{R+MMA} = 9 \times 10^5$ l/mol s. These values are about one order of magnitude higher than the rate constants of the reaction of benzoyl radicals with St and MMA (see Table 7), indicating that acetyl radicals are much more reactive than benzoyl radicals.

6. Azo Compounds

Azo compounds are very effective initiators for free radical polymerization of vinyl compounds.[66] They can be decomposed thermally or photolytically. With respect to the decomposition mechanism two possibilities are discussed: (1) simultaneous scission of the two C-N bonds according to Reaction 33:

$$R_1-N=N-R_2 \rightarrow R_1^\cdot + N_2 + R_2^\cdot \qquad (33)$$

and (2) stepwise cleavage according to Reaction 34:

$$R_1-N=N-R_2 \rightarrow R_1-N=N\cdot + R_2^\cdot \rightarrow R_1^\cdot + N_2 + R_2^\cdot \qquad (34)$$

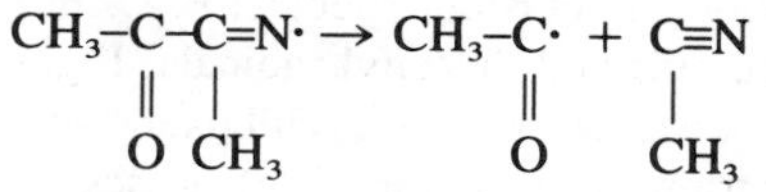

Evidence for concerted two bond cleavage according to Reaction 33 was obtained for the thermolysis of symmetrical azo compounds: 1,1'-diphenylazoethane,[67] α,α'-azocumene,[68,69] and DL-3,4-diethyl-3,4-dimethyldiazetine.[70] A stepwise decomposition mechanism according to Reaction 34 was evidenced in the case of cis-1,1'-azo-adamantane, a symmetrical compound.[71] A theoretical study on the thermolysis of trans-azoethane arrived at the conclusion that the initial step is *trans* → *cis* isomerization which is followed up by dissociation into diazenyl and ethyl radicals.[72] According to Engel,[73] the "azoalkane thermolysis" proceeds by a continuum of mechanisms between (1) and (2); the more unsymmetrical the azo compound the more unsymmetrically it cleaves.

The mechanism of the photolysis of azo compounds was studied only scarcely. Porter et al.[74] postulated a stepwise mechanism based on their work with unsymmetrical compounds such as Ph–N=N–C(CH₃)₂Ph and Ph₂CH–N=N–CH₂Ph. An investigation concerning the photofragmentation of azomethane in the gas phase showed that the fragmentation was complete within 2 ns. By coherent anti-Stokes Raman spectroscopy the nascent vibrational distribution appeared consistent with a (theoretically predicted) stepwise bond cleavage.[75]

Recently, a mechanistic flash photolysis study with α,α'-azocumenes in cyclohexane solution also yielded evidence for the stepwise decomposition of symmetrical azo compounds.[76] α,α'-Azocumene and 2,2'-di-*p*-tolyl-2,2'-azopropane have a UV absorption band with a maximum at 367 nm, but possess an absorption window between 270 and 340 nm. This fact permitted observation of transient species, e.g., in the case of α,α'-azocumene: 1-methyl-1-phenylethyl (α-cumyl) and diazenyl radicals. Figure 17 shows transient absorption spectra recorded with α,α'-azocumene in cyclohexane solution. The spectrum observed at the end of the flash (spectrum (1) in Figure 17a) possesses peaks at 272, 310, and 322 nm and was assigned to α-cumyl radical. Upon studying the spectral changes after the flash, it was found that the peaks at 272 and 322 nm grew in further (see Figure 17b) whereas the absorption between 270 and 305 nm decayed rapidly. The difference spectrum obtained by subtracting spectrum 2 from spectrum 1 in Figure 17a is shown in Figure 17c. It has a maximum at 285 nm and is assigned to α-cumyldiazenyl radical. The results indicate that α-cumyl radicals were formed in two modes, a fast mode occurring during the 20-ns flash and a slow mode after the flash. These findings strongly suggest the occurrence of the stepwise decomposition mechanism. The ratio of the optical densities pertaining to the two modes depended on the absorbed dose rate: the higher the absorbed dose per flash, the more dominant was the fast mode. This behavior was explained in terms of competition between unimolecular decomposition and dimerization of α-cumyldiazenyl radicals. The total mechanism is illustrated by Reactions 35 to 39:

$$Ph–C(CH_3)_2–N=N–C(CH_3)_2–Ph \rightarrow Ph–C(CH_3)_2–N=N\cdot + \cdot(CH_3)_2–Ph \qquad (35)$$

$$Ph–C(CH_3)_2–N=N\cdot \rightarrow Ph–C(CH_3)_2^\cdot + N_2 \qquad (36)$$

$$2Ph–C(CH_3)_2–N=N\cdot \rightarrow Ph–C(CH_3)_2–N=N–N=N–C(CH_3)_2–Ph \qquad (37)$$

$$Ph–C(CH_3)_2^\cdot + X \rightarrow product \qquad (38)$$

$$2Ph–C(CH_3)_2^\cdot \rightarrow product \qquad (39)$$

Figure 18 shows a plot of the ratio O.D.$_{max}$/O.D.$_0$ vs. the absorbed dose per flash which corresponds to the absorbed dose rate. Extrapolation of the experimental curve to (Dabs/flash) = 0 yields (O.D.$_{max}$/O.D.$_0$) = 2. At low absorbed dose rate α-cumyl radicals should be formed in about equal amounts by the two modes, i.e., during the flash according to Reaction 35 and after the flash according to Reaction 36, because dimerization according

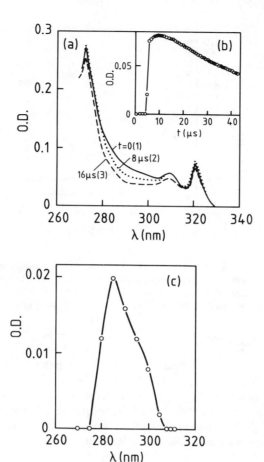

FIGURE 17.   Flash photolysis of α,α'-azocumene (AC) in Ar-saturated cyclohexane solution at room temperature. [AC] = 4.7 × 10⁻³ mol/l. D_{abs} = 7.9 × 10⁻⁵ einstein/l. (a): transient absorption spectra recorded at the end of the flash (t = 0) and 8 and 16 μs later, (b): kinetic trace demonstrating changes in the optical absorption at 322 nm, (c): difference spectrum obtained by substraction of spectrum 2 from spectrum 1 in (a). (From Sumiyoshi, T., Kamachi, M., Kuwae, Y., and Schnabel, W., *Bull. Chem. Soc. Jpn.*, 60, 77, 1987. With permission.)

to Reaction 37 cannot compete with unimolecular decomposition according to Reaction 36. Reaction 38 stands for the reaction of α-cumyl radicals with intact α,α'-azocumene, solvent, and impurities. On the basis of this mechanism the rate constants of decomposition and dimerization were estimated to be 1.1 × 10⁵ s⁻¹ and 5 × 10⁹ l/mol s, respectively.[76]

It should be pointed out that, in the case of α,α'-azocumene, the quantum yield of cleavage is relatively high: ϕ = 0.36. Since triplet sensitized and direct photolysis yielded almost the same value, a triplet mechanism was assumed for the C-N bond cleavage in azocumenes.[77]

B. HYDROGEN ABSTRACTION-TYPE INITIATORS

Hydrogen abstraction-type initiators belong to the class of initiators that form radicals via bimolecular reactions. The mechanism of hydrogen abstraction frequently involves electron transfer as the primary step. Provided the change in free energy, ΔG, is negative,

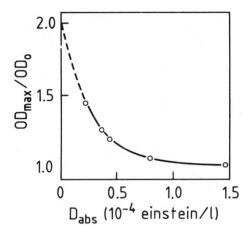

FIGURE 18.   Flash photolysis of α,α-azocumene in Ar-saturated cyclohexane solution. Dependence of the ratio O.D.$_{max}$/O.D.$_0$ on the absorbed dose per flash. (From Sumiyoshi, T., Kamachi, M., Kuwae, Y., and Schnabel, W., *Bull. Chem. Soc. Jpn.*, 60, 77, 1987. With permission.)

$$\Delta G = E_D^{ox} - E_A^{red} - E_{exc.st} - W \qquad (40)$$

where $E_{exc.st.}$ = energy of excited state, E_D^{ox} = oxidation potential of donor, E_A^{red} = redox potential of acceptor, $W = e_0^2/\epsilon a$ (Coulomb term, a = encounter distance), electron transfer is a highly probable reaction, i.e., the rate constant of electron transfer is rather high (approaching the encounter-controlled limit). Generally, electron transfer is considered to involve complex (exciplex) formation. Provided the reaction occurs in media of low polarity, charge separation is followed up by intracomplex proton transfer resulting in the formation of free radicals. In highly polar media, dissociation of complexes into free ions is possible.

Hydrogen abstraction, not involving electron transfer, is a comparably slow process. Respective rate constants are several orders of magnitude lower than those of electron transfer reactions.

With respect to photoinitiated polymerizations, most prominent acceptor compounds, regarding both electron and hydrogen transfer, are aromatic ketones of the benzophenone and thioxanthone families. Moreover, other aromatic ketones such as xanthones, 1,2-diketones (benzil), and 1,4- and 1,2-quinones are noteworthy. These compounds are prone to undergo hydrogen and/or electron transfer reactions because of their capability of forming triplet states in high quantum yields. For practical purposes, thioxanthones are favorable initiators. They have a strong absorption band in the near UV wavelength range and can be applied, therefore, in conjunction with certain pigments.

The mechanism of the photolysis of hydrogen abstraction type initiators has been studied with the aid of flash photolysis by various authors.[81-104] Some of this work, which is devoted to the identification of intermediates as well as to the kinetics of the reactions of excited initiator molecules and of free radicals with monomers will be reviewed here.

1. Benzophenone and Derivatives

Benzophenone (BP) reacts in its electronically excited triplet state effectively with aliphatic alcohols (e.g., 2-propanol), ethers (e.g., tetrahydrofuran), and amines (e.g., triethyl amine), according to Reaction 4 by hydrogen abstraction. Since BP triplets are formed with a quantum yield close to unity, the initiator efficiency of BP depends essentially on the rate constant of hydrogen abstraction k_{T+H} and the rate constants k_{R+M} of the reaction of radicals, formed in the hydrogen abstraction reaction, with monomers. Moreover, triplet quenching

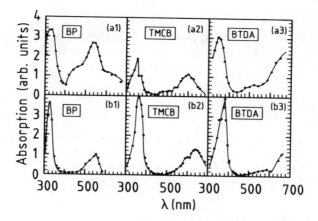

FIGURE 19. Flash photolysis of benzophenones in dilute Ar-saturated solution. $\lambda_{inc} = 347$ nm. Flash duration: 25 ns. $D_{abs} = 3.0 \times 10^{-5}$ einstein/l. [BP] $= 7.4 \times 10^{-4}$ mol/l; [TMCB] $= 6.0 \times 10^{-4}$ mol/l; [BTDA] $= 7.0 \times 10^{-4}$ mol/l. (a): T-T absorption spectra recorded in acetone solution at the end of the flash. (b): Ketyl radical spectra recorded in THF solution at the end of the flash. (From Kuhlmann, R. and Schnabel, W., *Angew. Makromol. Chem.*, 57, 195, 1977. With permission.)

by monomer is a crucial factor, which plays a detrimental role in various cases, e.g., in the case of styrene. In order to optimize the initiator efficiency many potential hydrogen donors were examined. Moreover, apart from BP, also BP derivatives were employed, e.g., 3,3′,4,4′-tetramethoxycarbonylbenzophenone (TMCB) and 3,3′,4,4′-benzophenone tetracarboxylic dianhydride (BTDA).[78,79]

(TMCB)

(BTDA)

More recently water-soluble BPs of the following structure were investigated.[80]

$$R = CH_2SO_3Na$$

$$R = CH_2N(CH_3)_3Cl$$

Flash photolysis experiments were carried out in order to determine the rate constants k_q, k_{T+H}, and k_{R+M}.[81-93] BP and its derivatives possess characteristic T-T-absorption spectra. Also, ketyl radicals formed by hydrogen abstraction are conveniently detected by absorption spectroscopy. Therefore, the kinetic parameters mentioned above could be easily measured by following formation or decay of the absorption of triplets and ketyl radicals, respectively.[81,82]

Figure 19 shows absorption spectra of triplets and of ketyl radicals of BP, TMCB, and BTDA recorded in acetone and tetrahydrofuran solution, respectively.

a. Triplet Quenching by Monomers

Rate constants of monomer quenching, k_q, are listed in Table 10. It can be seen that,

TABLE 10

Rate Constants k_q (l/mol s) of Quenching of Triplets of Benzophenones by Monomers at Room Temperature

Monomer	Initiator		
	BP[a]	TMCB[b]	BTDA[b]
Styrene	3.2×10^9	5.5×10^9	7.4×10^9
Vinylpyrrolidone	3.6×10^8	5.1×10^9	6.5×10^9
Vinylacetate	5.4×10^6	4.9×10^7	1.4×10^8
Methylmethacrylate	6.9×10^7	5.6×10^7	1.1×10^8
Acrylonitrile	3.4×10^7	9.4×10^6	5.0×10^7

[a]   Determined in benzene solution.
[b]   Determined in acetone solution.

From Kuhlmann, R. and Schnabel, W., *Angew. Makromol. Chem.*, 57, 195, 1977. With permission.

TABLE 11

Free Ion Yields Obtained at the Irradiation of Benzophenones in Different Solvents at λ_{inc} = 347 nm

Solvent	ϵ[a]	Ketone	$\phi(i)$[b]
Pentane-1-ol	14.4	TMCB	1.7×10^{-3}
Acetone	20.7	TMCB	2.5×10^{-2}
Acetonitrile	37.5	TMCB	9.2×10^{-2}
Acetone	20.7	BTDA	4.5×10^{-2}
Acetonitrile	37.5	BTDA	7.4×10^{-2}
Acetone	20.7	BP	1.0×10^{-2}
Acetonitrile	37.5	BP	1.5×10^{-2}

[a]   Dielectric constant of the solvent at 25°C.
[b]   Calculated on the basis of $\mu = 4 \times 10^{-4}$ cm^2 V^{-1} s^{-1}.

From Kuhlmann, R. and Schnabel, W., *J. Photochem.*, 7, 287, 1977. With permission.

because of high k_q values some monomers, such as styrene and vinylpyrrolidone, cannot be polymerized with the aid of BP initiators. Efficient energy transfer from initiator to monomer occurs in these cases because the triplet energies of the monomers are lower than those of the initiators. Other monomers, such as vinylacetate and MMA react relatively slowly with initiator triplets, probably because their triplet energies are higher than those of the initiators. In order to elucidate the reaction mechanism in these cases, photoconductivity experiments were carried out.[84] It was found that, upon irradiating BP and BP derivatives in the absence of monomer, a transient photocurrent was generated indicating the existence of exciplexes formed by triplet excited BP compounds and ground state solvent molecules. The quantum yield $\phi(i)$ of free ions increased with increasing dielectric constant ϵ of the solvent. Results are presented in Table 11, where it can be seen that the highest $\phi(i)$ values were obtained with acetonitrile solutions. Upon addition of unsaturated compounds $\phi(i)$ was reduced. If the triplet energy of the added compound was lower than that of the BP compound, e.g., in the case of styrene, $\phi(i)$ decreased steadily with increasing additive concentration. On

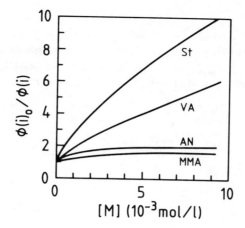

FIGURE 20. Flash photolysis of TMCB in acetone solution. Relative yield of free ions, determined by photocurrent measurements, as a function of monomer concentration. [TMCN] = 6×10^{-4} mol/l. λ_{inc} = 347 nm. (From Kuhlmann, R. and Schnabel, W., *J. Photochem.*, 7, 287, 1977. With permission.)

TABLE 12
Rate Constants k_{T+H} (10^6 l/mol s) of the Reaction of Triplets of Benzophenones of the Structure

R⟨○⟩-C⟨○⟩
‖
O

with Tetrahydrofuran Determined in Benzene Solution in the Absence of O_2[85]

R	H	Cl	OCH$_3$	F	COOH
	4.5	6.4	2.5	5.6	4.3
	3.0 [81]				

the other hand, if the triplet energy of the added compound was higher than that of the BP compound, as in the cases of MMA and AN, $\phi(i)$ decreased only slightly and leveled off with increasing additive concentration. It was, therefore, concluded that free ions were formed by interaction of the unsaturated compound with the triplet excited BP compound. This finding was taken as direct evidence for the formation of exciplexes of charge transfer character. Typical results are presented in Figure 20, where the relative ion yield is plotted vs. the monomer concentration. It can be seen that vinyl acetate takes a position intermediate between that of AN and MMA, on the one side, and that of St, on the other side, probably indicating the simultaneous occurrence of several deactivation reactions.

b. Reaction of Triplet with Hydrogen Donors

The radical-forming capability of the benzophenones is characterized by the rate constants k_{T+H} which is relatively high in the cases of tetrahydrofuran and 2-propanol: $k_{T+H} = 3 \times 10^6$ l/mol s (THF)[81] and $k_{T+H} = 1 \times 10^6$ l/mol s (2-propanol).[83] Table 12 presents additional rate constants of the reaction of triplets of benzophenones with THF.[85] Amines are much

TABLE 13

Rate Constants k_{T+H} (10^9 l/mol s) of the Reaction of Triplets of Thioxanthones and Benzophenone with Hydrogen Donors, Determined in Benzene Solution at Room Temperature in the Absence of O_2[94]

Amine[a]	Initiator				
	TX	CTX	MTX	ITX	BP
E4DB	6	5	4	4	7
BMA	8	6	6	6	2
TEA	2.5				1.9[86]
DMA		8			
DMB		0.85			
BC					2[86]
DABCO					0.56[86]

[a] E4DB: ethyl-4-(dimethylamino)-benzoate, BMA: bis-(2-hydroxy-ethyl)-methylamine, TEA: triethylamine, DMA: dimethylamine, DMB: 2-(dimethylaminoethyl)-benzoate, BC: methyl-4-amino-benzoate, and DABCO: 1,4-diazabicyclo[2.2.2]octane.

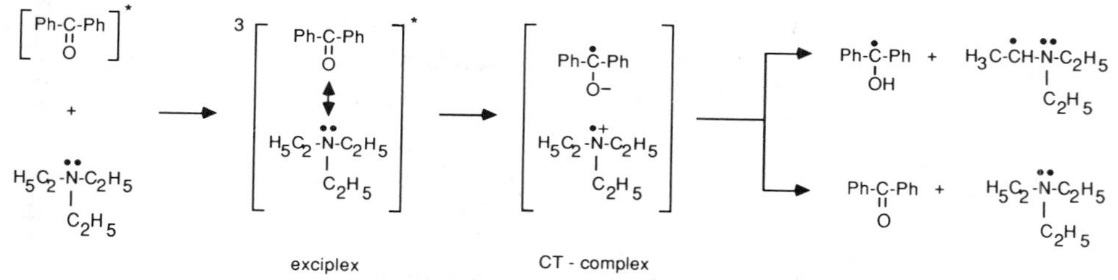

SCHEME VI

more reactive than 2-propanol and THF. Typical rate constants are presented in the last column of Table 13. Apart from DABCO, the rate constants are about 10^3 times higher than those of the reactions of 2-propanol or THF with BP triplets. The reaction mechanism involves electron transfer from the amine to the triplet state of BP forming a charge transfer complex as is illustrated in Scheme VI for the system BP/TEA. The exciplex is deactivated by proton transfer, leading to ketyl and amine radicals, or by spin inversion followed by back transfer of the electron resulting in the starting compounds: BP and amine. Evidence for the existence of the exciplex was obtained by picosecond laser flash photolysis of BP in acetonitrile solution in the presence and absence of triethylamine or diazabicyclo-octane.[88] BP triplets were formed within 10 ps after excitation. In the presence of amine, BP triplets were quenched with a half-life of about 10 ps forming the exciplex. Figure 21 shows the spectra of BP triplet and exciplex. The spectrum of the latter decayed with a half-life of about 15 ps leaving the spectrum of BP ketyl radical.

The efficiency of the reaction of amines with BP triplets is significantly dependent on the ionization potential of the former. Notably, a similar situation is met with thioxanthones. This aspect will be discussed, therefore, in more detail in Section 2.

c. Reaction of Ketyl Radical with Monomers

In some cases it was possible to determine the rate constant k_{R+M} of the reaction of the ketyl radical of BP with monomers.[81] The values obtained are presented in Table 7. They

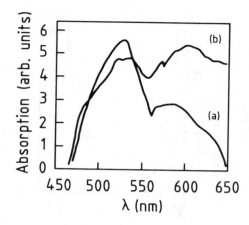

FIGURE 21. Transient absorption spectra of benzophenone triplet (a) and of benzophenone triplet/triethylamine exciplex (b) recorded 10 ps after excitation in acetonitrile solution. (a): [BP] = 0.1 mol/l. (b): [BP] = 0.1 mol/l; [TEA] = 3 mol/l. (From Shaefer, C. G. and Peters, K. S., *J. Am. Chem. Soc.*, 102, 5685, 1980. With permission.)

are rather low (ranging from 4×10^3 to 9×10^3 l/mol s). Obviously, not the ketyl radicals but the corresponding amine, alcohol, or ether radicals are capable of effectively initiating the polymerization of olefinic monomers.

2. Thioxanthone and Derivatives

Thioxanthone (TX) readily reacts in its electronically excited triplet state, similar to BP, with amines. However, it is rather unreactive towards aliphatic alcohols and THF. This is due to the fact that, in the case of TX, the lowest triplet state is a $\pi\pi^*$ state ($E_T = 274$ kJ/mol) whereas BP ($E_T = 288$ kJ/mol) has a lowest $n\pi^*$ triplet state. Only ketones with lowest $n\pi^*$ triplet states are readily reduced by alcohols and THF, whereas amines reduce quite effectively both $n\pi^*$ and $\pi\pi^*$ triplets.

The initiator efficiency of TX and TX derivatives is essentially determined by the rate constants k_q, k_{T+H}, and k_{R+M}. These rate constants could be determined conveniently because triplets and ketyl radicals of TX and its derivatives possess characteristic, rather strong absorption bands at $\lambda > 400$ nm. Figure 22 shows typical transient spectra recorded with 2-chlorothioxanthone (CTX) in cyclohexane solution.[94]

An outstanding property of (ground state) thioxanthones is their capability of absorbing rather strongly light in the near UV wavelength region. Table 14 presents extinction coefficients of thioxanthones of the following structure:[94]

	R	
	H	TX
	Cl	CTX
	CH$_3$	MTX
	CH(CH$_3$)$_2$	ITX

a. Triplet Quenching by Monomers

The lifetime of the triplets of TX and its derivatives was significantly reduced by olefinic

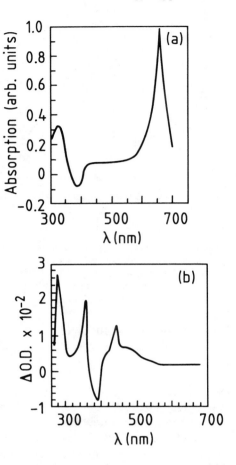

FIGURE 22.   Flash photolysis of thioxanthones in O_2-free dilute solution. $\lambda_{inc} = 347$ nm. (a) Absorption spectrum of MTX triplet recorded in benzene solution at the end of the 25-ns flash. [MTX] $= 4.0 \times 10^{-5}$. $D_{abs} = 7.6 \times 10^{-6}$ einstein/l. (b) Absorption spectrum of the ketyl radical of CTX recorded in cyclohexane solution 150 ns after the flash [BMA] $= 4 \times 10^{-2}$ mol/l. [CTX] $= 5 \times 10^{-5}$ mol/l. $D_{abs} = 2.8 \times 10^{-6}$ einstein/l. (From Amirzadeh, G. and Schnabel, W., *Makromol. Chem.*, 182, 2821, 1981. With permission.)

TABLE 14
Optical Properties of Thioxanthone
in Benzene Solution

Initiator	λ_{max} [nm]	$\epsilon\lambda_{(max)}$ [l/mol cm]	ϵ_{347nm} [l/mol cm]
TX	380	6.6×10^3	2.15×10^3
CTX	388	6.7×10^3	1.7×10^3
MTX	384	7.3×10^3	2.4×10^3
ITX	386	6.9×10^3	1.8×10^3

compounds. Upon recording the lifetime of the decay of the T-T absorption at 650 nm, the rate constants k_q compiled in Table 15 were obtained. These data are also presented in Figure 23, which shows that k_q depends on the chemical nature of the monomer. It decreases in the series St > VP > MMA > AN > BVE > VA. Styrene acts as a very effective triplet quencher, because its triplet energy is lower than that of the thioxanthones. Apart from

TABLE 15
Rate Constants k_q (l/mol s) of the Reaction of Triplets with Monomers

Initiator	Monomer[a]					
	St	MMA	AN	VAc	VP	BVE
TX	3×10^9	1.5×10^7	4×10^6	2×10^5	4×10^7	1×10^6
CTX	6×10^9	2×10^6	4×10^5	2×10^4	3×10^7	1×10^6
MTX	6×10^9	3×10^6	1×10^6	3×10^4	5×10^6	3×10^5
ITX	6×10^9	3×10^6	1×10^6	4×10^4	6×10^6	3×10^5
BP[b]	3×10^9	7×10^7	3×10^7	5×10^6	4×10^8	

[a] St: styrene, MMA: methyl methacrylate, AN: acrylonitrile, VAc: vinyl acetate, VP: *N*-vinyl-2-pyrrolidone, and BVE: butyl vinylether.
[b] BP: benzophenone.

styrene, all monomers interact more effectively with unsubstituted TX as compared with the substituted TX compounds. The relatively high k_q values found with unsubstituted TX can be explained in terms of sterical hindrance exerted by the substituent in 2-position. It retards the close approach of donor and acceptor necessary for the exchange mechanism to become operative. Notably, the systems CTX/VP and CTX/BVE exhibit a somewhat different behavior. This might be understood in terms of VP and BVE possessing electron-rich C=C double bonds, in contrast to MMA, VA, and AN. Moreover, the electron attracting power of the chlorine in CTX will cause an electron deficiency at the carbonyl group, thus enhancing its interaction with the double bond of the monomer.

b. *Reaction of Triplet with Hydrogen Donors*

The statement concerning quasi nonreactivity of TX triplets towards aliphatic alcohols made in Section 2 needs to be commented on. Actually, TX triplets were indeed found to undergo a rather slow reaction, resulting in the formation of ketyl radicals, upon irradiating TXs in neat alcohols and in neat cyclohexane.[94] On the basis of the triplet lifetimes listed in Table 16, rate constants $k_{T+H} \approx 10^3$ to 10^4 l/mol s are estimated for the reaction of aliphatic alcohols and cyclohexane with TX triplets.

On the other hand, many amines react very rapidly with thioxanthone triplets. Typical rate constants are listed in Table 13. The high reactivity is best explained in terms of an exciplex mechanism involving the formation of ion pairs by electron transfer. In a recent study with the system TX/dimethylanilin, devoted to the elucidation of this mechanism, ion pairs were evidenced by characteristic optical absorptions at 460 nm (dimethylaniline radical cation) and at 650 nm (TX anion).[101] Figure 24 shows the respective transient absorption spectrum recorded 136 ns after irradiating an acetonitrile solution with a flash of 337-nm light, produced by a nitrogen laser. That a charge-transfer intermediate is also formed in benzene solution was inferred from the fact that k_{T+H} is correlated with the ionization potentials of the amines. As can be seen from Table 17, k_{T+H} decreases with increasing ionization potential. It is interesting to note that a similar correlation has been found for BP.[89]

c. *Reaction of Ketyl Radical with Monomers*

Rate constants k_{R+M} of the reaction of thioxanthone ketyl radicals with olefinic monomer

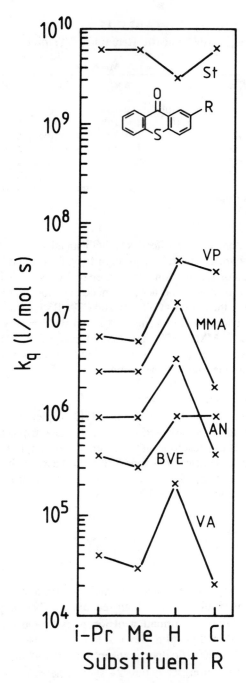

FIGURE 23.   Quenching of thioxanthone triplets by monomers. Influence
of the substituent R in position 2 on the rate constant k_q. iPr: $CH(CH_3)_2$,
Me: CH_3, Cl: chlorine. (From Amirzadeh, G. and Schnabel, W., *Makromol. Chem.*, 182, 2821, 1987. With permission.)

were determined in some cases in benzene solution by measuring the rate of decay of the
optical absorption of the ketyl radicals at λ = 435 nm.[94] The values obtained are listed in
Table 7. Actually, TX ketyl radicals are of low reactivity towards olefinic monomers. On
the basis of these rate constants it can be concluded that, in cases of practical importance,
ketyl radicals will be mostly deactivated by radical-radical reactions rather than by reactions

TABLE 16
Lifetimes of ^{3}MTX* in Various Argon-Saturated (Neat) Solvents at Room Temperature

Solvent	τ (μs)
Cyclohexane	10
2-propanol	12
1,2-dichloroethane	20
Acetonitrile	40
Ethanol	60
Methanol	68
Benzene	95

Note: Measurements were carried out at [MTX] = 4×10^{-5} mol/l and $D_{abs} = 2 \times 10^{-6}$ einstein/l.

From Amirzadeh, G. and Schnabel, W., *Makromol. Chem.*, 182, 2821, 1981. With permission.

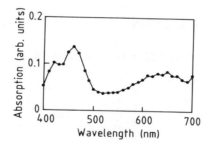

FIGURE 24.   Flash photolysis of the system thioxanthone/dimethylaniline in N_2-saturated acetonitrile solution at room temperature. Transient absorption spectrum recorded 136 ns after the flash. λ_{inc} = 337 nm. [TX] = 6.8×10^{-4} mol/l. [DMA] = 9.9×10^{-3}. (From Yates, S. F. and Schuster, G. B., *J. Org. Chem.*, 49, 3349, 1984. With permission.)

TABLE 17
Rate Constants k_{T+H} of the Reaction of Triplet TX with Various Amines and Gas Phase Ionization Potentials of Amines

Amine	k_{T+H} (l/mol s)		Ionization potential (eV)
	Benzene	**Acetonitrile**	
Dimethylaniline	9.1×10^9	5.3×10^9	7.14
Triethylamine	1.3×10^9		7.50
Dibutylamine	2.7×10^9		7.70
t-Butylamine	4.6×10^6	8.7×10^6	8.64

with monomer molecules.[15] The initiating species, in TX/amine systems, is the amine radical that is generated together with the ketyl radical in the reaction of TX triplets with amines. Because of the transparency of amine radicals in the wavelength region of experimental accessibility, reactions of amine radicals with monomers could not be examined directly.

3. 1-Phenyl-2-Acetoxy-2-Methylpropanone-1

In the previous sections attention was paid, to some extent, to the intermediacy of exciplexes or CT complexes in the reaction of ketones with amines. It was shown that direct evidence for the existence of these complexes could be obtained in polar media by the detection of radical ions.[84,88,102] In connection with this work it is interesting to note that spectroscopic kinetic evidence for the intermediacy of an exciplex in the reaction of 1-phenyl-2-acetoxy-2-methylpropanone-1 (PAMP) with tri-[2-hydroxyethyl]-amine (THEA) in benzene solution, i.e., in an unpolar medium, was obtained. The results will be reported in the following. Contrary to 1-phenyl-2-hydroxy-2-methylpropanone-1 (PHMP, see Section A.3) its acetic acid ester, PAMP, does not undergo α-cleavage. Flash photolysis work with PAMP in cyclohexane and benzene solution yielded evidence for a rather long-lived triplet state.[105] Figures 25a and b show the decay of the T-T absorption at 410 nm and of the phosphorescence at 440 nm in cyclohexane solution at room temperature. First-order plots are shown in Figure 25c. The triplet lifetime is about 2 μs. It was found that PAMP reacts rapidly with aliphatic amines such as (THEA) or di-[2-hydroxyethyl]-methyl amine (DHEMA) giving rise to the formation of ketyl and amine radicals:[105]

$$
\begin{bmatrix} \text{Ph–C–R} \\ \| \\ \text{O} \end{bmatrix}^{3*} + \; \begin{array}{c} \text{CH}_2\text{–CH}_2\text{OH} \\ / \\ \text{N – CH}_2\text{–CH}_2\text{–OH} \\ \backslash \\ \text{CH}_2\text{–CH}_2\text{OH} \end{array} \; \xrightarrow{h\nu} \; \text{PH–}\overset{\cdot}{\text{C}}\text{–R} + \; \begin{array}{c} \text{CH}_2\text{–CH}_2\text{OH} \\ / \\ \text{N – }\overset{\cdot}{\text{C}}\text{H–CH}_2\text{–OH} \\ \backslash \\ \text{CH}_2\text{–CH}_2\text{OH} \end{array}
$$

$$
\text{R: } -\underset{\underset{\underset{\|}{\text{O}}}{\overset{\overset{\text{CH}_3}{|}}{\underset{|}{\text{C–CH}_3}}}}{}\text{O–C–CH}_3
\tag{41}
$$

The following rate constants k_{T+H} (in liters per mole per second) were obtained: 1.0×10^9 (DHEMA) and 1.4×10^9 (THEA).[106] Actually, PAMP can be used as a photoinitiator of the hydrogen abstraction type for the polymerization of monomers that do not quench PAMP triplets effectively.[105] A detailed study of the mechanism of Reaction 41 revealed that ketyl radicals were formed much more slowly than triplets disappeared.[106] This is illustrated in Figure 26, where kinetic traces obtained with benzene solutions of PAMP (3.3 $\times 10^{-3}$ mol/l) in the absence and presence of THEA are presented. A comparison of traces a-1 and b-1 shows that THEA reacts rapidly with PAMP triplets. Trace a-2 depicts how, in the presence of amine, a new absorption is formed at $\lambda = 360$ nm after the decay of the absorption of the triplet. Obviously, the triplet decay rate is much faster than the build-up rate of the new absorption. Therefore, it has been concluded that a mechanism according to Reaction 4 applies (vide ante Section I.A.2), that is, PAMP triplets form an intermediate with THEA which decomposes, in benzene solution, into free radicals and this intermediate is identical with the triplet exciplex. The increase in the absorption of ketyl radicals in benzene or cyclohexane solution followed a first-order rate law with a rate constant k = 9

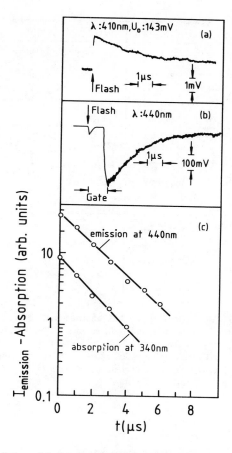

FIGURE 25. Irradiation of 1-phenyl-2-acetoxy-2-methylpropanone-1 (1.5×10^{-4} mol/l) in Ar-saturated cyclohexane solution at room temperature. $\lambda_{inc} = 265$ nm. Flash duration = 15 ns. $D_{abs} = 1.2 \times 10^{-5}$ einstein/l. (a): Kinetic trace depicting the decay of the optical absorption of triplets at $\lambda = 410$ nm. (b): Kinetic trace depicting the decay of the phosphorescence at $\lambda = 440$ nm. As indicated in the graph, the photomultiplier became fully operative after a certain "gate" time. (c): first-order plots of the decay of triplet absorption and phosphorescence after the flash.(From Eichler, J., Herz, C. P., and Schnabel, W., *Angew. Makromol. Chem.*, 91, 39, 1980. With permission.)

$\times 10^6$ s^{-1}. Actually, this is the rate constant of hydrogen transfer in the complex. Quite similar results were obtained with DHEMA.[106]

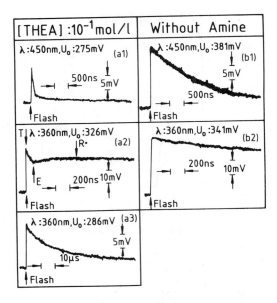

FIGURE 26.   Flash photolysis of PAMP in Ar-saturated benzene solution in the presence [(a-1) to (a-3)] and in the absence of THEA [(b-1) and (b-2)]. [PAMP] = 3.3×10^{-3} mol/l. Kinetic traces depicting formation and decay of the optical absorption at 360 and 450 nm. (From Salmassi, A., Eichler, J., Herz, C. P., and Schnabel, W., *Z. Naturforsch.*, 35a, 1273, 1980. With permission.)

REFERENCES

1. **Pappas, S. P., Ed.,** *UV Curing: Science and Technology,* Vol. 1 and 2, Technology Marketing, Norwalk, CT, 1978 and 1985.
2. **Guillett, J.,** *Polymer Photophysics and Photochemistry,* Cambridge University Press, New York, 1985.
3. **Timpe, H.-J. and Baumann, H.,** Photopolymersysteme — Prinzipien und Anwendungen, Deutscher Verlag für Grundstoffindustrie, Leipzig, 1987.
4. **Hageman, H. J.,** Photoinitiators for free radical polymerization, *Progr. Org. Coatings,* 13, 123, 1985.
5. **Baumann, H., Timpe, H.-J., and Böttcher, H.,** Initiatorsysteme für radikalinduzierte Photopolymerisation, *Z. Chem.,* 23, 197, 1983.
6. **Baumann, H., Timpe, H.-J., and Böttcher, H.,** Initiatorsysteme für ionische Photopolymerisation, *Z. Chem.,* 23, 394, 1983.
7. **Timpe, H. J.,** Photoinduzierte Polymerbildungsprozesse, Sitzungsberichte der Akademie der Wissenschaften der DDR, Akademie-Verlag, Berlin, 1987.
8. **Pappas, S. P.,** Photoinitiated radical polymerization, *J. Radiat. Curing,* 14, 6, 1987.
9. **Bassi, G. L.,** Photoinitiators of polymerization: recent developments and evolution, *J. Radiat. Curing,* 14, 26, 1987.
10. **Crivello, J. V.,** Photoinitiated polymerization by sulphonium salts, *Dev. Polym. Photochem.,* 2, 1, 1981.
11. **Crivello, J. V.,** Cationic polymerization — iodonium and sulfonium salt photoinitiators, *Adv. Polym. Sci.,* 62, 3, 1984.
12. **Ledwith, A.,** New initiation systems for cationic polymerization, *Makromol. Chem. Suppl.,* 3, 348, 1979.
13. **Ledwith, A., Al-Kass, and Hulme-Lowe, A.,** Mechanism for photoinitiation of cationic polymerization, in *Cationic Polymerization and Related Processes,* Goethals, E. J., Ed., Academic Press, London, 1984.
14. **Pappas, S. P.,** Photogeneration of acid. VI. A review of basic principles for resist imaging applications, *J. Imaging Technol.,* 11, 146, 1985.
15. **Schnabel, W.,** Mechanistic and kinetic aspects concerning the initiation of free radical polymerization by thioxanthones and acylphosphine oxides, *J. Radiat. Curing,* 13, 26, 1986.

16. **Crivello, J. V.,** Photoinitiated cationic polymerization, in *UV Curing: Science and Technology,* Vol. 1, Pappas, S. P., Ed., Technology Marketing Corp., Norwalk, CT, 1978, 23.

17. **Crivello, J. V. and Lam, J. W. H.,** Diaryliodonium salts. A new class of photoinitiators for cationic polymerization, *Macromolecules,* 10, 1307, 1977.

18. **Crivello, J. V., Lam, J. W. H., and Volante, C. N.,** Photoinitiated cationic polymerization using diaryliodonium salts, *J. Radiat. Curing,* 4, 2, 1977.

19. **Knapczyk, J. W., Lubinkowski, J. J., and McEwen, W. E.,** Photolysis of diphenyliodonium salt in alcohol solution, *Tetrahedron Lett.,* 3739, 1972.

20. **Beck, G.,** Fast Conductivity Techniques in Pulse Radiolysis, Proc. 6th Int. Congr. Rad. Res., Tokyo, 1979, 279.

21. **Janata, E.,** Pulse radiolysis conductivity measurements in aqueous solutions with nanosecond time resolution, *Rad. Phys. Chem.,* 19, 17, 1982.

22. **Kevan, L. and Schwartz, R. N., Eds.,** *Time Domain Electron Spin Resonance,* John Wiley & Sons, New York, 1979.

23. **Atkins, P. W., McLauchlan, K. A., and Percival, P. W.,** Electron spin-lattice relaxation times from the decay of ESR emission spectra, *Mol. Phys.,* 25, 281, 1973.

24. **De Boer, J. W. M., Chan Chung, T. Y. C., and Wan, J. K. S.,** Time-resolved CIDEP in the photo-reduction of quinones. A study of the spin lattice relaxation time of semiquinone radicals in solution, *Can. J. Chem.,* 57, 2971, 1979.

25. **Trifunac, A. D., Thurnauer, M. C., and Norris, J. R.,** Submicrosecond time-resolved EPR in laser photolysis, *Chem. Phys. Lett.,* 57, 471, 1978.

26. **Kuwata, K., Kominami, S., Hayashi, K., and Kobayashi, K.,** Time-resolved ESR spectroscopy in radiation chemistry and photochemistry, *Memoirs Inst. Sci. Ind. Res.,* 43, 59, 1986.

27. **Kuhlmann, R. and Schnabel, W.,** Flash photolysis investigation on primary processes of the sensitized polymerization of vinyl monomers. II. Experiments with benzoin and benzoin derivatives, *Polymer,* 18, 1163, 1977.

28. **Kuhlmann, R. and Schnabel, W.,** Flash photolysis investigation on primary processes of the sensitized polymerization of vinylmonomers. III. Photoreactions of benzoin compounds and stationary polymerization experiments, *Angew. Makromol. Chem.,* 70, 145, 1978.

29. **Lewis, F. D., Lauterbach, R. T., Heine, F.-G., Hartmann, W., and Rudolph, H. J.,** Photochemical α-cleavage of benzoin derivatives. Polar transition states for free radical formation, *J. Am. Chem. Soc.,* 97, 1519, 1975.

30. **Heine, H.-G., Hartmann, W., Kory, D. R., Magyar, J. G., Hoyle, C. E., McVey, J. K., and Lewis, F. D.,** Photochemical α-cleavage and free radical reactions of some deoxybenzoins, *J. Org. Chem.,* 39, 691, 1974.

31. **Lewis, F. D., Hoyle, C. E., and Magyar, J. G.,** Substituent effects on the photochemical α-cleavage of deoxybenzoin, *J. Org. Chem.,* 40, 488, 1975.

32. **Lewis, F. D. and Magyar, J. G.,** Cage effects in the photochemistry of (S)-(+)-2-phenylpropiophenone, *J. Am. Chem. Soc.,* 95, 5973, 1973.

33. **Amirzadeh, G., Kuhlmann, R., and Schnabel, W.,** Blitzphotolyse von Desoxybenzoinen in Lösung, *J. Photochem.,* 10, 133, 1979.

34. **Gehlhaus, J. and Kieser, M.,** Hydroxyalkylphenones useful as photosensitizers, *Ger. Offen.,* 2,722, 264, A1, 1978.

35. **Felder, L., Kirchmayr, R., and Hüsler, R.,** Photopolymerizable systems containing aromatic-aliphatic ketones and use of the ketones as photoinitiators, *Eur. Pat. Appl.* 3002, 1979.

36. **Hageman, H. J., Van der Maeden, F. P. B., and Jansen, P. C. G. M.,** Vinyl polymerization photoinitiated by benzoinmethylether, *Makromol. Chem.,* 180, 2531, 1979.

37. **Groenenboom, C. J., Hageman, H. J., Overeem, T., and Weber, A. J. M.,** Photoinitiators and photoinitiation. III. Comparison of the photodecomposition of α-methoxy- and α,α'-dimethoxydeoxybenzoin in 1,1-diphenylethylene a model substrate, *Makromol. Chem.,* 183, 281, 1982.

38. **Eichler, J., Herz, C. P., Naito, I., and Schnabel, W.,** Laser flash photolysis investigation of primary processes in the sensitized polymerization of vinyl monomers. IV. Experiments with hydroxy alkylphenones, *J. Photochem.,* 12, 225, 1980.

39. **Hugenberger, C., Lipscher, J., and Fischer, H.,** Self-termination of benzoyl radicals to ground and excited state benzil. Symmetry control of radical combination, *J. Phys. Chem.,* 84, 3467, 1980.

40. **Salmassi, A., Eichler, J., Herz, C. P., and Schnabel, W.,** On the photolysis of 1-phenyl-2-hydroxy-2-methyl-propanone-1 in the presence of methylmethacrylate, styrene and acrylonitrile: laser flash photolysis studies, *Polym. Photochem.,* 2, 209, 1982.

41. **Lechtken, P., Buethe, I., and Hesse, A.,** Acylphosphinoxidverbindungen und ihre Verwendungen, *Ger. Offen.,* 2830927, 1980; **Lechtken P., Buethe, I., Jacobi, M., and Trimborn, W.,** Acylphosphinoxidverbindungen, ihre Herstellung und Verwendung, *Ger. Offen.,* 2909994, 1980.

42. **Heine, H.-G., Rosenkranz, H. J., and Rudolph, H.,** Photopolymerisierbare Mischungen mit Aroyl-phosphonsäureestern als Photoinitiatoren, *Ger. Offen.,* 3023486, 1980.

43. **Jacobi, M. and Henne, A.,** Acylphosphine oxides: a new class of photoinitiators, *J. Radiat. Curing,* 10, 16, 1983.

44. **Sumiyoshi, T., Henne, A., Lechtken, P., and Schnabel, W.,** Optical absorption spectra of phosphonyl radicals, *Z. Naturforsch.,* 39a, 434, 1984.

45. **Sumiyoshi, T., Schnabel, W., Henne, A., and Lechtken, P.,** On the photolysis of acylphosphine oxides. I. Laser flash photolysis studies with 2,4,6-trimethylbenzoyldiphenylphosphine oxide, *Polymer,* 26, 141, 1985.

46. **Sumiyoshi, T., Schnabel, W., and Henne, A.,** Photolysis of acylphosphine oxides. II. The influence of methyl substitution in benzoyldiphenylphosphine oxides, *J. Photochem.,* 32, 119, 1986.

47. **Sumiyoshi, T., Schnabel, W., and Henne, A.,** Photolysis of acylphosphine oxides. III. Laser flash photolysis studies with pivaloyl compounds, *J. Photochem.,* 32, 191, 1986.

48. **Sumiyoshi, T. and Schnabel, W.,** On the reactivity of phosphonyl radicals towards olefinic compounds, *Makromol. Chem.,* 186, 1811, 1985.

49. **Sumiyoshi, T., Schnabel, W., and Henne, A.,** Laser flash photolysis of acyl phosphonic acid esters, *J. Photochem.,* 30, 63, 1985.

50. **Sumiyoshi, T., Weber, W., and Schnabel, W.,** Thiophosphonyl radicals. Photolytic generation and reactivity towards olefinic compounds, *Z. Naturforsch.,* 40a, 541, 1985.

51. **Schnabel, W. and Sumiyoshi, T.,** Initiation of free radical polymerisation by acylphosphine oxides and derivatives, in *New Trends in the Photochemistry of Polymers,* Allen, N. S. and Rabek, J. F., Eds., Elsevier Applied Science Publishers, London, 1985, 69.

52. **Sumiyoshi, T., Katayama, M., and Schnabel, W.,** Pulse radiolysis studies of energy transfer reactions from benzene excimer to phosphine oxides, *Chem. Lett. Chem. Soc., Jpn.,* 1647, 1985.

53. **Kamachi, M., Kuwata, K., Sumiyoshi, T., and Schnabel, W.,** ESR studies on the photodissociation of 2,4,6-trimethylbenzoyldiphenylphosphine oxide, *J. Chem. Soc. Perkin Trans.,* 2, 961, 1988; **Kuwata, K., Kominami, S., Hayashi, K., and Kobayashi, K.,** Time-resolved ESR spectroscopy in radiation chemistry and photochemistry, *Memoirs Inst. Sci. Ind. Res.,* 43, 59, 1986.

54. **Baxter, J. E., Davidson, R. S., Hageman, H. J., McLauchlan, K. A., and Stevens, D. G.,** The photo-induced cleavage of acylphosphine oxides, *J. Chem. Soc., Chem. Commun.,* 73, 1987.

55. **Baxter, J. E., Davidson, R. S., Hageman, H. J., and Overeem, T.,** Photoinitiators and photoinitiation. VII. The photo-induced α-cleavage of acylphosphine oxides. Trapping of primary radicals by a stable nitroxyl, *Makromol. Chem. Rapid Commun.,* 8, 311, 1987.

56. **Hayashi, H., Sakaguchi, Y., Kamachi, M., and Schnabel, W.,** Laser flash photolysis study of the magnetic field effect on the photodecomposition of (2,4,6-trimethylbenzoyl)diphenylphosphine oxide in micellar solution, *J. Phys. Chem.,* 91, 3936, 1987.

57. **Majima, T. and Schnabel, W.,** Flash photolysis of 2,4,6-trimethylbenzoyl-diphenylphosphine oxide and sulfide, unpublished results.

58. **Haag, R., Wirz, J., and Wagner, P. J.,** The photoenolization of 2-methylacetophenone and related compounds, *Helv. Chim. Acta,* 60, 2595, 1977.

59. **Kerr, C. M. L., Webster, K., and Williams, F.,** Electron spin resonance studies of γ-irradiated phosphite and phosphate esters. Identification of phosphinyl, phosphonyl, phosphoranyl and phosphine dimer cation radicals, *J. Phys. Chem.,* 79, 2650, 1975.

60. **Geoffroy, M. and Lucken, E. A. C.,** Electron spin resonance spectrum of x-irradiated single crystals of diphenylphosphine oxide, *Mol. Phys.,* 22, 257, 1971.

61. **Delzenne, G. A., Laridon, U., and Peeters, H.,** Photopolymerization initiated by O-acyloximes, *Eur. Polym. J.,* 6, 933, 1970.

62. **Hong, S. I., Kurosaki, T., and Okawara, M.,** Photopolymerization initiated by oxime derivatives, *J. Polym. Sci., Polym. Chem. Ed.,* 12, 2553, 1974.

63. **McGinniss, V. D.,** Acrylate systems for U.V. curing. I. Light sources and photoinitiators, *J. Radiat. Curing,* 2, 3, 1975; **McGinniss, V. D. and Ting, V. W.,** Acrylate systems for U.V. curing. II. Monomers and crosslinking resin systems, *Radiat. Curing,* 2, 12, 1975.

64. **Baas, P. and Cerfontain, H.,** Photochemistry of α-oxo-oximes. VII. Photolysis of α-oxo-oxime esters, *J. Chem. Soc. Perkin Trans. 2,* 1653, 1979.

65. **Amirzadeh, G.,** Mechanismen der photoinitiierten Polymerisation. Eine Laserblitz-photolytische Unter-suchung von Thioxanthonen und α-Keto-Oximestern, Dissertation, Technische Universität, Berlin, 1981.

66. **Odian, G.,** *Principles of Polymerization,* 2nd ed., John Wiley & Sons, New York, 1981, 194.

67. **Seltzer, S.,** The secondary α-deuterium isotope effect in the thermal decomposition of α-phenylethylazo-2-propanol, *J. Am. Chem. Soc.,* 85, 14, 1963.

68. **Neuman, R. C., Jr. and Amrich, M. J., Jr.,** Pressure effects on azocumene decomposition rates, efficiencies of radical production and semibenzene dimers, *J. Org. Chem.,* 45, 4629, 1980.

69. **Pryor, W. A. and Smith, K.,** The viscosity dependence of bond homolysis. A qualitative and semiquantitative test for cage return, *J. Am. Chem. Soc.,* 92, 5403, 1970.

70. **White, D. K. and Greene, F. D.,** Decomposition of meso- and DL-3,4-diethyl-3,4-dimethyldiazetine (a 1,2-diaza-1-cyclobutene), *J. Am. Chem. Soc.,* 100, 6760, 1978.

71. **Neuman, R. C., Jr. and Binegar, G. A.,** *cis*-Azoalkanes. Mechanisms of scission and isomerization, *J. Am. Chem. Soc.,* 105, 134, 1983.

72. **Dannenberg, J. J. and Rocklin, D.,** A theoretical study of the mechanism of thermal decomposition of azoalkanes and 1,1-diazenes, *J. Org. Chem.,* 47, 4529, 1982.

73. **Engel, S. P.,** Mechanism of the thermal and photochemical decomposition of azoalkanes, *Chem. Rev.,* 80, 99, 1980.

74. **Porter, N. A., Marnett, L. J., Lochmüller, C. H., Closs, G. L., and Shobataki, M.,** Application of chemically induced dynamic nuclear polarization to the study of the decomposition of unsymmetric azo compounds, *J. Am. Chem. Soc.,* 94, 3664, 1972.

75. **Holt, L. P., McCurdy, K. E., Adams, J. S., Burton, K. A., Weisman, R. B., and Engel, P. S.,** Direct studies of photodissociation of azomethane vapor using transient CARS spectroscopy, *J. Am. Chem. Soc.,* 107, 2180, 1985.

76. **Sumiyoshi, T., Kamachi, M., Kuwae, Y., and Schnabel, W.,** Laser flash photolysis of azocumenes. Direct observation of stepwise decomposition, *Bull. Chem. Soc. Jpn.,* 60, 77, 1987.

77. **Engel, P. S. and Steel, C.,** Photochemistry of aliphatic azo compounds in solution, *Acc. Chem. Res.,* 6, 275, 1973.

78. **Block, H., Ledwith, A., and Taylor, A. R.,** Polymerization of methylmethacrylate photosensitized by benzophenones, *Polymer,* 12, 271, 1971.

79. **Hutchinson, J. and Ledwith, A.,** Mechanisms and relative efficiencies in radical polymerization photoinitiated by benzoin, benzoinmethyl ether and benzil, *Polymer,* 14, 250, 1973.

80. **Ledwith, A.,** Photoinitiation by aromatic carbonyl compounds, *J. Oil Colour Chem. Assoc.,* 59, 157, 1976.

81. **Kuhlmann, R. and Schnabel, W.,** Laser flash photolysis investigations on primary processes of the sensitized polymerization of vinyl monomers. I. Experiments with benzophenones, *Polymer,* 17, 419, 1976.

82. **Kuhlmann, R. and Schnabel, W.,** On the primary processes of sensitized photopolymerization of vinyl monomers. Laser flash photolysis studies and stationary polymerization experiments, *Angew. Makromol. Chem.,* 57, 195, 1977.

83. **Beckett, A. and Porter, G.,** Primary photochemical processes in aromatic molecules. IX. Photochemistry of benzophenone in solution, *Trans. Faraday. Soc.,* 59, 2038, 1963.

84. **Kuhlmann, R. and Schnabel, W.,** Transient photocurrents induced by laser flash photolysis of aromatic carbonyl compounds in solution, *J. Photochem.,* 7, 287, 1977.

85. **Merlin, A., Lougnot, D.-J., and Fouassier, J.-P.,** Laser spectroscopy of substituted benzophenone used as photo-initiator of vinyl polymerization, *Polym. Bull.,* 2, 847, 1980.

86. **Merlin, A., Lougnot, D.-J., and Fouassier, J.-P.,** The benzophenone amine photoinitiator in vinyl polymerization, *Polym. Bull.,* 3, 1, 1980.

87. **Lougnot, D.-J., Jacques, P., and Fouassier, J.-P.,** The behavior of triplet benzophenone in micellar solutions: evidence for an exciplex mechanism?, *J. Photochem.,* 19, 59, 1982.

88. **Shaefer, C. G. and Peters, K. S.,** Picosecond dynamics of the photoreduction of benzophenone by triethylamine, *J. Am. Chem. Soc.,* 102, 7580, 1980.

89. **Guttenplan, J. B. and Cohen, S. G.,** Triplet energies, reduction potentials and ionization potentials in carbonyl-donor partial charge-transfer interactions. I, *J. Am. Chem. Soc.,* 94, 4040, 1972.

90. **Stone, P. G. and Cohen, S. G.,** Catalysis by aliphatic thiol of photoreduction of benzophenone by primary and secondary amines, *J. Am. Chem. Soc.,* 102, 5685, 1980.

91. **Stone, P. G. and Cohen, S. G.,** Effects of structure on rates and quantum yields in photoreduction of fluorenone by amines, catalysis and inhibition by thiols, *J. Am. Chem. Soc.,* 104, 3435, 1982.

92. **Griller, D., Howard, P. R., Marriot, P. R., and Scaiano, J. C.,** Absolute rate constants for the reaction of *tert*-butoxyl, *tert*-butylperoxyl, and benzophenone triplets with amines: the importance of the stereoelectronic effect, *J. Am. Chem. Soc.,* 103, 619, 1981.

93. **Davidson, R. S. and Goodin, J. W.,** The polymerization of acrylates using a combination of carbonyl compounds and an amine as a photoinitiator system, *Eur. Polym. J.,* 18, 597, 1982.

94. **Amirzadeh, G. and Schnabel, W.,** On the photoinitiation of free radical polymerization. Laser flash photolysis investigations on thioxanthones, *Makromol. Chem.,* 182, 2821, 1981.

95. **Allen, N. S., Catalina, F., Green, P. N., and Green, W. A.,** Photochemistry of thioxanthones. I. Spectroscopic and flash photolysis study on oil soluble structures, *Eur. Polym. J.,* 21, 841, 1985.

96. **Allen, N. S., Catalina, F., Green, P. N., and Green, W. A.,** Photochemistry of thioxanthones. II. A spectroscopic and flash photolysis study on water-soluble structures, *Eur. Polym. J.,* 22, 347, 1986.

97. **Allen, N. S., Catalina, F., Moghaddam, B., Green, P. N., and Green, W. A.,** Photochemistry of thioxanthones. III. Spectroscopic and flash photolysis study on hydroxy and methoxy derivatives, *Eur. Polym. J.,* 22, 691, 1986.

98. **Allen, N. S., Catalina, F., Green, P. N., and Green, W. A.,** Photochemistry of thioxanthones. IV. Spectroscopic and flash photolysis study on novel n-propoxy and methyl n-propoxy derivatives, *Eur. Polym. J.*, 22, 793, 1986.

99. **Allen, N. S., Catalina, F., Green, P. N., and Green, W. A.,** Photochemistry of thioxanthones. V. A polymerization, spectroscopic and flash photolysis study on novel water soluble methyl substituted 3-(9-oxo-9H-thioxanthene-2-yloxyl)-*N,N,N*-trimethyl-1-propanaminium salts, *Eur. Polym. J.*, 22, 871, 1986.

100. **Allen, N. S., Catalina, F., Green, P. N., and Green, W. A.,** Photochemistry of thioxanthones. VI. A polymerization, spectroscopic and flash photolysis study on novel water-soluble substituted 3-(9-oxo-9H-thioxanthene-2,3-γ-4-yloxy)-*N,N,N*-trimethyl-1-propanaminium salts, *J. Photochem.*, 36, 99, 1987.

101. **Yates, S. F. and Schuster, G. B.,** Photoreduction of triplet thioxanthone by amines: charge transfer generates radicals that initiate polymerization of olefins, *J. Org. Chem.*, 49, 3349, 1984.

102. **Fouassier, J. P., Lougnot, D. J., Paverne, A., and Wieder, F.,** Time-resolved laser-pumped dye laser spectroscopy: energy and electron transfer reactions, *Chem. Phys. Lett.*, 135, 30, 1987.

103. **Jacques, P., Lougnot, D. J., and Fouassier, J. P.,** Specific micellar effects in the temporal behavior of excited benzophenone: consequences upon the polymerization kinetics, in *Surfactants in Solution*, Vol. 2, Mittal, K. L. and Lindman, B., Eds., Plenum Press, New York, 1984, 1177.

104. **Scaiano, J. C. and Lougnot, D. J.,** Electrostatic and magnetic field effects on the behavior of radical pairs derived from ionic benzophenones, *J. Phys. Chem.*, 88, 3379, 1984.

105. **Eichler, J., Herz, C. P., and Schnabel, W.,** On the photolysis of hydroxyalkylphenones and O-substituted derivatives, *Angew. Makromol. Chem.*, 91, 39, 1980.

106. **Salmassi, A., Eichler, J., Herz, C. P., and Schnabel, W.,** Evidence for triplet exciplex formation in the reaction of amines with 1-phenyl-2-acetoxy-2-methyl-propanone-1, *Z. Naturforsch.*, 35a, 1273, 1980.

Chapter 6

LASER SPECTROSCOPY OF EXCITED STATE PROCESSES IN WATER-SOLUBLE PHOTOINITIATORS OF POLYMERIZATION

Daniel-Joseph Lougnot and Jean-Pierre Fouassier

TABLE OF CONTENTS

I. INTRODUCTION

In recent years, radical photopolymerization reactions have been increasingly used in the coating industry.[1] Typically, a monomer (M) or a prepolymer film is polymerized under exposure to UV light, a photoinitiator (I) being incorporated in order to absorb the luminous energy and to start the reaction:

4845/ch 6/str1-2

$$I \xrightarrow{h\nu} R^{\cdot} \xrightarrow{M} RM^{\cdot} \rightsquigarrow polymer$$

The intrinsic photochemical reactivity in the excited states of the photoinitiator governs the practical efficiency of the process. Useful relationships have been proposed to account for the differences in reactivity observed[2-4] in oil-soluble photoinitiators.

Several ionic water-soluble photoinitiators (WSPs) have appeared in the past few years.[5-7] They allow grafting reactions to be carried out in waterborne systems and, in this way, significantly reduce the homopolymerization (e.g., in grafting hydrophilic monomers onto cellulosic materials).[8] Presumably, they could be of interest if incorporated in UV-curable systems. On the other hand, they exhibit a very good efficiency in micelle photo-polymerization.[9-12] This chapter is devoted to a discussion of the excited state processes detected through time-resolved laser spectroscopy.

II. BASIC DEFINITIONS

The design of WSPs is generally derived from backbone of usual structures (benzo-phenone — BP,[8] thioxanthone — TX,[5] benzil — BZ,[10] hydroxy alkylketone — HAP[7]). Water solubility may be expected to some extent, due to the presence of polar substituents (morpholino group in TPMK[13] or hydroxy groups in HAP series[7]) but is more efficiently achieved by introducing ionic groups. More recently, a new system was proposed (phenacyl thiosulfate derivatives, PTS).[6] All these compounds are listed in Table 1.

The ground state absorption spectra of these compounds in water are very similar to those of the parent oil-soluble compounds (Figure 1). The presence of an alkylthio group in TPMK, however, strongly modifies the UV absorption spectrum because of a change in the spectroscopic nature of the lowest occupied molecular orbital of the carbonyl chrom-ophore.

The general behavior of the excited states involves either the usual electron and proton transfer processes in the presence of amines (e.g., for BPs, TXs, BZs — Scheme IA)[14,15] or cleavage photoreactions at the α-position (e.g., for HAPs,[16] TPMK[17] — Scheme IB) or at the C-S or S-S bond (presumably for PTS)[16] with respect to the carbonyl group.[8]

These compounds are quite efficient photoinitiators in the polymerization of acrylamide in water (Table 2). Some of them require the presence of an electron donor such as meth-yldiethanolamine (MDEA).

The design of new efficient systems presupposes a good knowledge of the processes involved in the excited states, which provides an insight into the various steps of the overall initiation mechanism.

III. EXCITED STATE PROCESSES IN WATER

A. GENERAL BEHAVIOR[15]

Laser excitation of WSPs (BP, TX, BZ) leads to transient absorptions in the near UV and visible wavelength range, ascribed to triplet states (Figure 2).

TABLE 1
Typical Structure of Photoinitiators Exhibiting Water Solubility

		R	
		$-CH_2SO_3^-$ Na^+	BP^-
		$-CH_2N^+$ $(CH_3)_3Cl^-$	BP^+
		$-OCH_2COOH$	TX^-
		$-O(CH_2)_3N^+$ $(CH_3)_3SO_3CH_3^-$	TX^+
		$-SCH_3$	TMPK
		$-CH_2SO_3^-Na^+$	BZ^-
		$-CH_2N(CH_3)_3Cl^-$	BZ^+
		$-OCH_2COO^-Na^+$	HAP^-
		$-OCH_2CH_2OH$	HAP
			PTS

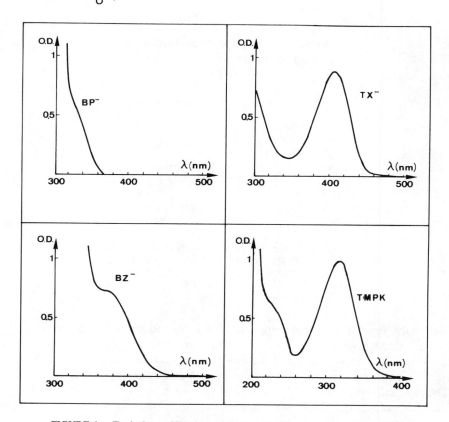

FIGURE 1. Typical near-UV absorption spectra of water-soluble initiators.

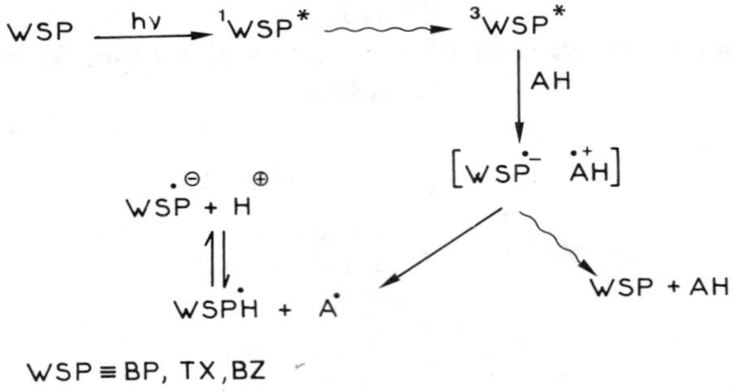

$$WSP \xrightarrow{h\nu} {}^{1}WSP^{*} \rightsquigarrow {}^{3}WSP^{*}$$

$$\downarrow AH$$

$$\left[WS\overset{\bullet}{P}{}^{-} \ \overset{\bullet}{A}\overset{+}{H} \right]$$

$$WS\overset{\ominus}{P} + \overset{\oplus}{H}$$

$$\Updownarrow$$

$$WSPH + \overset{\bullet}{A}$$

$$WSP + AH$$

$$WSP \equiv BP, TX, BZ$$

A

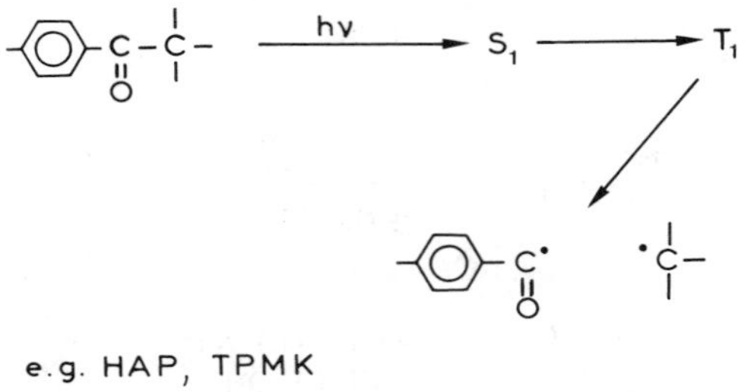

e.g. HAP, TPMK

B

SCHEME I

TABLE 2
Relative Rates of Polymerization of Acrylamide in the Presence of Various WSPs

	No amine	[MDEA] = 0.05 M
BP$^+$	0	194
BP$^-$	23	76
TX$^-$	21	136
TX$^+$	23	136
BZ$^-$	0	6
BZ$^+$	0	7
TPMK	37[a]	—
BP$^+$	—	430[a]

Note: [AA] = 0.7 M in water, I_0 = 5 × 10^{15} phot.cm^{-2} s^{-1}, OD = 0.1 at 366 nm, and λ = 366 nm.

[a] λ >300 nm.

    The triplet states of carbonyl compounds are known to interact with acrylic monomers and amines through an electron transfer mechanism involving a charge transfer complex (CTC). Depending on the nature of the two partners and the environment, the CTC may undergo back electron transfer, proton transfer, or generate radical ions.

    The interaction of ^{3}WSPs with amines and monomers is shown in Figure 3 and 4 where the rate of constants (k) of the triplet state decay are plotted against the concentration of

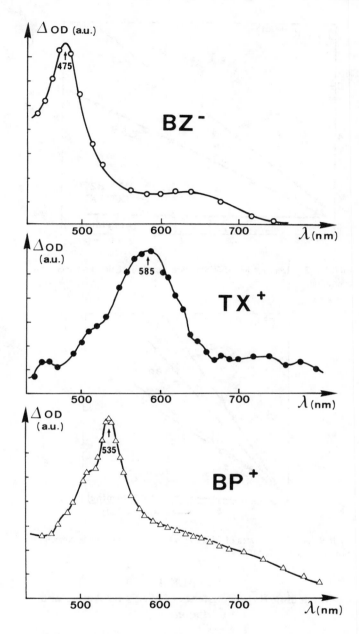

FIGURE 2. Transient absorption spectra of WSPs in their triplet state.

additive. While the triplets of BP derivatives are seen to interact strongly both with amines and monomers, those of TXs and BZs are only slightly quenched by monomers. On the other hand, electron transfer reactions are not very efficient in the case of TX$^-$ (Table 3).

In addition, the decay of the triplet state becomes entangled due to a complex relaxation process (Figure 5) involving several species, some of them growing and other ones decaying which were ascribed to ketyl radicals and radical ions originating from acid-base equilibria and to free ions.

B. QUANTUM YIELD OF THE ELECTRON TRANSFER PROCESS[15]

The quantum yield of the elementary step generating the CTC can be calculated from the experimental values of the rate constants corresponding to the bimolecular quenching

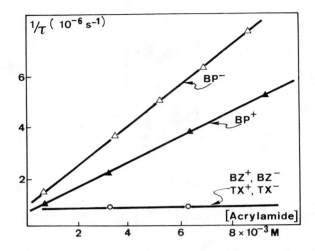

FIGURE 3.   Decay rates of WSPs as a function of the monomer concentration.

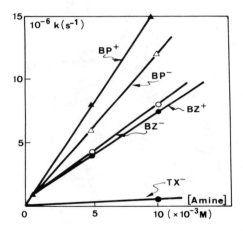

FIGURE 4.   Decay rates of WSPs as a function of the amine concentration. (The amine is MDEA.)

TABLE 3
Quenching Rate Constants of WSPs by Amine and Monomer

Initiator	k_Q ($\times 10^{-5} M^{-1} s^{-1}$)[a]	k_e ($\times 10^{-7} M^{-1} s^{-1}$)[b]
BP^+	4900	1.50
BP^-	8800	1.55
TX^+	<1	6.5
TX^-	0.8	3.2
BZ^+	0.5	38
BZ^-	0.6	45

[a]   Quenching by acrylamide.
[b]   Quenching by MDEA.

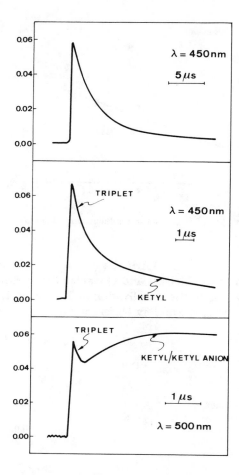

FIGURE 5. Transient signals observed with TX$^+$ in water.

by amine and monomer. It is representative of the efficiency of the electron transfer process. The data show that ϕ_{CT} approaches unity for BZ and TX derivatives, and is four to seven times lower for BPs. The rates of polymerization, however, do not really parallel the modification of ϕ_{CT} in this series (e.g., compare the results for BP$^+$ and TX$^-$ or BP$^+$ and BZ$^-$).[15]

C. QUENCHING OF THE SINGLET STATE[18]

Singlet quenching does not take place with BPs and is almost unimportant with usual oil-soluble TXs, e.g., chlorothioxanthone. On the contrary, TXs substituted by an alkoxy group, such as the water-soluble TXs studied here, exhibit higher yields of fluorescence and more efficient quenching by amines (Figure 6). The term $k_q\tau$ represents the product of the quenching rate constant by the lifetime of the first excited singlet state. In the TXs series, this lifetime (Table 4) lies in the nanosecond range and the calculated k_q value suggests that the process is not diffusion controlled.[18]

Thus, due to the high rate constant of intersystem crossing, the quenching of the singlet state becomes effective only at high amine concentration and it is even doubtful whether this state actually contributes to the initiation process. In fact, the CTC involved in this interaction is likely to undergo back electron transfer to the carbonyl ground state, as was demonstrated in the case of BP itself.

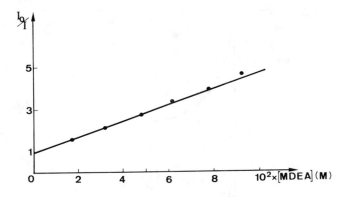

FIGURE 6.   Stern-Volmer plot of the fluorescence intensity of TX$^+$ as a function of [MDEA].

TABLE 4
Stern-Volmer Coefficients and Quenching Rate Constants Corresponding to the Deactivation of TX$^+$ First Excited Singlet State by Different Amines

Amine	K_{SV} (M^{-1})	k_Q^s ($\times 10^{-9} M^{-1} s^{-1}$)[a]
Triethylamine	44.0	3.0
Diethylamine	8.1	0.5
Triethanolamine	45.2	3.1
Methyldiethanolamine	34.6	2.4
Dimethylethanolamine	19.0	1.3
Ethanolamine	0.3	0.02

[a]   With $\tau = 14.7$ ns.[18]

D. MODELS FOR THE POLYMERIZATION INITIATION

The different water-soluble compounds behave according to several typical mechanisms.

In the case of BP structures,[19] the initiation process (Scheme IIA) usually involves electron and proton transfer from the cosynergist, thus yielding the amine-derived radical; BP$^+$ typically exemplifies this behavior. However, a plot of the logarithm of the rate of polymerization against that of the monomer concentration (Figure 7) shows that the expected dependence is not observed in the case of BP$^-$. In order to explain the initiation in the absence of an electron donor and the anomalous dependence of the rate of polymerization on the monomer concentration in the case of BP$^-$, another initiation pathway may be taken into account: the CTC formed between the triplet state and the monomer would decay not only through a back electron transfer but could also produce radical species capable of initiating a polymerization. It was also shown that amines can quench the CTC through generation of a three-center intermediate referred to as a triplex.

For TX derivatives (Scheme IIB), the initiation mechanism[15,18] involves the photoreduction of the triplet state by an electron donor, which competes with the quenching of its precursor (i.e., the singlet state) by the monomer molecules. The possible event of an initiation originating from this first excited singlet state[20] can be completely ruled out.[15,18]

As a consequence, the analytical expression of R$_p$ which accounts for the effect of the monomer and amine concentrations can be written as follows:

$$R_p \propto [M]\sqrt{\phi_i} \propto [M]\sqrt{\phi_{ST}\phi_{CT}\phi_{RM}}$$

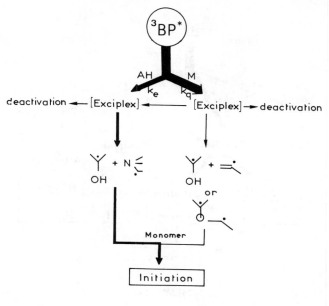

A

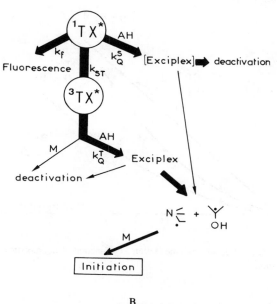

B

SCHEME II

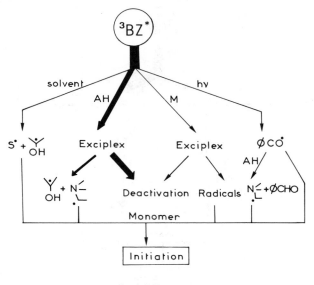

SCHEME   IIC

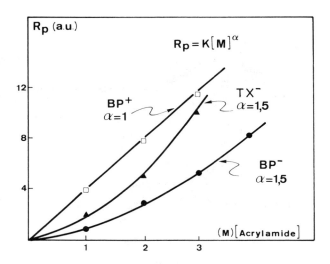

FIGURE 7.   Rates of polymerization as a function of the monomer concentration.

$$\phi_{ST} = \frac{k_{ST}}{k_{ST} + k_f + k_Q^S[Amine]}$$

$$\phi_{CT} = \frac{k_Q^T[Amine]}{k_Q^T[Amine] + \Sigma k \text{ (other processes)}}$$

$$\phi_{RM\cdot} = \frac{k_i[M]}{k_i[M] + \Sigma k \text{ (other processes)}}$$

Should the initiation efficiency $\phi_{RM\cdot}$ remain moderate (<0.1), the final expression of R_p becomes

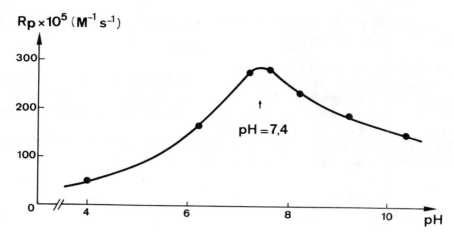

FIGURE 8.   Dependence of R_p as a function of the pH for TX^+ ([MDEA] $= 5 \times 10^{-2}$ *M*; [AA] $= 0.7$ *M*, $OD_{366} = 0.1$, and $I_0 = 5 \times 10^{15}$ phot.cm^{-2} s^{-1}).

$$R_p \propto [M]^{3/2} \sqrt{\frac{[\text{Amine}]}{A \,[\text{Amine}]^2 + B \,[\text{Amine}] + C}}$$

where A, B, and C stand for the products of rate constants involved the expressions of ϕ_{ST}, ϕ_{CT}, and ϕ_{RM}. The fair agreement[18] between the experimental data and the calculated curve obviously reflects the self-consistency of the kinetic model used for this calculation. In conclusion, this may be regarded as circumstantial evidence that the initiation mechanism involves mainly, if not exclusively, the triplet state of the TXs.

BZ structures, like BZ itself are known to undergo electron transfer reactions and are able to initiate the polymerization of styrene through charge transfer processes.[21] Apart from their being weakly quenched by acrylamide, water-soluble BZs photoinitiate polymerizations with insignificant yields in the absence of hydrogen donors. In the presence of an amine, although the usual mechanism involving an electron and proton transfer takes place, the rate of polymerization remains low because of the poor efficiency of the process generating the active radical from the exciplex (Scheme IIC). Under weak levels of excitation at 366 nm, the photofragmentation of BZ derivatives to benzoyl radicals is a very minor pathway (see Section G).

E. THE PROCESSES OCCURRING AFTER THE PRIMARY STEP OF THE ELECTRON TRANSFER: EFFECT OF THE pH

The effect of the pH on the polymerizable mixture and the activity of the initiating species was mentioned in many cases to account for some oddity in the polymerization kinetics of ionic initiators. A direct effect of the pH on the reducing power of the amine cosynergist was often put forward to account for the high sensitivity of the initiation efficiency to pH changes for pH values ranging from 7 to 9. Consideration was also given to the possibility of a direct participation either of the ketyl or the ketyl anion radical in the termination step.

By way of illustration, Figure 8 shows the influence of the pH on the conversion of a water-soluble TX.[18] The curve shows a maximum rate in the region of pH = 7.4. This characteristic shape is obviously due to the balance of two deactivating processes, whose pH dependencies are opposite. The protonation of the amine at the nitrogen in acid medium is certainly responsible for the lack of activity at low pH values. In the same way, the concentration of the ketyl anion radical possibly involved in the termination step of the photopolymerization may be expected to become crucially pH dependent when the detrimental effect combines with a pH increase.

F. MULTIPHOTONIC PROCESSES

A recent investigation conducted by Scaiano et al.[22] on the photophysical processes arising from the upper triplet levels of several carbonyl compounds provided evidence for some unusual biphotonic reactions which are reputedly forbidden in the corresponding lowest triplet state.

Thus, BZ excited in one of its upper triplet states was reported to spontaneously cleave to benzoyl radicals according to the following scheme:[22]

$$\text{Benzil} \xrightarrow[308\text{nm}]{h\nu_1} {}^1\text{Benzil*} \longrightarrow {}^3\text{Benzil*}$$

$$^3\text{Benzil*} \xrightarrow[490 \text{ nm}]{h\nu_2} {}^3\text{Benzil**} \longrightarrow 2\left\langle \bigcirc \right\rangle\!\!-\!\overset{\displaystyle |}{\underset{\displaystyle O}{\overset{\|}{C}}}\!\cdot$$

As benzoyl radicals are known to interact with acrylic monomers, this interesting observation raises questions as to the initiation mechanism prevailing in the case of the water-soluble BZ derivatives. It is worthy of note that, in addition to its first absorption band centered in the region of 450 nm, the lowest triplet of BZ exhibits also several other bands in the near UV. One may, therefore, safely assume that some benzoyl radicals, arising from the cleavage of an upper triplet state of BZs, take part in the photoinitiation mechanism.

Although no conclusive piece of evidence supporting this assumption has been reported as yet, the formation of benzoyl radicals (identified by their characteristic transient absorption while photolysing BZ in the submillisecond range by a polychromatic source)[23] is in agreement with a biphotonic cleavage of BZ. Indeed, such a process could well be one of the causes of the divergent statements and conflicting analyses which were introduced to account for the kinetics of photopolymerizations initiated by BZ derivatives.[15]

This type of biphotonic process, seemingly, does not affect the photochemistry of BPs and TXs.

G. GENERAL DISCUSSION

Inspection of the values of ϕ_{CT} and R_p obtained with WSPs reveals at a glance that the rates of polymerization do not parallel the efficiency of the electron transfer process. Indeed, they rather depend on the ability of the CTC to undergo proton transfer and on the nature of the radical species generated during the initiation process.

In the case of BPs (Scheme IIA), the active species is usually generated with a high efficiency, which is often close to unity. The only limiting process is the quenching by the monomer itself. Great improvements of the initiating efficiency should, therefore, not be expected from elaborate substituent effects, insofar as the polymerization of acrylamide is concerned. However, it could be interesting to carry out a kinetic study of the polymerization of several other water-soluble monomers with a view to discovering other structures that would be better suited to initiation by ionic BPs.

Regarding TXs (Scheme IIB) the approach of the problem is completely different since initiation efficiencies close to unity cannot be achieved with these compounds, in spite of the weakness of the detrimental interaction with acrylamide. The origin of this lack of activity was shown to result from a purely physical quenching of their singlet state by the cosynergist (i.e., the amine electron donor) and from a proton transfer following the electron transfer which is often difficult. A comprehensive study of the substituent effects on the rate constants of the photophysical process taking place in this type of compound (carried out on a set of 15 water-soluble derivatives of TX), has revealed that substitution in position 3 reduces to a great extent the singlet lifetime of TXs. Clearly, the consequence of this observation must be a reduction of the sensitivity of the triplet quantum yield to amine quenching, a property which is in line with an improvement of the initiating efficiency.

As for the availability of the proton from the amine moiety in the radical pair, this question cannot be answered in a simplified manner: first of all, this availability depends on the lability of α-hydrogens, a property which is governed by the structure of the amine, but it also depends on the degree of solvation of the CTC and of eventual intramolecular hydrogen bonding when the amine carries hydroxyles. Moreover, the basicity of the amine often exercises an indirect influence on the initiation mechanism by constraining the ketyl radical to convert to its basic form, a species which is suspected to prematurely interrupt the propagation of the polymer chain. Thus, the part played by the amine cosynergist is multiform and it is rather difficult to try to forecast the direction and the amplitude of the overall response in terms of R_p, of a system in which a given amine is substituted for another one. Therefore, this conclusion leaves the door open for complementary experimentation on TXs, especially with an aim at finding out the best cosynergist for each initiator monomer pair.

The question of the poor activity of BZs does not require such a careful approach; the compounds of this series, though weakly quenched by the monomer, do not lead to efficient initiation. This property is probably related to some reluctance to undergo proton transfer subsequently to electron transfer, on account of a fast back electron transfer process and of a stabilization of the free ion radicals by resonance. Thus, when BZs are used to photoinitiate the polymerization of acrylamide, this series of compounds never rises, in terms of quantum yield, above mediocrity in comparison with BPs or TXs.

IV. EXCITED STATE PROCESSES IN ORGANIZED MICROHETEROGENEOUS MEDIA

It is a very well accepted fact that "emulsion" polymerization leads to polymer chains with both high R_p and DP_n. In this kind of medium, thermal as well as photochemical initiation has been shown to induce very efficient polymerizations. Although less currently used, the photoactivated systems were dealt with in many specific studies; the role of the photoinitiator was specially emphasized[9,10,24] and the characteristics of the polymer chains were thoroughly investigated.[25]

In micellar environment, in addition to the usual photoprocesses described previously, the WSPs interact with the surfactant assembly and generate a triplet radical pair which leads either to a singlet pair by intersystem crossing or to free radicals by decorrelation and separation (Scheme III).[26-28]

The photophysical and photochemical behavior of WSPs (or of the corresponding oil-soluble structures) is expected to be influenced by different factors (e.g., substituents, charges, polarity, basicity, etc.) which govern the localization in the aggregate and the interactions with the surfactant or the guest molecules (monomer or amine).

Among these factors, the nature and position of the substituents, the high electrostatic field induced by the charges, the polarity of the bulk phase, as well as that of the microphase and the acid-base equilibria seem to be most important.

A. THE LOCALIZATION OF THE PARTNERS

Micellar structures are generally suspected of exercising some influence on the elementary reactions through the compartmentalization of the species, which goes hand in hand

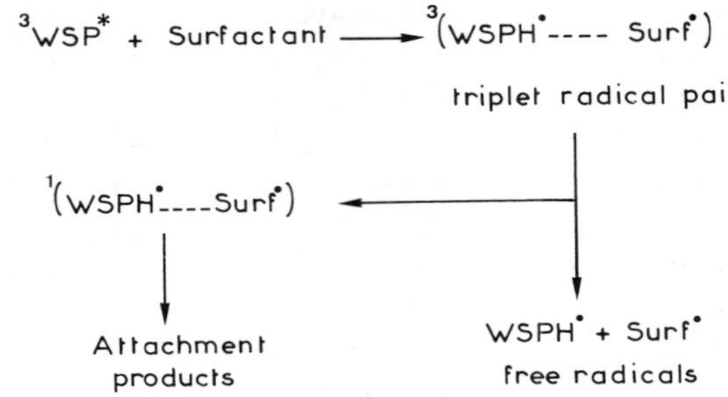

$$^3WSP^* + \text{Surfactant} \longrightarrow {}^3(WSPH^\bullet ---- Surf^\bullet)$$

triplet radical pair

$${}^1(WSPH^\bullet ---- Surf^\bullet) \longleftarrow$$

Attachment
products

WSPH$^\bullet$ + Surf$^\bullet$
free radicals

SCHEME   III

with microheterogeneity. Therefore, a full understanding of the process involved in such systems often implies a fair knowledge of the localization of the reactants with respect to the aggregate.

The approach which is resorted to in this study is twofold, and involves either the study of the solvatochromacy of the UV spectrum of WSPs or the study of the behavior of the excited states, with a view to assigning a micropolarity to the polymerization site. Information on the depth of this site in the aggregate is, thus, provided. Such an approach is certainly fraught with ambiguity, since the micellization process is regarded from a static point of view. A dynamic description taking into account the dynamic aspect of the microhetero-geneous structure would be probably much more accurate. However rough this approach may be, it is very useful to point out some correlations between localization and reactivity in the investigated series.[29-32]

By way of illustration, Figure 9A[31] shows the n $\rightarrow$ π^* transition of three BPs in SDS solution. As evident from this figure, the shape of their absorption in the 350-nm region is different, thus, indicating a variable polarity of their solubilization site in the aggregate. Unsubstituted BP which is weakly soluble in water is expected to exhibit a high binding constant to the micelle, and its mobility must be severely restricted by this interaction. On that account, the purely static approach mentioned above provides a simplified view of the location of ionic BPs which easily dissolve in water. The interaction must be rather weak, and a low rate constant of the micellization equilibrium should be expected. Figure 9B shows how useful is the empiric polarity parameter (E_T) introduced by Dimroth, which makes it possible to determine the depth of the solubilization site in the aggregate. When $\nu_{n\pi^*}$ is plotted against E_T for different solvents, a straight line is obtained. This linear correlation can be used to estimate the E_T of the solubilization site in sodium dodecyl sulfate (SDS) micelles.

At first sight, this study shows that, with respect to that of BP, the carbonyl group of BP^- is in a region of high polarity, very similar to bulk water. On the other hand, BP^+ should be considered as more or less encapsulated in micelles. This unexpected behavior could be the result of driving forces acting in opposite directions: the hydrophilic character resulting from the charge which attracts BP^+ to the bulk water phase, and the coulombic repulsion between the charge of this species and that one carried by the surfactant which opposes its demicellization.

Regarding the information derived from time-resolved absorption studies, it is worth mentioning that the degree of correlation between the triplet lifetime of a series of substituted BPs and the corresponding Hammett substituent constant is definitely improved by taking a localization parameter into consideration when dealing with micellar systems. This em-pirical parameter, which is derived from the amplitude of the solvatochromic shifts in a given set of solvent, images better than qualitatively the indirect influence exercised by a given substituent on the localization of the corresponding carbonyl compound and, hence, on its characteristic reactivity.[29]

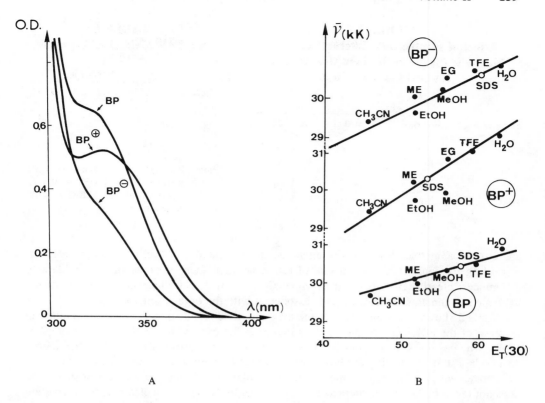

FIGURE 9.   (A) Absorption spectra of BPs in micellar solution [SDS] = 0.1 *M*. (B) Position of the absorption maximum of the n → π* transition in various solvents as a function of $E_T(30)$. (ME: methoxyethanol, EG: ethyleneglycol, and TFE: trifluoroethanol).

B. ELECTROSTATIC AND MAGNETIC FIELD EFFECTS

Since water solubility is generally imparted by grafting ionic groups onto the backbone of oil-soluble carbonyl compounds, the charge carried by the substituent must interact with that of ionic surfactants. Moreover, electrostatic effects can also take place between Gouy's layer and some transient species or end products arising from the photochemistry of ionic WSPs.

In the same manner, the behavior of radical pairs in organized systems, and especially in micelles, is known to be affected by electrostatic as well as magnetic field effects. It has been established that the decay of these radical pairs involves predominantly a competition between intersystem crossing and separation of the geminate biradical. Application of an intense magnetic field results in a splitting of the sublevels (due to Zeeman effect); while intercrossing through the subchannels from T_+ and T_- is shut off, that from T_0 remains unaffected.[33-35]

By way of illustration, four BPs were examined in SDS (anionic) and indodecyltrimethylammonium chloride (cationic) micelles. The series BP, BP^+, and BP^- was completed with $BP-C_{12}$ (dodecyl-4 benzophenone) which is expected to dissolve in the core of the micelle due to the long-chain alkyl group. The magnetic and electrostatic field effects are clearly demonstrated by the results shown in Table 5,[36] which shows the influence of a 2000 G field on the percent of separation and exit of a radical pair derived from BPs (by hydrogen abstraction from the alkyl chain of the surfactant). The location of ground state BPs in micelles can be inferred (as shown in the previous paragraph) from the position of the nπ* band in environments of various polarities. Thus, BP and $BP-C_{12}$ are resident in an environment whose polarity is comparable to that of methanol. Likewise, the ionic BPs

TABLE 5

Effect of Electrostatic Interactions and Magnetic Fields on the Behavior of Some Radical Pairs at 300 K

		% exit	
Surfactant	Substrate	H = 0 G	H = 2000 G
SDS	BP	7	60
	BP$^-$	45	62
	BP$^+$	6	41
	BP-C$_{12}$	2	14
DTAC	BP	25	78
	BP$^-$	22	51
	BP$^+$	61	81

TABLE 6

Magnetic Field H Effect on R$_p$

	Surfactant		
Initiator	SDS	CTAC	DTAC
BP	2.4	2.3	2.6
BP$^-$	1.1	1.5	1.6
BP$^+$	1.8	~1	~1

Note: Relative R$_p$ obtained as the ratio R$_p$(H)/ R$_p$(H = 0) — [SDS] = 0.5 *M*; [MMA] = 0.5 *M*; [TEA] = 0.005 *M*; and H = 800 G.

are also in a methanol-like environment when associated to micelles of opposite sign. Regarding the ionic BPs in a micelle of the same sign, these systems show considerable differences: clearly, BP$^+$ is more hydrophobic than BP$^-$ and whatever the charge of the aggregate, it remains, at least in part, associated with the organic microphase.

Table 5 shows that application of the magnetic field results in approximately a ninefold increase of the efficiency of the pair separation in the case of BP in SDS (i.e., from 7 to 60%), whereas this effect is much less pronounced in the case of BP$^-$ in SDS (i.e., from 45 to 62%), thus, reflecting the repulsive electrostatic interactions between BP$^-$ and SDS. This statement is supported by the system BP$^+$/SDS, where the percent exit at high field does not reach by far, its theoretical maximum, due to the attractive electrostatic interaction which has a tendency to hold back the BP moiety in the micellar system.[36] This analysis is in line with the conclusions of a preliminary investigation of the effect of an internal magnetic field on the polymerization of methyl methacrylate (MMA) in micelle. Table 6 shows[9] that application of a magnetic field to a system in which surfactant and initiator carry opposite charges results in a strong increase of the polymerization rate, whereas the systems carrying the same charges are almost unaffected.

C. QUENCHING OF THE TRIPLET STATE BY THE MONOMER

This general process, which involves a charge transfer from the olefin to the carbonyl, often plays a decisive part in the initiation mechanism, since it wastefully consumes a fraction of the precursors of the initiating species.[37] This process is generally less efficient than the quenching by amines or other compounds carrying an electron pair, due to the specific geometry which is required in the solvent cage to make the electron transfer possible. This transfer is, thus, directly affected by the accessibility of the carbonyl group. The detrimental character of this process is related to the fact that the CTC decays predominantly via back electron transfer to ground state products. In some cases, depending on the triplet energy of the carbonyl acceptor, an energy transfer can also take place.[38,39]

As a rule, the experimental values of the rate constant of the bimolecular quenching between the ketone and MMA (derived from a study of the triplet lifetime as a function of the monomer concentration) are significantly lower with BZ and TX structures than BPs (Table 7).[9,16,24] This unfavorable feature, however, is balanced in the case of BPs by a high efficiency of the radical generation (i.e., by reaction with the amine donor) as it was pointed out for the polymerization of acrylamide in homogeneous solution. Incidentally, it is worth mentioning that the Stern-Volmer plot of the triplet decay constant remains linear up to high monomer concentrations (of the order of the concentrations used in polymerization experiments), thus, excluding any static quenching[40,41] of the excited carbonyl by MMA.

In reverse micellar or microemulsion systems, the general behavior of WSPs is basically

TABLE 7
Quenching of ³WSP by Monomers
in SDS Micelles

Initiator	k_Q ($\times 10^{-6} M^{-1} s^{-1}$)	
	MMA	Styrene
BP	1500	190
BP⁻	1000	410
BP⁺	700	140
TX⁻	1	7
TX⁺	— [a]	77
BZ⁻	<0.1	3
CTX	1.4	95
DTX	8	— [a]

Note: ([SDS] = 0.5 *M*)—comparison with the deactivation of the triplet state of chlorothioxanthone (CTX) and dodecyl thioxanthone (DTX).

[a] Not measured.

the same, with a reactivity between the olefin and the carbonyl in its triplet state controlled by compartmentalization. In this case, the absolute value of the "bimolecular-like" quenching constants depends to a tremendous extent on the amount of water swelling the micelle, i.e., of the total area of the interface between the mesophase and the bulk phase (Figure 10).[42]

This finding may be regarded as evidence for the existence of a limiting step in the kinetics of the quenching process. This could correspond to the crossing of the interface, a process which occurs with increasing probability as the area of the interface increases. This limiting step could also be related to the partitioning of the quencher between the bulk organic phase and the water droplet.

D. QUENCHING BY THE ELECTRON DONOR

Addition of amines lead (just as in homogeneous solution) to a shortening of the triplet state of WSPs in micelles and to the generation of long-lived transient absorptions. However, contrary to what was reported for homogeneous solutions of WSP[15,19,43] or for the corresponding oil-soluble compounds,[44,45] the relaxation process of these long-lived species becomes somewhat intricate. Obviously, it results from superimposition of several absorptions corresponding to transient species, among which some ones are building up while other ones are decaying. Clearly, the decay process of these long-lived absorptions involves the generation of a correlated radical pair, the separation of the pair and the exit of free radicals out of the aggregate, and the acid-base equilibrium between the various stage of protonation of the transient species, inside as well as outside the micelle.[31] In addition, the absorption spectra of the transient species (triplet, ketyl BPH˙, and ketylanion BP˙⁻) are not basically different, so that there is no spectral window which could allow their contributions to be studied separately.

By way of example, the transient absorption spectra of ³BP*, BPH˙, and BP˙⁻ are reported in Figure 11;[31] Figure 12[31] shows the progressive modification of the spectrum taken in four-time windows along the transient signal recorded in the system BP/triethylamine (TEA) in SDS micelles. This experiment reveals that the shape of the transient absorption progressively changes from that of ³BP (in the initial stage of the decay) to that of the ketyl anion (at the very end of the decay). This observation is in line with the fact that some free

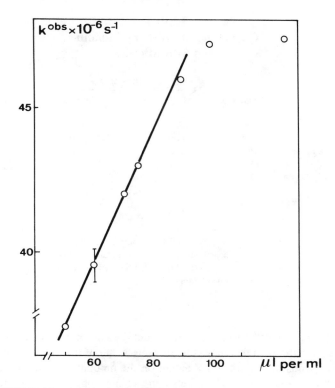

FIGURE 10.   Actual lifetime of BP$^-$ in reverse micelles of AOT as a function of the volume of water added in the presence of MMA; 12 < H$_2$O/AOT < 32, [MMA] = 0.18 M.

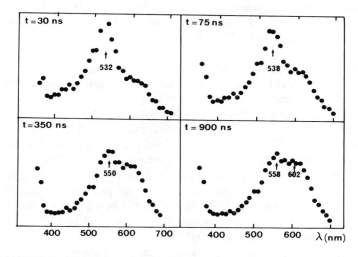

FIGURE 11.   Progressive change of the transient spectrum of BP in an SDS micellar solution along the decay of the transient absorption; [BP] = 1.1 × 10^{-2} M and [TEA] = 10^{-2} M.

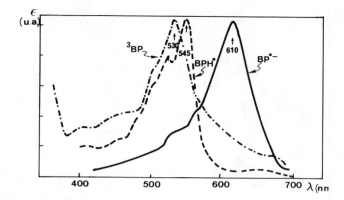

FIGURE 12. Spectrum of the triplet state of BP (^{3}BP), of the ketyl radical (BPH$^{\cdot}$) and of the corresponding ketyl anion (BP$^{\cdot-}$) in SDS 0.2 M.

TABLE 8
Quenching of ^{3}WSP and Oil-
Soluble Compounds CTX and
DTX by Amines in SDS Micelles
([SDS] = 0.5 M)

Initiator	k_e ($\times 10^{-9} M^{-1} s^{-1}$) MDEA	TEA
BP	1.5	4
BP$^-$	1.5	3
BP$^+$	0.4	6
TX$^-$	0.04	0.12
TX$^+$	0.02	—[a]
BZ$^-$	0.3	—[a]
CTX	0.34	0.71
DTX	0.42	0.44

[a] Not measured.

ketyl radicals formed inside the micelle escape from the macrocage into the bulk phase, where the pH is alkaline, due to the presence of TEA. The neutral ketyl radicals, then, equilibrate and convert to their anion form.

Two pieces of evidence favor this statement: (1) when the pH of the bulk phase is made more alkaline by adding NaOH, the conversion of the ketyl radical into its anion form (i.e., the deprotonation) is much faster and (2) when TEA is replaced by another amine of similar basicity, with the same ionization potential but with a much higher binding constant to the micelle[7] (e.g., dimethyldodecylamine, DMDA), much of the ketyl species escaping from the pseudophase remains under its neutral form.

Regarding the other WSPs of the BP, BZ, or TX series, their general behavior is basically the same. The only difference is the existence of an additional element in the decay of the transient species involved in the interaction between the excited carbonyl and the electron donor: the coulombic interaction between the charge carried by the ionic substituent and the transient dipole induced by the CT complex.

Some experimental values of quenching rate constants are shown in Table 8.[9,16,24] As a rule, the triplet of WSPs interacts very efficiently with amines in direct micelles. In a given series, the rate of this process depends greatly on the compartmentalization of the

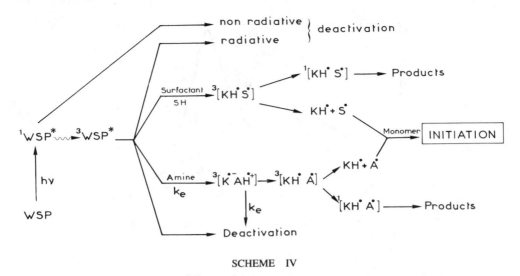

SCHEME IV

reactant one with respect to the other ones. MDEA, which is weakly bound to the micelle, interacts more efficiently with BP^- (which is also weakly bounded) than with BP^+ (which is kept in the pseudophase by the coulombic interaction). Conversely, BP^+ is quenched twice as fast than BP^- with TEA which is more closely bound to the micellar structure BP^+. The case of DMDA, which resides almost exclusively in the oil phase, is in line with this general statement: the anionic BPs or TXs which localize in the external shield of the SDS micelle are quenched about 25 times slower than their cationic homologues.

E. THE GENERAL MODEL

From a general point of view, the model accounting for the kinetic behavior of WSPs in micelle photopolymerization reactions does not basically differ from the one which is currently used to describe bulk or solution photopolymerization (Scheme IV). However, the strong interaction which takes place between the surfactant and ³WSPs (in direct as well as reverse micelles) plays in some cases a determining part in the initiation mechanism.

The macroscopic rates of polymerization of MMA and styrene in the presence of a series of WSPs are shown in Table 9.[9,16,24] Again, BP, CTX, and DMPA were tested under the same experimental conditions for comparison purposes. These rates of polymerization depend on two factors: the rate of photoinitiation (i.e., the event which initiate the chain) and the rate of propagation and termination of the growing chain. Clearly, the latter is affected by the nature of the aggregate and by the trapping of some radicals species, usually involved in the termination process, at the interface of the micellar structure. Thus, the effect of compartmentalizing the reactive system in a microheterogeneous structure is twofold, since the rate and the degree of polymerization alike are dependent on both the initiation and the propagation-termination steps.

The rate of photoinitiation R_i can be explicited as the product of several factors, each of them characterizing the efficiency of an elementary step of the overall initiation yield:

$$R_i = I_{abs}\phi_{ST}\phi_{CT}\phi_{RP}\phi_{A.}\beta$$

with ϕ_{ST} = quantum yield of intersystem crossing, ϕ_{CT} = yield of the step generating the micellized CTC, ϕ_{RP} = yield of triplet radical pair, $\phi_{A.}$ = yield of free initiating radical, and β = efficiency factor characterizing the reactivity of the amine-derived radical towards the olefin.

In the case of water-soluble BPs, if a parallel is drawn between the R_ps measured in

TABLE 9
Rates of Polymerization of MMA and Styrene
in SDS Solution

Initiator	Relative R_p	
	[MMA] = 0.5 *M*	**[Styrene] = 0.5 *M***
BP	50	37
BP$^-$	45	150
BP$^+$	250	162
TX$^-$	70	156
TX$^+$	135	165
BZ$^-$	45	171
DMPA	95[a]	74[a]
CTX	80	40

Note: ([SDS] = 0.5 *M*); I_0 = 5 × 10^{15} phot cm^{-2} s^{-1}; OD = 0.1; [MDEA] = 0.05 *M* — DMPA (dimethoxyphenylacetophenone) and CTX are taken as reference.

[a]   Without amine.

homogeneous solution (vide supra) and in SDS micelles, and the corresponding ϕ_{CT} values, the conclusion is that R_p increases significantly when going to microheterogeneous systems, although ϕ_{CT} remains quite as low in both media. The relative R_ps reported in Table 9 show a ratio of 2.6 between BP$^-$ and DMPA, an initiator which is known to cleave almost quantitatively to active species. Since the radicals A· are formed in solution with a yield (ϕ_A.) approaching unity, the product $\phi_{CT} \times \phi_A$. cannot exceed this limit. Therefore, the experimental observations cannot be accounted for by taking into consideration only the increase of ϕ_{CT}. It is, thus, asserted that the higher rates reported in micellar polymerization should be ascribed to chemical effects of compartmentalization (as in the case of thermally initiated polymerization) rather than to a specific enhancement of the photoinitiation efficiency.

Regarding the difference of reactivity in the series, they are certainly related to the introduction of an additional step in the initiation mechanism, i.e., the crossing of the micellar interface, on both sides of which the reactants are dissolved. In some cases, due to strong coulombic interactions, this new process may become a limiting step in the initiation kinetics and control at least partly the efficiency of this process.

In the case of water-soluble TX, the question of the initiation efficiency can be approached with the same basic ideas. The only difference lies in the very efficient quenching of the singlet state of TXs by electron-rich olefins, a process which is detrimental to the initiation efficiency. As compared with BPs, the performances of TXs are not significantly improved when used in micellar systems. Again, the determinant factor affecting the overall efficiency is not the electron transfer reaction, since its ϕ_{CT} approaches unity. The explanation must be sought for elsewhere, in some subeffects of micellization on all the elementary steps of the initiation process.

REFERENCES

1. **Roffey, C. G.**, *Photopolymerization of Surface Coatings*, John Wiley & Sons, New York, 1982.
2. **Schnabel, W. and Sumiyoshi, T.**, Initiation of free radical polymerization by acylphosphine oxides and derivatives, in *New Trends in the Photochemistry of Polymer*, Allen, N. S. and Rabek, J. F., Eds., Elsevier, London, 1985.

3. **Fouassier, J. P.,** Excited state properties of photoinitiators: lasers and their applications, in *Photopolymerization Science and Technology,* Allen, N. S., Ed., Elsevier, London, in press.

4. **Fouassier, J. P., Jacques, P., Lougnot, D. J., and Pilot, T.,** Lasers, photoinitiators and monomers: a fashionable formulation, *Polym. Photochem.,* 5, 57, 1984.

5. **Green, P. N.,** A novel series of water soluble thioxanthone photoinitiators, *Polym. Paint Resins,* 175, 246, 1985.

6. **Fratelli Lamberti,** European Patent 192 967, 1986.

7. **Merck,** Ger. Off. 3505998, 1986.

8. **Bottom, R. A., Guthrie, J. T., and Green, P. N.,** The photochemically induced grafting of 2-hydroxyethyl acrylate onto regenerated cellulose films from aqueous solutions, *Polym. Photochem.,* 6, 111, 1985.

9. **Fouassier, J. P. and Lougnot, D. J.,** Ionic photoinitiators for radical polymerization in direct micelles: the role of the excited states, *J. Appl. Polym. Sci.,* 32, 6209, 1986.

10. **Bonamy, A., Fouassier, J. P., Green, P. N., and Lougnot, D. J.,** A novel and efficient water-soluble photoinitiator of polymerization, *J. Polym. Sci., Polym. Lett. Ed.,* 20, 315, 1982.

11. **Rivière, D. and Fouassier, J. P.,** A comparative investigation of the photoinitiation processes in bulk and micelle polymerization. I. Steady state experiments, *Polym. Photochem.,* 3, 29, 1983.

12. **Fouassier, J. P. and Lougnot, D. J.,** A comparative investigation of the photoinitiation processes in bulk and micelle polymerization. II. Excited states dynamics, *Polym. Photochem.,* 3, 79, 1983.

13. **Rutsch, W., Berner, G., Kirchmayer, R., Husler, R., Rist, G., and Buhler, N.,** New photoinitiators for pigmented systems, in *Proc. Radcure,* Atlanta 1984, Tech. Paper FC 84 252, SME ed., Dearborn, MI, 1984.

14. **Fouassier, J. P. and Lougnot, D. J.,** Reactivity of water-soluble photoinitiators, *Radcure Baltimore,* Tech. Paper Fc86812, SME ed., Dearborn, MI, 1986.

15. **Lougnot, D. J. and Fouassier, J. P.,** Comparative reactivity of water-soluble photoinitiators as viewed in terms of excited states processes, *J. Polym. Sci., Polym. Chem. Ed.,* 26, 1021, 1988.

16. **Fouassier, J. P. and Lougnot, D. J.,** unpublished data.

17. **Fouassier, J. P., Lougnot, D. J., Payerne, A., and Wieder, F.,** Time resolved laser pumped dye laser spectroscopy: energy and electron transfer in ketones, *Chem. Phys. Lett.,* 135, 30, 1987.

18. **Lougnot, D. J., Turck, C., and Fouassier, J. P.,** Water-soluble polymerization initiators based on the thioxanthone structure: a spectroscopic and laser photolysis study, *Macromolecules,* 22, 108, 1989.

19. **Fouassier, J. P., Lougnot, D. J., Zuchowicz, I., Green, P. N., Timpe, H. J., Kronfeld, K. P., and Mueller, U.,** Photoinitiation mechanism of acrylamide polymerization in the presence of water-soluble benzophenones, *J. Photochem.,* 347, 36, 1987.

20. **Allen, N. S., Catalina, F., Green, P. N., and Green, W. A.,** Photochemistry of thioxanthones. VI. A polymerization, spectroscopic and flash photolysis study on novel water-soluble substituted 3-(9-oxo-9*H*-thioxanthene-2,3-γ-4-yloxy)-*N,N,N*-trimethyl-1-propanaminium salts, *J. Photochem.,* 36, 99, 1987.

21. **Encinas, M. V. and Lissi, E.,** Polymerization photoinitiated by carbonyl compounds. VI. Mechanism of benzil photoinitiation, *J. Polym. Sci., Polym. Chem. Ed.,* 22, 2469, 1984.

22. **Scaiano, J. C. and Johnston, L. J.,** Photochemistry of reaction intermediates, Proc. XI IUPAC Symp. Photochemistry, Lisbon, 1986.

23. **Allen, N. S., Catalina, F., Green, P. N., and Green, W. A.,** Photochemistry of carbonyl photoinitiators: photopolymerization, flash photolysis and spectroscopic study, *Eur. Polym. J.,* 22, 49, 1986.

24. **Fouassier, J. P. and Lougnot, D. J.,** Thioxanthone derivatives as photoinitiators in micelle photopolymerization, *J. Appl. Polym. Sci.,* 34, 477, 1987.

25. **Candow, F. and Holtzscherer, C. H.,** Microlatex inverses: étude de leur formation et de leurs propriétés, *J. Chim. Phys.,* 82, 691, 1985.

26. **Scaiano, J. C. and Abuin, E. B.,** Absolute measurements of the rates of radical exit and of radical-pair intersystem crossing in anionic micelles, *Chem. Phys. Lett.,* 81, 209, 1981.

27. **Braun, A. M., Krieg, M., Turro, N. J., Aikawa, M., Gould, I. R., Graff, G. A., and Lee, P. C.,** Photochemical process of benzophenone in microheterogeneous systems, *J. Am. Chem. Soc.,* 103, 7312, 1981.

28. **Jacques, P., Lougnot, D. J., and Fouassier, J. P.,** Specific micellar effects in the temporal behaviour of excited benzophenone. Consequences upon the polymerization kinetics, in *Surfactants in Solution,* Vol. 2, Mittal, K. L. and Lindman, B., Eds., Plenum Press, New York, 1984, 1177.

29. **Jacques, P., Lougnot, D. J., Fouassier, J. P., and Scaiano, J. C.,** Location effects upon the kinetic behaviour of benzophenone in micellar solution, *Chem. Phys. Lett.,* 127, 469, 1986.

30. **Fendler, J. H., Fendler, E. J., Infante, G. A., Shih, P. S., and Patterson, L. K.,** Absorption and proton magnetic resonance spectroscopic investigation of the environment of acetophenone and benzophenone in aqueous micellar solutions, *J. Am. Chem. Soc.,* 97, 89, 1975.

31. **Lougnot, D. J., Floret-David, D., Jacques, P., and Fouassier, J. P.,** Effets électrostatiques sur le comportement des états excités de benzophénones chargées en solution micellaire, *J. Chim. Phys.,* 82, 505, 1985.

32. **Fouassier, J. P., Lougnot, D. J., and Zuchowicz, I.,** Reverse micelle radical photopolymerization of acrylamide: excited states of ionic indicators, *Eur. Polym. J.*, 22, 933, 1986.
33. **Turro, N. J. and Kraeutler, B.,** Magnetic field and magnetic isotope effects in organic photochemical reactions. A novel probe of reaction mechanisms and a method for enrichment of magnetic isotopes, *Acc. Chem. Res.*, 13, 369, 1980.
34. **Scaiano, J. C., Abuin, E. B., and Stewart, L. C.,** Photochemistry of benzophenone in micelles. Formation and decay of radical pairs, *J. Am. Chem. Soc.*, 104, 5673, 1982.
35. **Turro, N. J., Chow, M. F., Chung, C. J., and Tung, C. H.,** An efficient high conversion photoinduced emulsion polymerization. Magnetic field effects on polymerization efficiency and polymer molecular weight, *J. Am. Chem. Soc.*, 102, 7391, 1980.
36. **Scaiano, J. C. and Lougnot, D. J.,** Electrostatic and magnetic field effects on the behaviour of radical pairs derived from ionic benzophenones, *J. Phys. Chem.*, 88, 3379, 1984.
37. **Loutfy, R. O., Dogra, S. K., and Yip, R. W.,** The interaction between the excited states of ketones and olefines: the role of triplet exciplexes, *Can. J. Chem.*, 57, 342, 1979.
38. **Gorner, H.,** Triplet states of phenylethylenes in solution. Energies lifetimes and absorption spectra of 1,1-diphenyl-, triphenyl- and tetraphenylethylene triplets, *J. Phys. Chem.*, 86, 2028, 1982.
39. **Morrison, H., Tisdale, V., Wagner, P. J., and Liu, K. C.,** Intramolecular charge transfer quenching in excited vinyl phenyl ketones, *J. Am. Chem. Soc.*, 97, 7189, 1975.
40. **Frank, A. J., Graetzel, M., and Kozak, J. J.,** On the reduction of dimensionality in radical decay kinetics induced by micellar systems, *J. Am. Chem. Soc.*, 98, 3317, 1974.
41. **Infelta, P. T., Graetzel, M., and Thomas, J. K.,** Luminescence decay of hydrophobic molecules solubilized in aqueous micellar systems. A kinetic model, *J. Phys. Chem.*, 78, 190, 1974.
42. **Lougnot, D. J. and Scaiano, J. C.,** A water-soluble benzophenone in reverse micelles kinetics and spectroscopy, *J. Photochem.*, 26, 119, 1984.
43. **Lougnot, D. J., Jacques, P., Fouassier, J. P., Casal, H. L., Kim-Thuan, N., and Scaiano, J. C.,** New functionalized water-soluble benzophenones: a laser flash photolysis study, *Can. J. Chem.*, 63, 3001, 1985.
44. **Schnabel, W. and Amirzadeh, G.,** On the photoinitiation of free radical polymerization. Laser flash photolysis investigations on thioxanthone derivative, *Makromol. Chem.*, 182, 282, 1981.
45. **Yates, S. F. and Schuster, G. B.,** Photoreduction of triplet thioxanthone by amines: charge transfer generates radicals that initiate polymerization of olefins, *J. Org. Chem.*, 49, 3349, 1984.

Chapter 7

LASER SPECTROSCOPY OF PHOTORESIST MATERIALS

Jean Faure and Jacques Delaire

TABLE OF CONTENTS

I. INTRODUCTION

In the processing of high-density electronic circuits, microlithographic methods involve the irradiation of an organic material (photoresist). In a "positive photoresist", the light induces chemical changes in a polymer which render the exposed part more soluble than the unexposed one. During the solubilization process, the polymer is removed selectively from the exposed part of the silicon wafer. This allows the etching process to take place on the unprotected area.[1]

The most common positive photoresists consists of a photoactive compound — a mixture of substituted ortho 2-diazonaphthalenones — dissolved in Novolak-type resin, a formol-metacresol polymer.[2,3] The photochemical decomposition mechanism of the so-called AZ class of photoresists has been investigated by UV, IR, and ^{13}C NMR, after steady-state irradiation at low temperature of the solid[2,3] or of glassy solutions.[4] Surprisingly, except for two reports in the literature on flash photolysis of 2-diazonaphthalenones in solution,[5,6] there was no laser photolysis study of these compounds until the last 2 years.

We present here the results of the identification of transients by laser spectroscopy. At almost the same time, two papers have appeared in the literature. One of them[7] contains our own results which will be developed and extended here, the other one, by Tanigaki and Ebbesen,[8] contains very similar experimental evidence, but a totally different interpretation. In the second part, we will present the generally accepted mechanism of the photolysis of α-diazoketones, known as Wolff rearrangement. In the third part, we will develop our own experimental results and compare them with data from the literature. In the last part, we will discuss the nature of the intermediates put into evidence by our results as well as those of Tanigaki and Ebbesen.[8]

II. MECHANISTIC CONSIDERATIONS ON THE PHOTOCHEMISTRY OF α-DIAZOKETONES

A. THE WOLFF REARRANGEMENT

The Wolff rearrangement (see Scheme I) is an important key to the knowledge of the photochemistry of α-diazoketones. Many studies have explained the different steps of the reaction and the nature of the intermediates formed after photonic excitation of the initial diazoketone. An important mechanistic question in the Wolff rearrangement centers on whether dinitrogen loss and 1,2-migration occur in a concerted manner, or whether the rearrangement is actually a two-step process which first proceeds through an α-ketocarbene in either a singlet or a triplet state.[9] From conformational studies, Kaplan and co-workers[10] proposed that a concerted rearrangement could occur from the s-Z conformation of α-diazoketone. By comparing product ratios obtained in direct and triplet-sensitized photolysis of these compounds, Tomioka[11] proposed a general scheme (see Scheme II) accounting for a concerted reaction for the s-Z conformation and a biradical step for the s-E conformation.

By picosecond laser spectroscopy, it has been shown that excitation of diphenyldiazo-methane leads to diphenylcarbene in a few picoseconds.[12] α-Ketocarbene has been very recently identified in strained α-diazoketones dissolved in argon matrix at 10 to 15 K by UV-visible, IR, and electron spin resonance spectroscopy.[9] At this time, there is no evidence of occurrence of ketocarbene at room temperature. Because of the rapidity of the conversation process of ketocarbene to ketene, the lifetime of ketocarbene is very short ($<10^{-12}$ s).

Furthermore, it has also been proposed that other intermediates such as oxirenes can be formed. Numerous theoretical calculations predict oxirene to be an energy minimum with a low barrier (2 to 8 kcal/mol) to rearrangement to ketene.[13-17] Singlet formyl carbene rearranges to ketene with no barrier. However, despite considerable efforts, there is no direct experimental evidence for oxirene formation.[9]

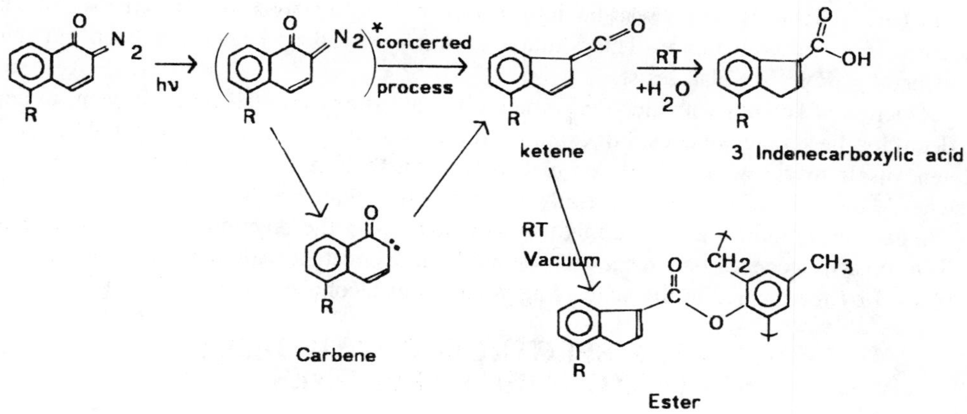

SCHEME I. Photolysis of diazonaphthalenones in a Novolak-type resin.

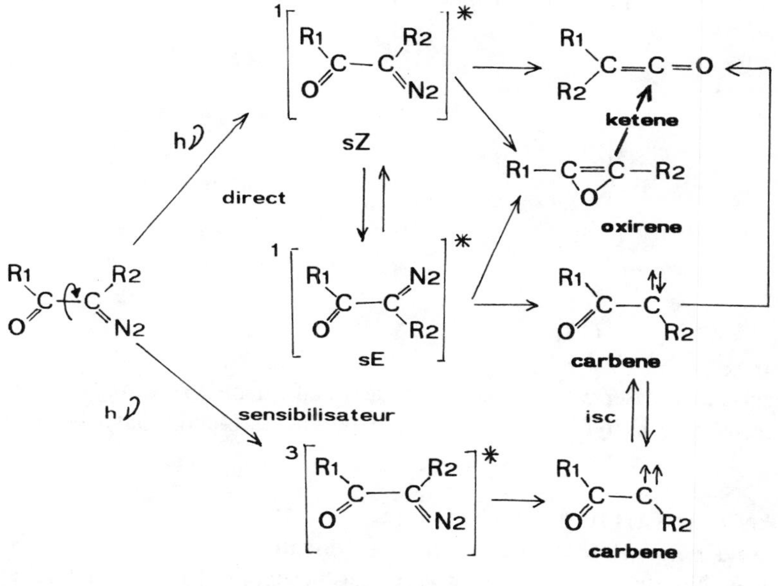

SCHEME II. Photolysis of diazoketones.

B. REACTIVITY OF KETENES

From theoretical calculations it is possible to foresee both addition and cycloaddition reactions on ketenes.[18,19] We will limit our interest here on addition reactions of ketenes on molecules bearing labile OH groups like water and alcohols. The reaction of ketene with water which gives a carboxylic acid is responsible for the increased solubility of the photoresist after irradiation.

Kinetics of addition of water on dimethylketene and diphenylketene in solution in ether[20,21] is slow enough to be measured by conventional spectrophotometry. It has been found that the hydration reaction is second order in water for dimethylketene and third order in water for diphenylketene. This has been interpreted as a consequence of the reaction of ketene with a water dimer in the first case, and a water trimer in the second case. Cyclic transition states were proposed as a consequence.[20]

Concerning hydration of the indene-ketene produced by photolysis of diazonaphthalenones (see Scheme I), there was no information before our work on the kinetics of the reaction at room temperature. The addition of water on indene-ketene has been put into evidence by IR in Novolak resist.[3]

Reaction of ketenes with alcohols, which gives a carboxylic ester, has also been studied in detail for the simpler ketenes. For example, dimethylketene reacts with an alcohol trimer.[20] Ketene itself reacts with a dimer of methanol.[22] There is not a general behavior for all ketenes, but arylketenes[18] are generally more reactive than alkylketenes with water and alcohols. Steric hindrance of alcohols tends to slow down the reaction.[20,23] An increase of solvent polarity generally causes a decrease in the rate constant with alcohols which can be explained by the change in the state of aggregation of alcohols in these solvents.[20,24]

III. LASER SPECTROSCOPY OF 1-[2H]2-DIAZONAPHTHALENONES

Two model compounds were studied in solution. 1-[2H]2-diazonaphthalenone-5-(4'-methylphenylsulfonate (compound **1**) is soluble in 2-ethoxyethyl acetate, dioxan, acetone, benzene, or toluene. 1-[2H]2-diazonaphthalenone-5-sodium sulfonate (compound **2**) is soluble in water.

The nanosecond laser was a passively mode-loked Quantel YGH 48 Nd:YAG ($\lambda = 353$ nm, 250 mJ/cm^2, 3-ns pulse width).[25] The experimental details have been given previously.[7,25]

A. ORGANIC SOLUTIONS
1. Compound 1 in 2-Ethoxyethyl Acetate or Dioxan

After excitation by a laser pulse, a very fast bleaching of the aerated solution of **1** in ethoxyethyl acetate (EEA) (2.5×10^{-4} M) occurs; the transient differential spectra are shown in Figure 1. At the end of the pulse, besides the bleaching between 330 and 450 nm, there is an absorption without a maximum below 330 nm. This absorption decays slowly and the decay kinetics are related with the quality of the solvent, in particular with traces of water which could be absorbed during the preparation of the solutions. In a solution of **1** in vacuum-distilled dioxan, the decay of this absorption was very slow, occurring over tens of milliseconds. The addition of water accelerated this decay (see below). As the absorption below 330 nm decays, there is an increase in absorption between 330 and 400 nm and both are correlated and depend on the water content of the solution. The differential spectrum exhibits a bathochromic shift and decays over tens of milliseconds to give a stable, weak absorption below 350 nm.

On the insert of Figure 1, we have shown the initial absorption spectrum of the solution and its time evolution after the laser pulse, calculated according to the transient spectra of Figure 1. It shows that the bleaching of the solution is weak and that above 380 nm the initial spectrum decays during the pulse and does not change in shape afterwards. From this

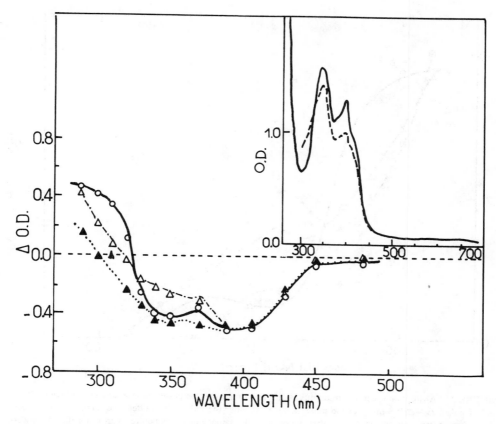

FIGURE 1.   Transient absorption spectra after laser photolysis of compound **1** ($2.5 \times 10^{-4}\,M$) in EEA: o—o end of pulse; △—△ 0.8 ms after the pulse; ▲—▲ 30 ms after the pulse. Insert: absorption spectra of the initial solution (path length = 5 mm) before the pulse (full line) and immediately after the pulse (dashed line).

observation, we were able to deduce at every time the neat absorption of the transients by simply subtracting at every wavelength the optical density due to the unbleached initial product. These transient spectra at three characteristic times are shown in Figure 2, and can be attributed to three different species, two transients A and B, and a stable product C.

The decay kinetics of species A is always first order, with a rate constant which depends strongly on the presence of traces of water. In freshly distilled EEA, the first-order rate constant k_A equal $2.1 \times 10^4\,s^{-1}$. Outgassing the solution has no influence on it. The absorption of B is stronger if water is deliberately added (see Figure 2). Species B also decays by a first-order kinetics, with a rate constant $k_B = 4.35 \times 10^3\,s^{-1}$; this constant is less sensitive than k_A to the presence of water, although there is a slight increase of k_B when the water concentration is increased. It is also insensitive to oxygen.

When methanol is added in a solution of compound **1** in EEA, the transient spectra observed after the laser pulse are very similar with those observed in the presence of water, as shown in Figure 3. The end-of-pulse spectrum is the same. After some time delay depending upon the methanol concentration, there is a bathochromic shift in the absorbance, which can lead to a new maximum around 340 nm. This intermediate (analogous to B species for water) decays in less than 1 ms to give a stable species.

2. Compound 1 in Benzene, Toluene, or Cyclohexane

As these solvents are hydrophobic, they can be obtained free from water after distillation over sodium. As a consequence, transient A has a much longer lifetime in these solvents.

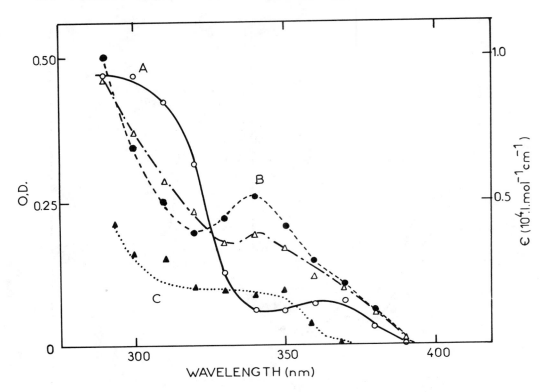

FIGURE 2.   Absorption spectra of transients at different times after the pulse (same solution as in Figure 1): ○—○ end of pulse; △—△ 0.8 ms after the pulse; ..... 30 ms after the pulse. The dotted line represents the transient absorption spectrum measured 0.3 ms after the pulse in the same solution as before, with water (0.1 *M*) added.

Without any additive, transient A disappears without any new intermediate in seconds to give a weak and stable absorption (see Figure 4). When excess methanol is added, transient A disappears faster, but without any bathochromic shift, to give the final weak absorption. The same effect is observed when traces of water are added to a benzene solution (the solubility of water in benzene is limited to $4 \times 10^{-2} M$). Thus, transient B is only stabilized in polar solvents.

B. WATER SOLUTIONS

The transient spectra of **2** in water ($2 \times 10^{-4} M$) are shown in Figure 5. The spectrum taken 1 μs after the end of the pulse is completely different from that of compound **1**. This spectrum has a maximum at 350 nm, and decays without changing in shape, on a millisecond time scale, leaving a stable absorption after one tenth of a second. By the same method as before, we could subtract the absorption of the unbleached diazide and obtain the initial and final spectra, named B′ and C′ in Figure 6. From the comparison of Figures 2 and 6, it appears that spectrum B′ looks like spectrum B and spectrum C′ resembles spectrum C. The transient B′ decays with first-order kinetics ($k_{B'} = 3.16 \times 10^3$ s^{-1}) on a time scale very similar to that of species B.

The same study was conducted recently by Tanigaki and Ebbesen,[8] who used a N$_2$ laser (pulse width 8 ns at 337 nm). They observed the same absorption spectrum as our spectrum B′ 3 μs after the end of their pulse. In addition, they observed a different absorption spectrum at the end of their pulse, which changes to spectrum B′ in half a microsecond. With the detection system that we used for this part of the study (Gould digital oscilloscope, bandwidth 20 MHz), we were not able to observe this fast transformation, all the more as it is near

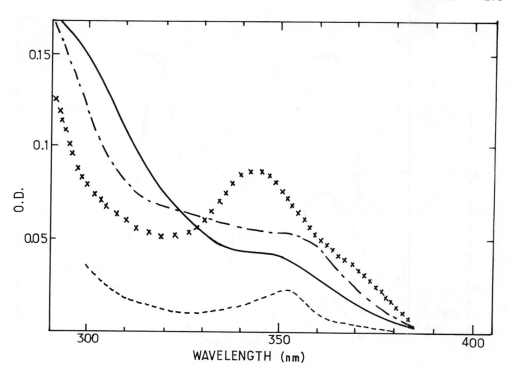

FIGURE 3. Absorption spectra of transients after laser photolysis of compound **1** in EEA added with methanol: —— end of pulse; —·—·— t = 25 μs ([CH₃OH] = 0.073 *M*); --- t = 0.55 ms ([CH₃OH] = 0.073 *M*); × × × × t = 0.7 μs ([CH₃OH] = 0.6 *M*).

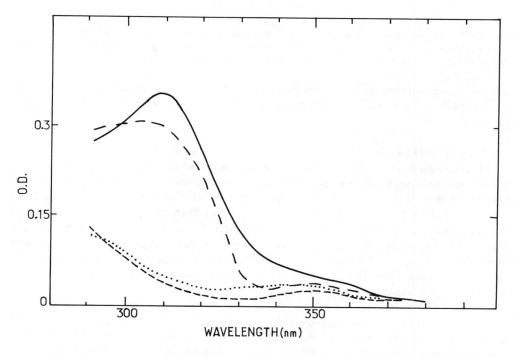

FIGURE 4. Absorption spectra of transients after laser photolysis of compound **1** in toluene (2.5 × 10⁻⁴ *M*): —— end-of-pulse; — — — end-of-pulse ([CH₃OH] = 4.2 × 10⁻² *M*); ---- t = 1.4 s (pure solvent); ..... t = 0.45 ms ([CH₃OH] = 4.2 × 10⁻² *M*).

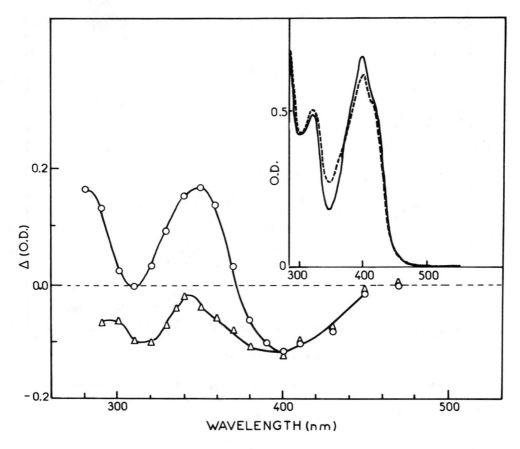

FIGURE 5.   Transient absorption spectra after laser photolysis of compound **2** (2.2×10^{-4} *M*) in water: ○—○
1 μs after the end of the pulse; △—△ 50 ms after the pulse. Insert: absorption spectra of the initial solution (path
length = 5 mm) before the pulse (full line) and immediately after the pulse (dashed line).

our excitation wavelength. It is interesting to note that the end-of-pulse spectrum observed
by Tanigaki and Ebbesen is very similar with our spectrum A observed in nonaqueous
solvents.

C. REVERSED MICELLES

The formation of reversed micelles in an hydrophobic solvent (cyclohexan, isooctan)
allows the introduction of large quantities of water molecules which are in strong mutual
interaction in the vicinity of the polar heads of the micelle.[26] Compound **1** was solubilized
in presence of AOT reversed micelles (AOT = sodium diethylhexylsulfosuccinate) formed
in isooctane and water. The ratio W (W = [H_2O]/[AOT]) was kept between 6 and 12, inside
the critical domain where reversed micelles exist.[27] At the end of the pulse, a transient
similar to A is observed which reacts very efficiently with the water included inside the
micelle and gives a transient similar to B absorbing around 350 nm. Let us recall that in
benzene, such a transient was not observed.

If the ratio W is kept constant, the size of reversed micelles remains constant and their
number varies. In such a situation, the decay of species A was found to be linear with water
concentration, or with the micelle concentration. We found that the reaction of transient A
with the micelles was diffusion limited. Concerning the disappearance of transient B, we
found it to be very fast (k = 5000 s^{-1}), in comparison to the decay of B in EEA.

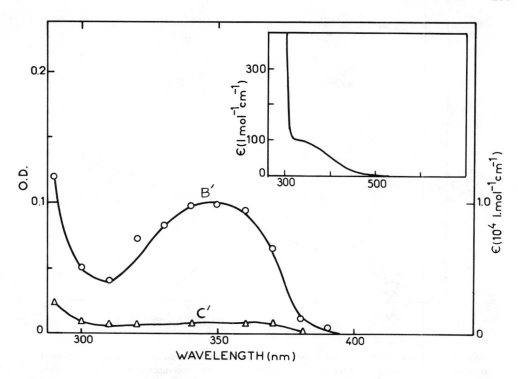

FIGURE 6.   Absorption spectra of transients at different times after the pulse (same solution as in Figure 5): ○—○ 1 μs after the end of the pulse; △—△ 50 ms after the pulse. Insert: absorption spectrum of a solution of 3-indenecarboxylic acid in water.

D. QUANTUM YIELD DETERMINATIONS

All determinations are based on the following hypothesis: the initial compound is bleached in a subnanosecond time range to give only transient A in the case of compound **1**, or transient B′ in the case of compound **2**, i.e., we suppose that, during this short time span, there are no competitive reactions parallel to the production of the observed transient. By this assumption, from the bleaching of the absorbance at wavelengths higher than 390 nm and from the energy deposited within the irradiated volume, we have determined $\Phi_A = 0.37 \pm 0.04$ for compound **1** in EEA and $\Phi_B = 0.48 \pm 0.05$ for compound 2 in water. These bleaching yields under laser excitation compare well with those determined under steady-state illumination for 2-diazonaphthalenone 5-sulfonic acid in water ($\Phi = 0.47$[28]) or in ethanol ($\Phi = 0.66$[28]).

As a consequence, extinction coefficients of A and B′ have been found simply by comparing optical density at wavelengths higher than 390 nm, where there is only bleaching, and below 390 nm where there is absorption and bleaching. These coefficients are reported in Figures 2 and 6 (right ordinate). For example, $\epsilon_A (300) = 9500 \; 1 \; M^{-1} \; cm^{-1}$ at 300 nm and $\epsilon_{B'} (300) = 10,000 \; 1 \; M^{-1} \; cm^{-1}$ at 350 nm. Regarding species B produced by partial hydrolysis of species A in "pure" EEA, since the reaction is not complete, the right-hand ordinate scale does not give the correct extinction coefficient. In the presence of an excess of added water ([H_2O] = $10^{-1} \; M$) the transient spectrum of B is higher (see Figure 2). By supposing that in this case the transformation A→B is complete, an extinction coefficient $\epsilon_B (350) = 5000 \; 1 \; mol^{-1} \; cm^{-1}$ can be estimated at the maximum (350 nm). If we consider that B changes to C, and B′ to C′, the extinction coefficients of C and C′ can also be deduced, as shown in Figure 6 for C′.

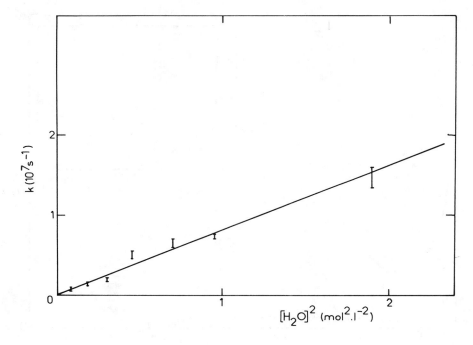

FIGURE 7.   Pseudo-first-order rate constant for the decay of the ketene vs. the square of water concentration. The decay of ketene has been observed at 310 nm for compound **1** (2.5 × 10^{-4} *M*) in EEA.

TABLE 1
Reactivity of Indene-Ketene with
Water and Methanol

	Solvent	
Reactant	**EEA**	**Benzene**
Water	8.5 × 10^6	3.75 × 10^3
Methanol	2.4 × 10^7	4.07 × 10^6

Note: Rate constants are in l^2 mol^{-2} s^{-1}.

E. REACTION OF TRANSIENT A WITH WATER AND METHANOL

Transient A reacts efficiently with traces of water and, as its lifetime is long, it is possible to study the reaction for water concentrations from 10^{-6} to 1 *M*. At molar concentrations of water in EEA, the decay of A observed at 300 nm lasts 100 ns. Throughout this concentration range, the decay obeys first-order kinetics. As shown in Figure 7, the pseudo-first-order rate constant is linear vs. the square of H$_2$O concentration. From the slope of the linear plot, a rate constant of 8.5 × 10^6 l^2 mol^{-2} s^{-2} is obtained. In benzene, the reaction is still second order in water, but is much slower. As seen before, transient A reacts very efficiently with methanol. In Table 1, the values of rate constants of transient A with water and methanol in two solvents are gathered.

F. REACTIVITY OF TRANSIENTS WITH THE NOVOLAK POLYMER

In EEA, addition of large quantities of Novolak polymer (formol-metacresol polymer) does not lead to any neat increase of the decay rate of transients. However, this polymer has free OH groups, which could react on transient A as water or methanol does. The reaction probably occurs in polymer films where the local concentration of OH groups is very high.

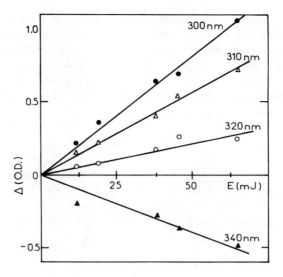

FIGURE 8. Initial optical density at the end of the pulse vs. laser energy in millijoules, for a solution of compound **1** ($2 \times 10^{-4} M$ in EEA). The different wavelengths are indicated on the left.

G. INFLUENCE OF LASER INTENSITY ON ABSORPTION OF TRANSIENT A

When laser intensity is varied from 5 to 50 mJ/cm^2, the optical density of transients measured immediately after the pulse in a solution of compound **1** in EEA increases linearly with the intensity (see Figure 8). Hence biphotonic processes are not involved in the formation of A. Moreover, since A absorbs slightly at 350 nm (see Figure 2), photobleaching of A could occur at high intensities and would result in a downward bending of the curves of Figure 8 at high intensities. The absence of such a bending indicates that photobleaching of A is negligible under our experimental conditions.

IV. IDENTIFICATION OF INTERMEDIATES

The analysis of the experimental results can be summarized as follows: the laser photolysis of compound **1** in every solvent leads, via a monophotonic process, to a transient A formed on a subnanosecond time scale. In the absence of water, the lifetime of A is rather long; but A reacts strongly with water to give B, which transforms into C. If water is used as the solvent for compound **2**, a transient A′ is formed which transforms into a species B′ in less than 1 μs. B′ has the same characteristics as B. The stable species C′ and C can be identified as 3-indenecarboxylic acid (see Scheme I) whose absorption spectrum is given in the insert of Figure 6.

In their interpretation of the data, Tanigaki and Ebbesen[8] identified our transient B′ as being the ketene, as did Nakamura et al.[5] previously. We think that their identification is wrong. On the contrary, we think that there is much evidence identifying A (or A′), the precursor of B (or B′), as the ketene. The first reason is that the absorption spectrum of A is very similar to the one observed by Hacker and Turro[4] after continuous irradiation at 77 K of 1- and 2-diazonaphthalenones in a glass of 3-methylpentane-cyclohexane: the UV absorption spectrum with maxima at 310 and 250 nm was attributed to the ketene, in agreement with the IR absorption spectrum observed in the same conditions. When a sulfonyl group was substituted in the 5-position on the naphthalene ring, the ketene absorption spectrum had maxima at 317 and 264 nm. Although the absorption spectrum of A has only a shoulder at 300 nm, we think that the shape of the spectrum and the extinction coefficient are in agreement with the photolysis work at 77 K.

The second experimental argument for the previous assignment is the reactivity of transient A with water and methanol: we found that the reaction is always second order in water or in methanol. Such kinetics have already been observed for normal ketenes (see above). In water as a solvent, the lifetime of the indene-ketene is found to be very short[8] in agreement with what is expected for the reaction of ketene with the solvent.

In presence of excess of added water or methanol (XOH; X = CH$_3$ or H), the pseudo-first-order rate constant measured for the ketene is

$$k_{obs} = k_A + k[XOH]^2 \tag{1}$$

In this equation, k_A is the pseudo-first-order rate constant observed for the decay of ketene in pure solvent, and k is the measured rate constant of ketene with XOH listed in Table 1. The mechanism which has been proposed involves the reaction of ketene with a XOH dimer, which prevails over the reaction with a monomer as the concentration of XOH increases:

$$2\ XOH \rightleftharpoons (XOH)_2 \tag{2}$$

The equilibrium (2) lies well to the left. The dimer then reacts with the ketene to form the following complex:

$$(3)$$

$$(4)$$

Reaction 3 is probably the slowest step in the mechanism. Thus, if we assume that equilibrium (2) is fast. with an equilibrium constant K_2, we deduce that

$$k = k_3\ [(XOH)_2] = k_3\ K_2\ [XOH]^2 \tag{5}$$

We recently studied dimer formation of water and methanol in different solvents by IR spectroscopy.[29] In EEA, hydrogen bonding between the solvent and water or methanol hinders the observation of dimers. In benzene, methanol forms dimers and polymers. OH stretching band of monomers is observed at 3600 cm^{-1}, and that of dimers at 3500 cm^{-1} From the intensities of these bands, the dimerization equilibrium constant K_2 was estimated to be 0.65 ± 0.1 l mol^{-1} at 25°C. Taking Equation 5 into account would give k_3 = 6.3 × 10^6 l mol^{-1} s^{-1}, which is lower than the diffusion limit. Water in benzene forms dimers

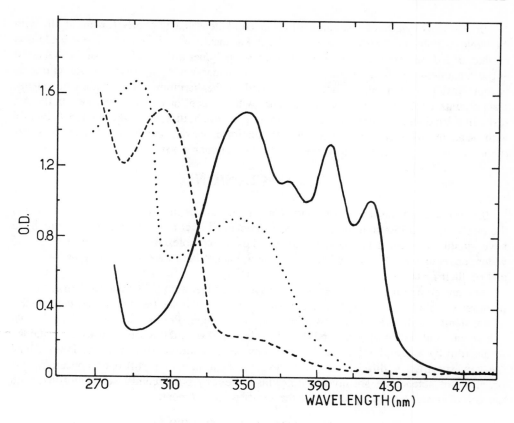

FIGURE 9.   Absorption spectra of compound **1** in MTHF glass (added with water 0.1 *M*): —— before irradiation at 77 K: --- after irradiation with a xenon arc at 77 K; .... after warming the sample at 200 K.

in very low yield.[30] This result explains why reaction of indene-ketene on water is inefficient in benzene (see Table 1).

In the literature, the bracketted cyclic complex was supposed to be a transition state.[18,20] Our results show that the long lifetime transient (B, or B′ if the solvent is water) could be due to this complex. The bathochromic shift of the lowest transition of the ketene is in agreement with a transition of CTTS character. This complex decomposes slowly to give the 3-indenecarboxylic acid, by a first-order process. The slight enhancement of the rate constant with addition of water in EEA is presumably an effect of the polarity. It is interesting to note that the lifetime of B species in EEA added with water (<0.1 *M*) is comparable with the lifetime of B′ in pure water: in no way could we explain this result if B (and B′) were a ketene. With benzene as a solvent, this complex has not been observed, and the ketene gives the final stable product (acid or ester) without intermediate. In reversed micelles in isooctane, this complex was observed again, but with a very short lifetime.

In our search for further experimental evidence on the structure of B species, we made the absorption spectra of compound **1** in an MTHF glass added with water (0.1 mol l⁻¹). After irradiation by continuous UV light (75 W Xenon arc) at 77 K, the absorption spectrum found was similar to the one depicted by Hacker and Turro[4] as shown in Figure 9. During warming of the sample, there is a change in the absorption spectrum, which gives a new one with absorption bands at 290 and 350 nm, the last peak being similar with B (or B′) absorption spectrum. This new species is stable at 200 K and finally disappears to give an absorption spectrum attributed to the indenecarboxylic acid. These experimental conditions would be appropriate to other spectroscopic investigations (IR, NMR).

In this work, we have also shown that indene-ketene reacts poorly (or not at all) with the Novolak polymer. This is in agreement with Pacanski,[3] who did not observe any addition product of indene-ketene on the formol-metacresol polymer after irradiation of the AZ resist at room temperature. It comes out that reaction of indene-ketene with water (even in traces) is more efficient than other reactions proposed in the literature: dimerization of ketene, reaction with unbleached diazonaphthalenone, with oxygen, or with the polymer. In the AZ resist film irradiated at room temperature, there is no doubt that indene-ketene reacts mainly with water absorbed in the polymer.[31] If irradiation is carried out in the absence of water (under vacuum), ketene probably reacts with the polymer matrix itself.

V. CONCLUSION

Laser photolysis has been proved to be a good technique in order to investigate in details the mechanism. Different transients have been detected after a nanosecond pulse, which we have identified as being the indene-ketene and cyclic complex formed between ketene and water (or methanol). These studies bring a new insight both on spectroscopy of transients and on their reactivity.

We are going on these studies in two ways: first, the study of transients formed in AZ photoresists films by laser spectroscopy should reveal whether the nature and the kinetics of transients are different from those observed in solution. Preliminary results have shown that ketene is also formed at the end of a 10-ns laser pulse, and it reacts with water contained in the film by a complex kinetics in two steps, as in solution. However, the intermediate transient spectrum is different. Second, as several questions are still open concerning the precursor of ketene, we are investigating this system by picosecond laser photolysis, with the aim of observing the dynamics of the Wolff rearrangement.

ACKNOWLEDGMENTS

The authors would like to thank F. Renou and M. Soreau who did most of the experiments. They also acknowledge the IBM France Company for a grant, and specially Mr. Mayeux from this company for his continuous interest in the work. The model compounds **1** and **2** were kindly given by the PCAS company.

REFERENCES

1. For recent reviews on microlithography, see a. **Bowden, M. J.,** *Materials for Microlithography,* (ACS Symp. Ser., Vol. 39), Thompson, L. F., Wilson, C. G., and Fréchet, J. M. J., Eds., American Chemical Society, Washington, D.C., 1984, 266; b. **Serre, B., Schue, F., Montginoul, C., and Giral, L.,** Le point sur les résines utilisables en microlithographie, *Actual. Chim.,* 11, 27, 1985.
2. **Pacansky, J. and Johnson, D.,** Photochemical studies on a substituted naphthalene-2,1 diazooxide, *J. Electrochem. Soc.,* 124, 862, 1977.
3a. **Pacansky, J. and Lyerla, J.,** Photochemical Decomposition Mechanisms for AZ-Type Photoresists, *IBM J. Res. Dev.,* 23, 42, 1979.
3b. **Pacansky, J.,** Recent advances in the photodecomposition mechanisms of diazo-oxides, *Polym. Eng. Sci.,* 20, 1049, 1980.
4. **Hacker, N. P. and Turro, N. J.,** Photochemical generation of ketenes. An ultra-violet and infra-red spectroscopy study of 1- and 2-diazonaphthalenones at 77 K, *Tetrahedron Lett.,* 23, 1771, 1982.
5. **Nakamura, K., Udagawa, S., and Honda, K.,** Studies of photosensitive resins. V. The photo-decomposition mechanism of *o*-naphtoquinone diazides, *Chem. Lett.,* 763, 1972.
6. **Bolsing, F. and Spanuth, E.,** Impulsphotolytishe unter suchungen zum mechanisms der photochemischen Wolff-Umlagerung, *Z. Naturforsch. B, Anorg. Chem., Org. Chem.,* 31, 1391, 1976.

7. **Delaire, J. A., Faure, J., Hassine-Renou, F., Soreau, M., and Mayeux, A.,** Laser photolysis of 1-[2H]-2-diazonaphthalenone derivatives, *N. J. Chem.*, 11, 15, 1987.

8. **Tanigaki, K. and Ebbesen, T. W.,** Dynamics of the Wolff rearrangement: spectroscopic evidence of oxirene intermediate, *J. Am. Chem. Soc.*, 109, 5883, 1987.

9. **McMahon, R. J., Chapman, O. L., Hayes, R. A., Hess, T. C., and Krimmer, H. P.,** Mechanistic studies on the Wolff rearrangement: the chemistry and spectroscopy of some α-ketocarbenes, *J. Am. Chem. Soc.*, 107, 7597, 1985.

10. **Kaplan, F. and Mitchell, M. L.,** Conformational control of decomposition pathways of α-diazoketones, *Tetrahedron Lett.*, 9, 759, 1979.

11. **Tomioka, H., Okuno, H., and Izawa, Y.,** Mechanism of the photochemical Wolff rearrangement, *J. Org. Chem.*, 45, 5278, 1980.

12. **Sitzmann, E. V. and Eisenthal, K. B.,** Studies of the chemical intermediate diphenylcarbene, in *Applications of Picosecond Spectroscopy to Chemistry,* (NATO ASI Ser., Ser. C, Vol. 127), Eisenthal, K. B., Ed., D. Reidel, Dordrecht, Netherlands, 1983, 41.

13. **Lewars, E. G.,** Oxirenes, *Chem. Rev.*, 83, 519, 1983.

14. **Tanaka, H. and Yoshimine, M.,** An ab initio study on ketene, hydroxyacetylene, formylmethylene, oxyrene, and their rearrangement paths, *J. Am. Chem. Soc.*, 102, 7655, 1980.

15. **Bouma, W. J., Nobes, R. H., Radom, L., and Woodward, C. E.,** On the existence of stable structural isomers of ketene. A theoretical study of the C_2H_2O potential energy surface, *J. Org. Chem.*, 47, 1869, 1982.

16. **Strausz, O. P., Gosavi, R. K., and Cunning, H. E.,** Ab initio molecular orbital calculations on oxiranylidene and ethynol, *J. Chem. Phys.*, 67, 3057, 1977.

17. **Dyktra, C. E.,** An ab initio study of the energies and structures of ketene, oxirene, and ethynol, *J. Chem. Phys.*, 68, 4244, 1978.

18. **Blake, P.,** *Kinetics and Mechanisms in the Chemistry of Ketenes, Allenes and Related Compounds,* Patai, S., Ed., John Wiley & Sons, New York, 1980, 309.

19. **Kagan, H. B.,** Les cétènes, *Ann. Chim.*, 10, 203, 1965.

20. **Lillford, P. J. and Satchell, D. P. N.,** Acylation: the kinetics and mechanism of the spontaneous and acid-catalysed reactions of dimethylketene with water and alcohols, *J. Chem. Soc. B, Phys. Org.*, 889, 1968.

21. **Poon, N. L. and Satchell, D. P. N.,** A comparison of the mechanisms of hydrolysis of diphenylketene and dimethylketene in diethylether solution, *J. Chem. Soc. Perkin Trans. 2*, 1381, 1983.

22. **Samtleben, R. and Pracejus, H.,** Saüreamide als bifunktionelle katalysatoren fur acylierung mit ketenen, *Tetrahedron Lett.*, 2189, 1970.

23. **Brady, W. T., Vaughn, W. L., and Hoff, E. F.,** Halogenated ketenes. Dihaloketene reactivities in an acylation reaction, *J. Org. Chem.*, 34, 843, 1969.

24. **Briody, J. M., Lillford, P. J., and Satchell, D. P. N.,** Acylation: the kinetics and mechanism of the addition of carboxylic acids to ketenes in diethyl ether and in dichlorobenzene solution, *J. Chem. Soc. B, Phys. Org.*, 84, 885, 1968.

25. **Delaire, J. A., Castella, M., Faure, J., Vanderauwera, P., and De Shryver, F. C.,** Inter- and intra-molecular charge transfer to cation-radicals of alkyl-esters of 2-anthracene carboxylic acid studied by nanosecond laser photolysis, *Nouv. J. Chim.*, 8, 231, 1984.

26. **Zulauf, M. and Eike, H. F.,** Inverted micelles and microemulsions in the ternary system H_2O/aerosol-OT/isooctane as studied by photon correlation spectroscopy, *J. Phys. Chem.*, 83, 480, 1979.

27. **Eike, H. F., Kubik, R., Hasse, R., and Zschokke, I.,** The water-in-oil microemulsion phenomenon: its understanding and predictability from basic concepts, in *Surfactants in Solution V. III.,* Mittal, K. L. and Lindman, B., Eds., Plenum Press, New York, 1984, 1533.

28. **Rehak, V., Poskocil, J., Majer, J., and Dvoracek, I.,** Photolysis of quinonediazides in alcohols and in water, *Czech. Chem. Commun.*, 44, 756, 1979.

29. **Soreau, M.,** Etude du Mécanisme de Photodégradation de Sensibilisateurs de Photorésists Positifs par Spectroscopie Laser, Thesis, Université de Paris-Sud, Orsay, France, 1987.

30. **Masterton, W. L. and Gendrano, M. C.,** Henry's law studies of solutions of water in organic solvents, *J. Phys. Chem.*, 70, 2895, 1966.

31. **Weill, A., Paniez, P., and Dechenaux, E.,** Etude calorimétrique de la rétention de l'eau par les résines novolaques, *Angew. Makromol. Chem.*, 122, 101, 1984.

Chapter 8

PRIMARY PHOTOPHYSICAL AND PHOTOCHEMICAL PROCESSES OF DYES IN POLYMER SOLUTIONS AND FILMS

Prashant V. Kamat and Marye Anne Fox

TABLE OF CONTENTS

I. INTRODUCTION

During the last few years photoactive polymeric systems have seen a phenomenal growth because of their applications in designing photoresists, xerography, photocuring of paints and resins, photodegradation and photostabilization of commercially used polymer products, and developing polymer modified electrodes.[1] Efforts are being made by photochemists and polymer chemists to understand the details of photophysical and photochemical processes in polymers. Fluorescence probes are often being used to investigate the static and dynamic behavior of these microheterogeneous systems. The majority of these investigations are designed to probe the polymer chain motions, small molecule-polymer coil interactions, and intercoil energy transfer processes.[2] For example, the fluorescence emission of intramolecular excimers could provide information on the rate of conformational transition and, thus, on the dynamic behavior of the polymer chain. Other interests in photoactive polymeric systems include developing polymeric photocatalysts and designing antenna polymers to mimic the function of light harvesting pigments.[3]

These photoactive polymer systems can be broadly classified into two categories: (1) polymers containing chromophores attached to the backbone of the polymer and (2) polymers acting as hosts to photoactive guest molecules. Two types of polymers of the former category have been synthesized: those in which the photoactive group is incorporated either as a part of the repeat unit or as an isolated chromophore attached to the end group. When excited electronically, the pendent chromophores often achieve a coplanar sandwich-like geometry to form an excimer which exhibits a new structureless band at lower energies.[4] Such an excimer formation defines the configurational and conformational motion of the polymer chain as well as the efficiency of energy migration along the chain. Excimer emission can also be used to investigate end to end cyclization,[6] compatibility in polymer blends,[7] and electropolymerization mechanisms.[8] Bimolecular excited state processes (such as excimer or exciplex formation, energy transfer, or electron transfer) have been studied in detail by several researchers.[4] Since excellent reviews related to this work can be found in the literature,[1,9] we will focus on the study of polymeric systems containing dye or other light-sensitive organic molecules as guests. The influence of the polymeric environment (both as films and in solution) on the excited state behavior of dye molecules will be described here.

II. PHOTOPHYSICS OF DYES IN POLYMERS

A. BINDING OF DYE MOLECULES TO THE POLYMER CHAIN AND ITS INFLUENCE ON THE ABSORPTION PROPERTIES

Ionic polymers (such as Nafion, a perfluorosulfonate ion exchange polymer, poly(4-vinylpyridine) (acidified), or polylysine contain either negative or positive charges all along the polymer chain, facilitating electrostatic binding of the oppositely charged dye molecules. The electrostatic forces caused by high local charges and the hydrophobic environment surrounding the dye molecules have important consequences on the excited state properties of the dye and its chemical interaction with the polymeric host. Any interaction between the dye molecule and the polymer can be probed with electronic absorption or emission spectra as the energetics of the ground and excited state are altered. These spectral changes include displacement or broadening of the absorption and emission spectra, changes in the extinction coefficient of absorption, appearance of the new absorption and emission bands, and changes in the excited state lifetimes.

The absorption spectra of thionine perchlorate in 50/50 v/v% ethanol/water mixture at different Nafion concentrations are shown in Figure 1. At very low concentrations of Nafion (0.025%) a decrease in the absorption ascribed to the monomeric form of the dye (602 nm) was observed with a corresponding increase in absorption around ~560 nm caused by dye

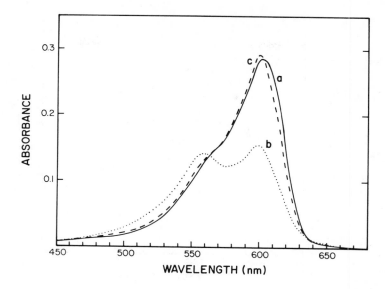

FIGURE 1. Absorption spectra of 5 μM thionine in ethanol at different Nafion (Aldrich) concentrations: (a) (——), 0%, (b) (....) 0.025%, and (c) (----) 1.25%. (Absorption spectra were recorded with a Cary 219 spectrophotometer.)

aggregation. This indicated that intra- and intermolecular aggregation between the dye molecules which are bound to Nafion are favored at very low polymer concentrations. However, at higher Nafion concentrations (1.25%), the dye molecules again become monomeric, exhibiting an absorption maximum around 599 nm which is blue shifted compared to the absorption maximum in neat solution. Such a shift in the absorption maximum matched the observed trend of the shift of the absorption peak with decreasing polarity.

In an earlier observation, a red shift in the absorption spectrum was noted for some anionic dyes, (croconate violet, croconate blue, rose bengal, and erythrosin B), upon introducing them into poly(4-vinylpyridine) solutions or films.[10-12] The hydrophobic environment was found to alter the absorption characteristics of the dye. Similar red-shift in the absorption bands of rose bengal bound to poly(styrene-co-vinylbenzylchloride) copolymer has also been reported.[13] Heterogeneous adsorption of dyes to free standing films of surface sulfonated polystyrene with cationic dyes such as acridine orange, methyl violet, and rhodiamin B has been investigated by Gibson and Bailey.[14] From the observed changes in absorbance, the binding capacity of the dye as well as the efficiency of displacement by another cationic dye could be estimated. The ability of azo dyes to form complexes with polyvinylpyrrolidone has been described by Takagishi et al.[15] to indicate the hydrophobic interaction between the dye and the polymer. The adsorption of a photochromic azo dye onto a styrene-divinylbenzene copolymer has also been studied.[16] The molecular mobility of the polymer matrix has been shown to influence the separation between the absorption and emission bands of fluorescent dyes.[17] Doping the rigid polymer matrix with molecules of high mobility enhanced the observed Stokes shift.

B. EMISSION SPECTRA AND EXCITED STATE LIFETIMES

The emission yield and lifetime of the excited dye can provide useful information on the influence of the polymeric environment on the energetics and the stability of the excited state. For example, the effect of the addition of Nafion to a solution of thionine is shown in Figure 2. The emission maximum of the polymer microencaged dye was at 624 nm, which is blue shifted when compared to the emission maximum of thionine in 50/50 v/v%

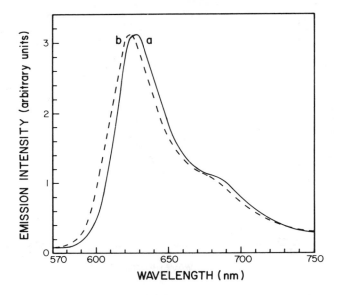

FIGURE 2. Emission spectra of 5 μM thionine in ethanol at different Nafion (Aldrich) concentrations: (a) (———), 0%, and (b) (---) 1.25% (Fluorescence emission was monitored with an SLM 8000 photon counting spectrofluorometer with excitation at 565 nm).

TABLE 1
Fluorescence Emission Yield and Fluorescence Lifetimes of Dyes in Organic Solvents and Polymer Solutions

Dye	Solvent	Dye (μM)	Poly(4-vinylpyridine) (mM)	ϕ_f	τ_f (ps)	Ref.
Croconate violet	Ethanol	10	—	0.002	3	10
	Ethanol	10	10	0.017	28	10
	Glycerol	10	—	0.026	33	10
Rose bengal	Ethanol	10	—	0.11	660	18, 25
	Ethanol	10	10	0.22	1105	18
	Acetonitrile	10	—	—	2380	26
Erythrosin B	Ethanol	10	—	0.08	570	18, 25
	Ethanol	10	10	0.15	1023	18

ethanol/water. Although no marked changes in the fluorescence yield could be observed for thionine, a pronounced effect of polymer microencagement has been observed for several dyes (Table 1). Up to a tenfold increase in fluorescence emission yield was observed for oxocarbon and xanthene dyes introduced into poly(4-vinylpyridine) solutions.[10,18] Environmental changes due to increased microviscosity and hydrophobic interactions were found to be responsible for the enhanced fluorescence yield.

1. Effect of Microviscosity

The excited state stability of oxocarbon dyes is dependent on the viscosity of the medium. For example, when the medium was changed from ethanol ($\eta = 1.2$ cp) to glycerol ($\eta = 954$ cp), the fluorescence yield of croconate violet increased dramatically (by an order of magnitude).[10] Since polymer segments effectively increase the microviscosity around the dye, they can facilitate stabilization of the excited singlet state.

Various relaxation processes can be proposed for the deactivation of the excited singlet states (Reactions 1 to 4),

$$dye(S_0) \xrightarrow{h\nu} {}^1dye^*(S_1) \tag{1}$$

$$^1dye^*(S_1) \xrightarrow{k_f} dye(S_0) + h\nu' \tag{2}$$

$$^1dye^*(S_1) \xrightarrow{k_{isc}} {}^3dye^*(T_1) + heat \tag{3}$$

$$^1dye^*(S_1) \xrightarrow{k_{nr}} dye(S_0) + heat \tag{4}$$

where k_f, k_{isc}, k_{nr} refer, respectively, to the rate constants for fluorescence decay, intersystem crossing, and nonradiative decay. The latter would include contribution from internal conversion, intramolecular quenching, and quenching caused by the interaction with the solvent. From Reactions 1 to 4, one can express the fluorescence quantum yield as

$$\phi_f = \frac{k_f}{k_f + k_{isc} + k_{nr}} \tag{5}$$

For croconate dyes one could consider ϕ_f to be equal to k_f/k_{nr} since $k_{nr} >> k_f$ and $k_{nr} >> k_{isc}$. Any decrease in k_{nr} would thus result in an increase in the fluorescence quantum yield. This analysis indicated that the increased microviscosity encountered inside the polymer cage decreased k_{nr}, thereby stabilizing the singlet excited state. Similarly several polymethine and cyanine dyes have been shown to possess viscosity-dependent radiation relaxation rates.[19,20] Among several other possible viscosity-dependent effects are the inhibition of solute-solvent or solvent-solvent hydrogen bonding or the interference of solvent within the path of a rotating pendent group.

Viscosity-dependent emission properties facilitate determination of microviscosity in polymeric systems. For example, Lee and Meisel[21] have used the excimer formation of pyrene in a *t*-butanol-swollen Nafion film to estimate the microviscosity of the polymer. Fluorescence studies of the solubilization and dye binding in hypercoiled poly(methacrylic acid) have been performed using 9-methylanthracene and rhodamine B as probes.[22] Measurements of the fluorescence lifetimes and rotational correlation times indicate that the probe molecules are rigidly held in the polymer matrix so that the transition dipole is fixed in each of the clusters. Emission properties of pyrene have been utilized to determine the interpolymer interactions of polyacrylic acid, poly(vinylamine hydrochloride), poly(1-aminoacrylic acid), and poly(1-acetylaminoacrylic acid).[23] A unique upper-state emission associated with the dimeric dye has also been reported recently for crystal violet bound to poly(acrylic acid).[24]

2. Effect of the Hydrophobic Environment

The photophysical properties of xanthene dyes are greatly influenced by the proton donating power of the medium.[25-27] As can be seen from Table 1, the lifetime of the excited singlet of rose bengal is considerably enhanced when the medium is changed from water to acetonitrile.

The energetics of the excited states of these dyes are influenced by hydrogen bonding. Usually solvent stabilization (ΔE) is more significant for the excited singlet state than for the diradical like triplet and is more important for the ionic ground state of charged dyes

TABLE 2
Fluorescence Lifetimes of Dyes in Polymer Films

Dye	Polymer	[Polymer]/[dye]	Fluorescence lifetime (ns)	Ref.
Rose bengal	Poly(4-vinylpyridine)	0	0.015	18
		100	0.054	18
		1000	0.802	18
		10000	1.820	18
Erythrosin B	Poly(4–vinylpyridine)	0	0.015	18
		100	0.054	18
		1000	0.504	18
		10000	1.350	18
Pyrene	Nafion	—	415	21
3,4,9,10-Perylenetetra carboxylic dianhydride	Poly(butyl methacrylate)	—	4,12	28

than for the excited singlet, i.e., $\Delta E(T_1) < \Delta E (S_1) < \Delta E (S_0)$. The absolute magnitude of this solvent stabilization will increase with solvent polarity: this implies that the S_1-T_1 energy gap will decrease as the polarity of the solvent increases.

When these dyes are microencaged with polymer, the hydrophobic backbone of the polymer displaces solvent, inhibiting solute-solvent interactions. This results in an increase of S_1-T_1 energy gap to stabilize the excited singlet state. Since k_{nr} for xanthene dyes is very small ($k_{nr} << k_f$ or k_{isc}), any decrease in k_{isc} would enhance the fluorescence yield and the lifetime of the dye.

C. PHOTOPHYSICAL BEHAVIOR OF DYES IN POLYMER FILMS

Though considerable efforts are being made to understand the photophysical behavior of the polymer with photoactive pendent groups, relatively little is known about the influence of a polymer on the behavior of photoactive guest molecules. Since dye-incorporated polymeric systems have a myriad of practical applications, it is important to study the photophysics of dyes in polymer matrices.

The fluorescence lifetimes of xanthene dyes exhibit an increase of nearly 2 orders of magnitude when dispersed in poly-(4-vinylpyridine) films (Table 2). Dispersal of the dye in a polymer facilitates the separation of dye molecules so that aggregation, contact quenching, and excited state-annihilation are inhibited (Figure 3). Further support for this argument can be gathered from the fact that the excited singlet lifetimes remain unchanged upon increasing the film thickness but maintaining a constant [dye]/[polymer] ratio. This clearly highlighted the role of polymer in stabilizing the excited singlet state. Construction of thicker films could facilitate the incorporation of larger amounts of dye without inducing dye aggregation.

This means that one could control the excited state behavior of the dye by controlling the [dye]/[polymer] ratio. An excimer-like emission has been observed recently for a perylene dye derivative.[28] A blue shift in the emission maximum was observed when this dye was dispersed in poly(butylmethacrylate) films. Excimer formation has also been observed for pyrene incorporated in Nafion films.[21]

Several other techniques are also being employed to better define the photophysical processes occurring in polymer matrices. A total internal reflection technique can be coupled with a subnanosecond fluorescence spectroscopy to elucidate the excited state behavior of the polymer surface. The importance of this technique has been demonstrated by Masuhara et al.[29] for 0.01- to 0.4-μm thick polystyrene films doped with *p*-bis[2-(5-phenylene oxazolyl)]benzene) and *N*-ethyl-carbazole. A femtosecond three-phase scattering technique has also been employed to observe dephasing of the lowest electronic transition of crystal violet in poly(methyl methacrylate).[30]

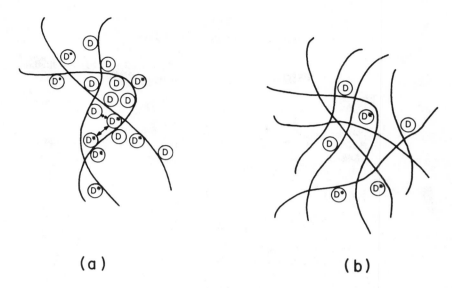

(a) (b)

FIGURE 3. Influence of polymer microencagement on the interaction between the ground state and excited state molecules (a) at low polymer/dye ratio and (b) at high polymer/dye ratio.

D. TRIPLET STATE BEHAVIOR OF DYES

The longer lived triplet excited state can play an important role in influencing the course of a photochemical reaction. The triplet state can be characterized by recording either the phosphorescence or T-T absorption spectra. The polymeric matrix surrounding the dye molecule often enhances the efficiency of phosphorescence. The various modes of radiationless decay are decreased in the polymer, as the high microviscosity restricts the mobility of the guest.

Triplet energy traps in copolymers of 1-vinylnaphthalene with styrene and with methyl methacrylate have been investigated in a glassy matrix at 77 K.[31] A delayed fluorescence caused by triplet-triplet annihilation within a single polymer segment was observed in these studies. A strong heavy atom-induced room temperature phosphorescence has been observed for pyrene incorporated into swollen Nafion films.[21]

The luminescence of benzophenone in polystyrene has been used to determine the effects of temperature and matrices on the behavior of triplet state.[32] The triplet state lifetime of anthracene in the absence of dynamic quenching have been measured in lucite solutions.[33] The triplet-singlet radiationless transition was found to dominate the deactivation of the triplet state.

The triplet lifetimes of several dyes in polymer solutions and films are summarized in Table 3. The polymeric environment enhanced the lifetime of the triplet in all these cases. For example, rose bengal triplet, which has a lifetime of 90 μs in ethanol, exhibited a lifetime of 205 μs in a poly(4-vinyl-pyridine) solution and 1.4 ms in a poly(styrene-vinylbenzyl-chloride) film.[18,34] It is evident that the polymeric environment, which alters the viscosity, polarity, and physical site-site interactions, facilitates triplet stabilization.

Excited state annihilation is an important process in determining the excited state lifetime and the photostability of dyes.[35] Such a process often proceeds with a net electron transfer in the quenching event, resulting in the generation of radical-ion pairs.[36]

$$\text{dye} *(T_1) + \text{dye} (S_0) \xrightarrow{k_q} \text{dye}^{+\cdot} + \text{dye}^{-\cdot} \longrightarrow \text{products}$$

$$2 \text{ dye} (S_0)$$

(6)

TABLE 3
Triplet State Properties of Dyes

Dye	Medium	T-T Absorption maximum (nm)	Triplet lifetime (μs)	Ref.
Rose bengal	-/Ethanol	600	90	18, 34
	Poly(4-vinylpyridine)/ethanol	610	205	18
	Poly(styrene-vinylbenzyl chloride)	—	1400	34
Erythrosin B	-/Ethanol	590	76	18
	Poly(4-vinylpyrydine)/ethanol	600	139	18
Croconate violet	-/Ethanol	538	69	11
	Poly(4-vinylpyrydine)/ethanol	542	134	11
Croconate blue	-/Ethanol	605	24	11
	Poly(4-vinylpyridine)/ethanol			11
Benzophenone	-/Benzene	—	5.4	32
	Poly(α,β,β-trifluorostyrene)	—	1000	32
	Poly(methylmethacrylate)	—	1000	32
	Poly(α-methylstyrene)	—	700	32
	Polystyrene	—	650	32
Anthracene	Lucite	—	13000	33

However, self-quenching can be retarded by encaging these dyes with a polymeric domain. For example, we have shown that for xanthene dyes the rate constant for the self-quenching can be decreased by an order of magnitude upon microencaging with poly-(4-vinylpyridine).[18] For rose bengal, the values of k_q were $2.7 \times 10^8 \ M^{-1} \ s^{-1}$ in ethanol and $8.5 \times 10^7 \ M^{-1} \ s^{-1}$ in 0.01 M poly-(4-vinylpyridine)/ethanol solution. For erythrosin B, the values of k_q were $2.9 \times 10^8 \ M^{-1} \ s^{-1}$ in ethanol and $8.5 \times 10^7 \ M^{-1} \ s^{-1}$ in 0.01 M poly(4-vinylpyridine)/ethanol solution. Strong hydrophobic interactions between the dye molecule and the hydrocarbon backbone of the polymer limit the diffusion of the dye molecule to the segmental diffusion of the polymer. Thus, the encapsulation with the polymer decreases the efficiency of self-quenching and stabilizes the excited state. On the other hand, if the molar polymer-to-dye ratio decreases to a level such that the dye molecules begin to aggregate within a polymer packet, enhanced self-quenching might lead to a decreased excited state lifetime.

III. PHOTOCHEMISTRY OF DYES IN POLYMERS

As described earlier, photochemistry in polymers can be considered from two different perspectives: that of the polymeric host and that of added or dissolved guest solute molecule.[37] The type and degree of interaction between the dye molecule and the polymer determine the course of a photochemical reaction. Details of the bimolecular processes involving formation of exciplexes and excimers have been presented elsewhere.[1,9,37] Some important aspects of energy and electron transfer processes in polymeric host systems will be considered here.

A. ENERGY TRANSFER PROCESSES

The transport of singlet and triplet energies in polymers has been the topic of many recent investigations. These studies not only elucidate the excited state behavior of the dye molecules in disordered systems but also provide information concerning the photodegradation of polymers, polymer photoconductivity, and photon harvesting systems. Energy transfer is also an important process in green plants, where chlorophyll functions as an "antenna pigment" by transferring the absorbed light energy to the reaction centers through nonradiative interactions.[38] Efforts are also being made to investigate different synthetic polymers which can mimic the function of the light harvesting pigment.[3,39,40]

Energy transfer by exciton migration was found to occur from singlet and triplet states of aromatic polymers in rigid solutions.[41] Trapping of singlet excitons in films and rigid solutions can be recognized from the emission spectra, while triplet excitons can be detected by quenching or by delayed fluorescence emission derived from T-T annihilation. The mobility of triplet excitons in poly(2-vinylnaphthalene) has been studied by Webber[42] in a quenching study with biacetyl. A detailed account of intermolecular energy transfer in polymers with radiative and nonradiative mechanisms can be found elsewhere.[43]

Triplet-triplet energy transfer between an excited donor (D) and an acceptor (A) can be performed in polymer solutions in a similar manner as that used in the homogeneous solution (Reaction 7).

$$^3D^* + A \rightarrow D + {}^3A^* \tag{7}$$

One of the two (D or A) or both species may be bound to the polymer in order to observe this energy transfer. For example, we have shown that triplet energy can be transferred from excited 9,10-dibromoanthracene to croconate dyes bound to poly(4-vinyl-pyridine). Laser flash photolysis experiments indicate that energy transfer in polymer solutions occurred at a slower rate than in neat ethanol. Segmental diffusion of the polymer chain (to which the dye molecules are physically attached) could well contribute to the effective diffusion of the dye molecule in ethanol. However, if both of the species, D and A, are bound to the polymer, a large enhancement in energy transfer can be anticipated. Shirai et al.[44] have studied excitation energy transfer between dyes bound to different polymers. They have also succeeded in inducing *cis-trans* isomerization of azobenzene and other derivatives in polymeric systems using triplet methylene blue as the sensitizer.

B. SINGLET OXYGEN PRODUCTION

A photodynamic study of singlet oxygen is important because of its damage-initiating properties in biological systems as well as in the degradation of polymers.[45] Considerable attention has been drawn in recent years to the lifetime, kinetic behavior, and the reaction mechanisms involving $O_2(^1\Delta_g)$ in heterogeneous media. Singlet oxygen can be conveniently generated in a polymer through photosensitization (Reaction 8).

$$^3S^* + O_2(^3\Sigma_g) \rightarrow O_2(^1\Delta_g) \tag{8}$$

$$O_2(^1\Delta_g) \rightarrow O_2(^3\Sigma_g) + h\nu \tag{9}$$

The emission of $O_2(^1\Delta_g)$ in the IR region (1270 nm) provides a convenient way to monitor its formation and reaction rate. Direct evidence for the formation of singlet oxygen in polymeric films via steady-state irradiations has been reported by Byteva et al.[46] Lee and Rodgers[47] have employed 2-acetonaphthene as a photosensitizer to produce singlet oxygen in Nafion powders. The singlet oxygen, which exhibited a lifetime ranging from 55 μs (H$_2$O swollen) to 320 μs (vacuum dried), was shown to be quenched with Ni^{2+}, Cu^{2+}, and $Co(NH_3)_6^{3+}$ which were also adsorbed on Nafion. The mobility of solvent pools within the polymer matrix and the binding of cations to the sulfonate headgroup of the polymer were found to influence the singlet oxygen quenching.

A time-resolved study of the reactions of singlet oxygen in a solid organic polymeric glass has been reported recently by Ogilby et al.[48] In a polymethyl methacrylate glass, a lifetime of ~100 μs has been observed for singlet oxygen when generated by sensitization by acridine, fluoranthene, 1,12-benzoperylene, and 2'-acetonaphthone incorporated in the

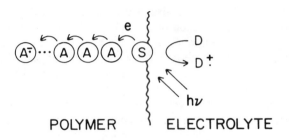

FIGURE 4.   Schematic representation of the photoinduced charge separation in a polymer system.

polymer matrix. The longer lifetime of singlet oxygen was found to facilitate its encounter with the sensitizer or a polymer-derived intermediate during the course of its diffusion in the host polymer.

Neckers and co-workers[49,50] have studied a series of polymer-based sensitizers for the formation of singlet oxygen. At low concentrations of the sensitizer (e.g., rose bengal), the quantum yield of $O_2(^1\Delta_g)$ formation increased with the sensitizer concentration, but decreased at high sensitizer concentrations (30 units of polymer per sensitizer molecule was found to be the optimum condition for $O_2(^1\Delta_g)$ production).

C. ELECTRON TRANSFER REACTIONS

The interest in the problems associated with polymer photodegradation and photochemical conversion and storage of light energy has necessitated the study of electron transfer reactions in polymers. One problem often encountered in such studies is the competition between reverse electron transfer and the separation of primary redox pairs as generated upon photoexcitation of the sensitizer. Studies in organized assemblies have indicated that charge separation can be greatly influenced by the effective control of the electron transport system, i.e., molecular diffusion of an electron mediator or migration of charge by electron hopping.[51] One such possible method of achieving charge separation is described in Figure 4. The redox reaction between the sensitizer (S) and the acceptor (A) is stabilized by electron migration to the neighboring acceptor species and by the regeneration of the sensitizer(s) with another redox couple. Similar stabilization of photochemical electron transfer products has already been demonstrated in Nafion film.[52] Attempts have also been made to employ polyelectrolytes,[52b] micelles,[53] vesicles,[54] and microemulsions[55] as model systems to investigate the influence of the microenvironment on photoinduced electron transfer reactions.

Transient absorption spectroscopy is a convenient technique to probe the various photochemical events occurring in the picosecond-millisecond time domain.[56,57] A holographic grating relaxation technique has been used to study the photochemistry of methyl red in polymeric hosts such as poly(methyl methacrylate) and polystyrene.[58a] Diffuse reflectance laser flash photolysis has also been employed to study the primary photochemical processes in dyed fabrics and polymers.[58b] Weir and Scaiano[58c] have examined the photochemistry of xanthone and of benzophenone in Nafion membranes in their acid and sodium exchanged forms. Electron transfer is an important primary step following the excitation of the sensitizer, as it polarizes the surrounding environment and triggers the chemical reaction:

$$S* (S_1 \text{ or } T_1) + D \xrightarrow{k_q} (S^{\mp\cdot} ... D^{\pm\cdot}) \begin{array}{c} \xrightarrow{k_{et}} S^{\mp\cdot} + D^{\pm\cdot} \\ \searrow^{k_n} \\ S + D \end{array} \tag{10}$$

Quenching of the excited state (S*) leads to the formation of a radical-ion pair, from which the dissociation to product occurs. The scheme presented in Reaction 10 is similar to the small molecule electron donor-acceptor systems in homogeneous solution. Environmental conditions such as micropolarity, microviscosity, segmental diffusion, and site-to-site interactions greatly influence the energetic and kinetic aspects of electron transfer process in the polymer. Masuhara[56] has discussed for polymers the theoretical aspects of Marcus theory which describes electron transfer.

Aqueous nylon which contains pendent viologen and quaternized amino groups has been employed by Nosaka et al.[59] to demonstrate the effect of an ionic polymer on photoinduced electron transfer from zinc porphyrin to viologen. The cationic and/or hydrophobic environment was found to retard back electron transfer, thereby increasing the efficiency of net electron transfer. The photocatalytic activities of several porphyrins trapped in polymer matrices have been examined as a function of the composition of the polymer matrix and of the reaction medium.[60] Environmental effects on the photoredox reactions of thionine-Fe(II) in polyethylene glycol solutions have been studied via the flash photolysis.[61] Photoinduced intramolecular charge transfer processes has also been studied in the pico- to nanosecond time regime to follow the sequential formation of high-energy intermediates such as exciplexes, radical-ion pairs, and triplets in an A-$(CH_2)_n$-D type polymer.

In order to understand the dynamics of electron transfer process in polymer systems, it is necessary to understand the electronic structure of the polymer ion radicals which in turn are closely related to the most stable ground state configurational and conformational structures of the polymers.[58] Radical anions and cations of several aromatic compounds in stretched polymer films have been characterized by Hiratsuka et al.[63] Polarization spectra of radical ions of acridine, phenazine, and azanaphthalene in polyethylene and poly(vinylchloride) films have been recorded to compare with those of the parent hydrocarbon radical ions.

IV. APPLICATIONS

A. PHOTOELECTROCHEMICAL CELLS

Polymer film-coated conducting or semiconducting electrodes have been successfully employed in many recent investigations.[64] Dye-loaded polymer electrodes can be important in extending the absorptive range of large bandgap semiconductors. Photoelectrochemical effects observed with several oxocarbon dyes incorporated into poly(4-vinylpyridine) films have been discussed earlier.[12] The photoelectrochemical response of an n-SnO_2 optically transparent electrode at different modification stages are presented in Table 4. Distinctively higher photocurrents and photovoltages were observed with the dye-loaded polymer films. Supersensitizers, such as benzoquinone (BQ)/hydroquinone in H_2O, stabilize the observed photocurrents by regenerating the sensitizer at the electrode surface. Similar sensitization of the n-SnO_2 and n-TiO_2 electrodes have also been achieved with polymer-coated films containing chlorophyll analogues.[65] The response of SnO_2/PVP/Cu-chlorophyllin to illumination and its action spectrum are shown in Figure 5. The photoaction spectrum exhibited maxima at 405 and 605 nm which coincided with the absorption bands of the pigment. This confirmed the expectation that the observed sensitization effects were initiated by the excitation of chlorophyllin in the polymer.

The generation of photoelectrochemical effects at a dye-loaded polymer electrode can be broadly classified into two categories: (1) photosensitization involving electron injection into the conduction band of the semiconductor and (2) photogalvanic process involving a redox reaction between the excited dye and a redox couple. The scheme described in Figure 6 highlights the principles of these two mechanisms. Factors such as the energy levels of the conduction and valence bands of the semiconductor and the redox potential of the excited

TABLE 4
Photoelectrochemical Effect at Dye-Loaded Polymer Electrodes

Electrode	Redox Couple	Photovoltage[a] ΔV (mV)	Photocurrent[b] Δi (μA)	Ref.
SnO₂/	—	−5	−0.01	12
SnO₂/PVP	—	−15	−0.06	12
SnO₂/croconate violet (ads)	—	10	0.05	12
SnO₂/PVP/croconate violet	—	50	3.0	12
SnO₂/PVP/croconate violet	BQ/H₂Q	55	6.0	12
SnO₂/PVP/croconate violet	Fe(III)/Fe(II)	15	0.5	12
SnO₂/PVP/chlorophyllin	BQ/H₂Q	100	0.7	65
SnO₂/PVP/chlorophyllin	Fe(III)/Fe(II)	45	0.4	65
SnO₂/PVP/rose bengal	Fe(III)/Fe(II)	−140	−3.6	66

Note: Dye incorporated poly(4-vinylpyrydine) film coated transparent SnO₂ electrodes in 0.1 *M* CF₃COONa (pH 3).

[a] Difference in open circuit photovoltage.
[b] Difference in short circuit photocurrent.

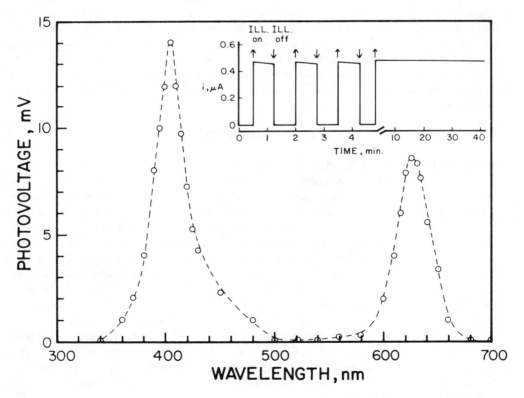

FIGURE 5.   Photoaction spectrum of SnO₂/poly(4-vinylpyridine/Cu-chlorophyllin electrode in 0.1 *M* CF₃COONa (pH 3) with Pt-counter electrode. Insert is a response of the polymer modified electrode to illumination.

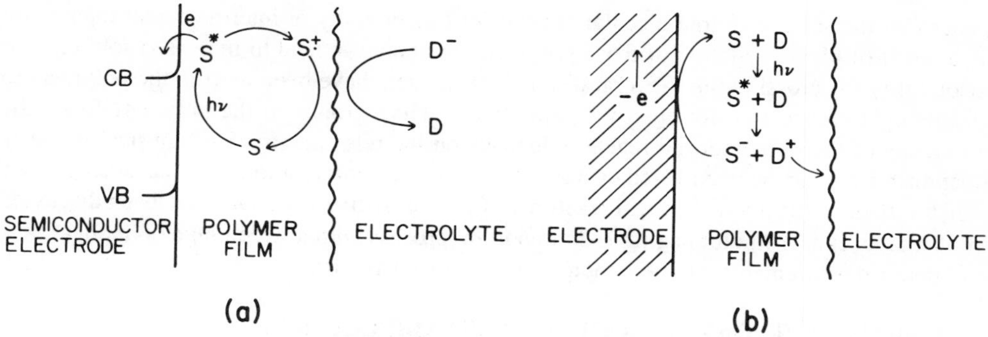

FIGURE 6. Mechanism of (a) photosensitization and (b) photogalvanic effect at a dye loaded polymer film coated electrode.

sensitizer and of the redox couple govern the mechanism for inducing photoelectrochemical effects. For semiconductors with low dopant concentrations, if the energy level of the conduction band lies below the oxidation potential of the excited sensitizer, electron injection into the conduction band can be anticipated. However, photogalvanic effects can be observed if the redox couple present in the polymer film can quench the excited state of the sensitizer (usually triplet excited state) and undergo electron transfer to yield an electroactive reduced (or oxidized) product.

The photogalvanic response of a poly(4-vinylpyridine) film loaded with rose bengal has been reported by us.[66] The dye-loaded polymer electrode, when immersed in an Fe(II)/Fe(III) solution and irradiated with visible light, results in the generation of a photogalvanic effect. The excited dye in the polymer film is reduced upon reaction with Fe(II) and the dye radical anion undergoes heterogeneous electron transfer at the electrode surface (Reaction 11 and 12)

$$RB + Fe(II) \xrightarrow{h\nu} RB* + Fe(II) \longrightarrow RB^{\cdot-} + Fe(III) \qquad (11)$$

external circuit

$$RB^{\cdot-} \xrightarrow{-e} RB \qquad\qquad Fe(III) \xrightarrow{+e} Fe(II) \qquad (12)$$

More than an order of magnitude enhancement was observed when the dye was incorporated into the polymer film.[66] Several conceivable explanations for such an enhancement include the possibility of attaining a high dye concentration at the electrode surface (up to 0.1 M) without aggregation, an extended lifetime of the excited state of the dye in the polymer matrix, the stabilization of mechanistically important radical-ions within the charged film, the promotion of mediated charge transfer through the film, and/or the spatial arrangement of useful concentrations of the dye under conditions where self-quenching can be limited. Macromolecular thionine films have also been employed to study the beneficial effect of polymer environment in the generation of the photogalvanic effect.[67,68]

B. IMAGING PROCESSES

Imaging with photoreactive polymers has drawn considerable attention toward the goal of developing photoresists to meet the demands of the electronic industry.[69] Photosensitive polymers, when applied as films to substrates such as semiconductor wafers and subjected to light irradiation through a mask, provide the necessary doping, etching, or plating of a pattern on the surface. Intramolecular changes of functionality or intermolecular interaction as a result of photoexcitation of these polymeric systems can lead to increased solubility or insolubility via crosslinking. Several different sensitizers have been used in the polymers to absorb light and to transfer energy to the photoreactive groups in the polymer. Since the efficiency of the photoinduced reaction depends on the relevant structure-physical property relationships, it can be maximized by a suitable structural modification. Various other aspects which influence the photochemical reactivity of photoresistive materials have been discussed by Williams et al.[70] Photoimaging is an important process with a wide range of applications, and major developments can be anticipated in the near future.

C. PHOTODEGRADATION AND PHOTOSTABILIZATION

Extensive work has been carried out to study the mechanism and kinetics of photodegradation of polymers, details of which can be found elsewhere.[71,72] The ketone group in synthetic polymers (e.g., polyphenylvinyl ketone or polymethylvinyl ketone) is known to play an important role, as it absorbs near UV and initiates photooxidation. The first chemical step in photodegradation is usually a homolytic bond scission to form the free radicals via Norrish type I or II reactions.[71]

Degradation of polymers can also be achieved by incorporating sensitizers as guest molecules into polymeric films to initiate oxidation by absorbing light. Aromatic ketones are useful in initiating the photooxidation of several polymers.[73,74] For example, photosensitized degradation of poly(α-methylstyrene) by benzophenone has been carried out by Ikeda et al.[74] (Reactions 13 to 15).

$$BP \xrightarrow{h\nu} {}^1BP^* \longrightarrow {}^3BP^* \tag{13}$$

$$ {}^3BP^* + PH \rightarrow {}^{\cdot}BPH + P^{\cdot} \tag{14}$$

$$P^{\cdot} \rightarrow P_m + P_n^{\cdot} \tag{15}$$

Short-lived radical species generated by hydrogen abstraction from a main chain methylene group of the polymer by excited benzophenone triplet have been characterized by spin trapping. Laser flash photolysis could also be an excellent technique for establishing the mechanism of photodegradation.[75,76]

Light-initiated autooxidation of the polymers has also been observed in air. A variety of components such as oxygen, ozone, oxides of nitrogen, and sulfur dioxide can initiate the oxidation process. Such polymer degradation can be prevented either by screening the polymer from UV damage or by quenching the photo-excited state.[77] Many nickel complexes have been used as UV stabilizers since they have the ability to quench excited states, in particular, carbonyl triplets and singlet oxygen. Multicomponent systems of antioxidants and UV stabilizers which reinforce one another by complementary mechanisms could bring about the necessary stabilization in the polymer. Increasing demand for light-stable polymers in practical applications necessitates a deeper understanding of the problem of photodegradation and photostabilization.

D. LUMINESCENT SOLAR CONCENTRATORS

A luminescent solar concentrator (LSC) consists of a transparent sheet of polymer such as PMMA (polymethylmethacrylate) impregnated with a highly fluorescent guest dye molecule. The purpose of LSC is to convert part of the incident solar irradiation into an emission with a high concentration factor, which is suitable for the operation of a photovoltaic cell. The light pipe trapping luminescence obtained with the LSC avoids the problem associated with the sunlight-tracking to achieve high flux gains and to minimize the cost of photovoltaic converters. Photophysical properties of coumarin, rhodamine, and oxazine dyes have been studied in detail by Zewail and Batchelder[78,79] as a means to evaluate their performance in the LSC. The photostability of dyes and the energy loss associated with Stokes shifts could be the limiting factors in attaining higher efficiency. A detailed discussion of the theoretical and experimental concerns associated of LSC can be found in References 78 to 82.

ACKNOWLEDGMENT

The research work, both at Notre Dame Radiation Laboratory and at The University of Texas at Austin was supported by the Office of Basic Energy Sciences of the U.S. Department of Energy. This is Document No. SR-111 from the Notre Dame Radiation Laboratory.

REFERENCES

1. **Guillet, J.**, *Polymer Photophysics and Photochemistry*, Press Syndicate of the University of Cambridge, 1985.
2a. **Holden, D. A. and Guillet, J. E.**, Singlet electronic energy transfer in polymers containing naphthalene and anthracene chromophores, *Macromolecules*, 13, 289, 1980.
2b. **Aspler, J. S., Hoyle, C. E., and Guillet, J. E.**, Singlet energy transfer in a 1-naphthylmethacrylate-9-vinylanthracene copolymer, *Macromolecules*, 11, 925, 1978.
2c. **Hargreaves, J. S. and Webber, S. E.**, Excited state properties of poly(2-vinylnaphthalene) containing pyrene groups, *Macromolecules*, 15, 424, 1982.
3a. **Guillet, J. E., Sherren, J., Gharapetian, H. M., and MacInnis, W. K.**, Prospects for solar synthesis. I. A new method for singlet oxygen reactions using natural sunlight, *J. Photochem.*, 25, 501, 1984.
3b. **Holden, D. A., Ren, X.-X., and Guillet, J. E.**, Studies of the antenna effect in polymer molecules. VII. Singlet and triplet energy migration and transfer in 2-vinylnaphthalene-phenyl vinyl ketone copolymers, *Macromolecules*, 17, 1500, 1984.
4a. **Phillips, D., Roberts, A. J., and Soutar, I.**, Transient decay studies of photophysical processes in aromatic polymers. II. Investigation of intramolecular excimer formation in copolymers of 1-vinyl naphthalene and methyl acrylate, *Polymer*, 22, 427, 1981.
4b. **Gupta, A., Liang, R., Mocanin, J., Kliger, D., Goldenbeck, R., Horwitz, J., and Miskowski, V. M.**, Internal conversion in poly(1-vinylnaphthalene). IV. Photochemical processes in polymeric systems, *Eur. Polym. J.*, 17, 485, 1981.
4c. **De Schryver, F. C., Demeyer, K., Van der Auweraer, M., and Quanten, E.**, Excimer formation of poly- and dichromophoric molecules in solution, *Ann. N.Y. Acad. Sci.*, 93, 366, 1981.
4d. **Ng, D. and Guillet, J. E.**, Interpretation of the excimer kinetics of poly(1-vinylcarbazole) and 1,3-dicarbazolyl-propane in dilute solutions, *Macromolecules*, 14, 405, 1981.
4e. **Ghiggino, K. P., Archibald, D. A., and Thistlewaite, P. J.**, Picosecond fluorescence studies of poly(N-vinylcarbazole), *J. Polym. Sci., Polym. Lett. Ed.*, 18, 673, 1980.
4f. **Morawetz, H.**, Some applications of fluorimetry to synthetic polymer studies, *Science*, 203, 405, 1979.
5. **Johnson, G. E. and Good, T. A.**, Molecular weight dependent emission properties of poly(N-vinylcarbazole) in polymer blends, *Macromolecules*, 15, 409, 1982.
6a. **Winnik, M. A., Redpath, T., and Richards, D. H.**, The dynamics of end-to-end cyclization in polystyrene probed by pyrene excimer formations, *Macromolecules*, 13, 328, 1980.
6b. **Winnik, M. A., Redpath, T., Paton, K., and Danhelka, J.**, Cyclization dynamics of polymers. X. Synthesis, fractionation and fluorescent spectroscopy of pyrene end-capped polystyrenes, *Polymer*, 25, 91, 1984.
7. **Semerak, S. N. and Frank, C. W.**, Excimer fluorescence as a molecular probe of blend miscibility. III. Effect of molecular weight of the host matrix, *Macromolecules*, 14, 443, 1981.
8. **Kamat, P. V.**, Fluorescence emission as a probe to investigate electrochemical polymerization of 9-vinylanthracene, *Anal. Chem.*, 59, 1636, 1987.
9. **Kalyansundaram, K.**, *Photochemistry in Microheterogeneous Systems*, Academic Press, New York, 1987, 255.
10. **Kamat, P. V. and Fox, M. A.**, Enhanced fluorescence emission of croconate violet in ethanol containing poly(4-vinylpyridine), *Chem. Phys. Lett.*, 92, 595, 1982.

11. **Kamat, P. V. and Fox, M. A.,** Triplet state properties of croconate dyes in homogeneous and polymer containing solutions, *J. Photochem.,* 24, 285, 1984.

12. **Kamat, P. V., Fox, M. A., and Fatiadi, A. J.,** Dye loaded polymer electrodes. II. Photoelectrochemical sensitization by croconate violet in polymer films, *J. Am. Chem. Soc.,* 106, 1191, 1984.

13. **Paczkowski, J. and Neckers, D. C.,** Polymer-based sensitizers for the formation of singlet oxygen. New studies of polymeric derivatives of rose bengal, *Macromolecules,* 18, 1245, 1985.

14. **Gibson, H. W. and Bailey, F. C.,** Chemical modification of polymers. XVII. Dyeing of sulfonated polystyrene films by ion-exchange with cationic dyes, *Polymer,* 22, 1068, 1981.

15a. **Takagishi, T. and Kuroki, N.,** Interaction of polyvinyl pyrrolidone with methyl orange and its homologs in aqueous solution: thermodynamics of the binding equilibria and their temperature dependence, *J. Polym. Sci., Polym. Chem. Ed.,* 11, 1889, 1973.

15b. **Takagishi, T., Nakagami, K., Imajo, K., and Kuroki, N.,** Interaction of polyvinylpyrrolidone with methyl orange and its homologs in aqueous solution over the temperature range 60—90°C, *J. Polym. Sci. Polym. Chem. Ed.,* 14, 923, 1976.

16. **Negishi, N., Ishihara, K., and Shinohara, I.,** Adsorption of photochromic azo dye onto styrene-divinylbenzene copolymer, *J. Polym. Sci. Polym. Lett. Ed.,* 19, 593, 1981.

17. **Sah, R. E., Baur, G., and Kelker, H.,** Influence of the solvent matrix on the overlapping of the absorption and emission bands of solute fluorescent dyes, *Appl. Phys.,* 23, 369, 1980.

18. **Kamat, P. V. and Fox, M. A.,** Photochemistry and photophysics of xanthene dyes in polymer solutions and films, *J. Phys. Chem.,* 88, 2297, 1984.

19. **Winkworth, A. C., Osborne, A. D., and Porter, G.,** Viscosity dependent internal rotation in polymethine dyes measured by picosecond fluorescence spectroscopy, in *Picosecond Phenomena III, (Springer Ser. Chem. Phys.,* Vol. 23), Eisenthal, K. B., Hochstrasser, R. M., Kaiser, W., and Laubereau, A., Eds., Springer-Verlag, New York, 1982, 228.

20. **Sundström, V. and Gillbro, T.,** Viscosity dependent radiationless relaxation rate of cyanine dyes. A picosecond laser spectroscopic study, *Chem. Phys. Lett.,* 61, 257, 1981.

21. **Lee, P. C. and Meisel, D.,** Photophysical studies of pyrene incorporated in Nafion membranes, *Photochem. Photobiol.,* 41, 21, 1985.

22. **Snare, M. J., Tan, K. L., and Treloar, F. E.,** Fluorescence studies of solubilization and dye binding in hypercoiled poly(methacrylic acid): a connected cluster model, *Macromol. Sci. Chem.,* A17, 189, 1982.

23. **Arora, K. S. and Turro, N. J.,** Photophysical investigations of interpolymer interactions in solutions of a pyrene substituted poly(acrylic acid), poly(vinylamine), poly(1-amino-acrylic acid), and poly(1-acetylaminoacrylic acid), *J. Polym. Sci., Polym. Physics, Ed.,* 25, 243, 1987.

24. **Jones, G., II, Goswami, K., and Halpern, A. M.,** Upper state emission for crystal violet under conditions of specific binding to poly(acrylic acid), *Nouv. J. Chim.,* 9, 647, 1985.

25. **Fleming, G. R., Knight, A. W. E., Morriv, J. M., Morrison, R. J. S., and Robinson, G. W.,** Picosecond fluorescence studies of xanthene dyes, *J. Am. Chem. Soc.,* 99, 4306, 1977.

26. **Cramer, L. E. and Spears, K. G.,** Hydrogen bond strengths from solvent-dependent lifetimes of rose bengal dye, *J. Am. Chem. Soc.,* 100, 224, 1978.

27. **Yu, W., Pellegrino, F., Grant, M., and Alfano, R. R.,** Subnanosecond fluorescence quenching of dye molecules in solution, *J. Chem. Phys.,* 67, 1766, 1977.

28. **Ford, W. E. and Kamat, P. V.,** Photochemistry of 3,4,9,10-perylenetetracarboxylic dianhydride dyes. III. Singlet and triplet excited-state properties of the bis(2,5-di-*tert*-butylphenyl)imide derivative, *J. Phys. Chem.,* 91, 6373, 1987.

29. **Masuhara, H., Mataga, N., Tazuke, S., Murao, T., and Yamazaki, I.,** Time-resolved internal reflection fluorescence spectroscopy of polymer films, *Chem. Phys. Lett.,* 100, 415, 1983.

30. **DeSilvestri, S., Weiner, A. M., Fujimoto, J. G., and Ippen, E. P.,** Femtosecond dephasing studies of dye molecules in a polymer host, *Chem. Phys. Lett.,* 112, 195, 1984.

31. **Fox, R. B., Price, T. R., and Cozzens, R. F.,** Photophysical processes in polymers. III. Properties of triplet energy traps in 1-vinylnaphthalene copolymers, *J. Chem. Phys.,* 54, 79, 1971.

32. **Jones, P. F. and Callowayu, A. R.,** Temperature effects on the luminescence of benzophenone in polymers, *J. Am. Chem. Soc.,* 92, 4997, 1970.

33. **Melhuish, W. H. and Hardwick, R.,** Lifetime of triplet state of anthracene in lucite, *Trans. Faraday. Soc.,* 58, 1908, 1962.

34. **Packnowski, J. and Neckers, D. C.,** Photochemical properties of rose bengal. XI. Fundamental studies in heterogeneous energy transfer, *Macromolecules,* 18, 2412, 1985.

35. **Kamat, P. V. and Lichtin, N. N.,** Properties of the triplet state of N,N,N',N'-tetraethyloxonine, *Isr. J. Chem.,* 22, 113, 1982.

36. **Kamat, P. V. and Lichtin, N. N.,** Electron transfer in the quenching of protonated triplet thionine and methylene blue by ground state thionine, *J. Photochem.,* 18, 197, 1982.

37. **Farid, S., Martic, P. A., Daly, R. C., Thompson, D. R., Specht, D. P., Hartman, S. E., and Williams, J. L. R.,** Selected aspects of photochemistry in polymer media, *Pure Appl. Chem.,* 51, 241, 1979.

38. **Porter, G. and Archer, M. D.,** In vitro photosynthesis, *Interdiscip. Sci. Rev.,* 1, 119, 1976.
39. **Guillet, J. E.,** Studies of energy transfer and molecular mobility in polymer photochemistry, *Pure Appl. Chem.,* 49, 249, 1977.
40. **Kim, N. and Webber, S. E.,** Photophysics of films of poly(2-vinylnaphthalene) doped with pyrene and tetracyanobenzene, *Macromolecules,* 15, 430, 1982.
41. **Klöpffer, W.,** Energy transfer and trapping in rigid solutions of aromatic polymers, *Spectrosc. Lett.,* 11, 863, 1978.
42. **Webber, S. E.,** Quenching of triplet excitons in poly(2-vinylnaphthalene) by biacetyl, *J. Photochem.,* 9, 269, 1978.
43. **Owen, E. D.,** Intermolecular energy transfer in polymers in *Developments in Polymer Photochemistry, Part 1,* Allen, N. S., Ed., Applied Science Publishers, Essex, England, 1980, 1.
44a. **Shirai, M., Ohyabu, M., and Tanaka, M.,** Interaction between dyes and polyelectrolytes. XII. Excitation energy transfer between bound dyes, *J. Polym. Sci., Polym. Chem. Ed.,* 190, 1847, 1981.
44b. **Shirai, M., Ohyabu, M., Ono, Y., and Tanaka, M.,** Interaction between dyes and polyelectrolytes. XIII. Effect of dye aggregation on the excitation energy transfer between bound dyes, *J. Polym. Sci., Polym. Chem. Ed.,* 19, 555, 1982.
44c. **Shirai, M., Ono, Y., and Tanaka, M.,** Interaction between dyes and polyelectrolytes. XIV. Triplet energy transfer in a cationic dye-polyanions system, *Makromol. Chem.,* 184, 153, 1983.
45a. **Ranby, B. and Rabek, J.,** *Singlet Oxygen: Reactions with Organic Compounds and Polymers,* John Wiley & Sons, New York, 1978.
45b. **Wasserman, H. H. and Murray, R. W.,** *Singlet Oxygen,* Academic Press, New York, 1979.
46. **Byteva, I. M., Gurinovich, G. P., Golomb, O. L., and Karpov, V. V.,** Luminescence of singlet oxygen in polymer films, *Opt. Spectrosk. (USSR),* 56, 923, 1984.
47. **Lee, P. C. and Rodgers, M. A. J.,** Kinetic properties of singlet oxygen in a polymeric microheterogeneous system, *J. Phys. Chem.,* 88, 4385, 1984.
48. **Ogilby, P. R., Iu, K.-K., and Clough, R. L.,** The photosensitized production of singlet molecular oxygen ($^1\Delta_gO_2$) in a solid organic polymer glass: a direct time resolved study, *J. Am. Chem. Soc.,* 109, 4746, 1987.
49. **Paczkowski, J. and Neckers, D. C.,** Polymer-based sensitizers for the formation of singlet oxygen, in *Organic Phototransformations in Nonhomogeneous Media,* Fox, M. A., Ed., (ACS Symp. Ser., Vol. 278), American Chemical Society, Washington, D.C., 1985, 223.
50. **Neckers, D. C.,** Heterogeneous sensitizers based on Merrifield copolymer beads, in *Reactive Polymers,* Vol. 3, Elsevier, Amsterdam, 1985, 277.
51. **Matsuo, T.,** Photoinduced electron transfer reaction in membranes and polymer systems for solar energy conversion, *Pure Appl. Chem.,* 54, 1693, 1982.
52a. **Gopidas, K. R. and Kamat, P. V.,** Photochemistry in polymers. Photoinduced electron transfer between Phenosafranin and triethylamine in perfluorosulfonate membrane, *Macromolecules,* submitted.
52b. **Sassoon, R. E.,** Improved energy storage in a two-polyelectrolyte system, *J. Am. Chem. Soc.,* 107, 6133, 1985.
53a. **Brugger, P.-A., Infelta, P. P., Braun, A. M., and Grätzel, M.,** Photoredox reactions in functional micellar assemblies. Use of amphiphillic redox relays to achieve light energy conversion and charge separation, *J. Phys. Chem.,* 103, 320, 1981.
53b. **Brugger, P.-A., Grätzel, M., Guarr, T., and McLendon, G. J.,** "Zwitterion" mediator/quenchers. Columbic minimization of the back-reaction in photocatalysis, *J. Phys. Chem.,* 86, 944, 1982.
54a. **Calvin, M.,** Simulating photosynthetic quantum conversion, *Acc. Chem. Res.,* 11, 369, 1978.
54b. **Fendler, J. H.,** Surfactant vesicles as membrane mimetic agents: characterization and utilization, *Acc. Chem. Res.,* 13, 7, 1980.
54c. **Ford, W. E. and Tollin, G.,** Chlorophyll photosensitized vectorial electron transport across phospholipid vesicle bilayers: kinetics and mechanism, *Photochem. Photobiol.,* 38, 429, 1983.
55a. **Kiwi, J. and Grätzel, M.,** Dynamics of light-induced redox processes in microemulsion systems, *J. Am. Chem. Soc.,* 100, 6314, 1978.
55b. **Atik, S. and Thomas, J. K.,** Photoinduced reactions in polymerized microemulsions, *J. Am. Chem. Soc.,* 105, 4515, 1983.
56. **Masuhara, H.,** An introduction to transient absorption spectroscopy and nonlinear photochemical behavior of polymer systems, in *Photophysical and Photochemical Tools in Polymer Science,* Winnik, M. A., Ed., D. Reidel, Dordrecht, Netherlands, 1986, 43 and 65.
57. **Schnabel, W.,** Laser flash photolysis study of the dynamics of polymers in solution, in *Photophysics of Synthetic Polymers,* Phillips, D. and Roberts, A. J., Eds., Science Reviews, England, 1982, 55.
58a. **Gong, S. S., Christensen, D., Zhang, J., and Wang, C. H.,** Holographic method for the investigation of the photochemical processes of the methyl red in poly(methylmethacrylate) and polystyrene, *J. Phys. Chem.,* 91, 4504, 1987.
58b. **Wilkinson, F., Willsher, C. J., and Pritchard, R. B.,** Laser flash photolysis of dyed fabrics and polymers. I. Rose bengal as a photosensitizing dye, *Eur. Polym. J.,* 21, 333, 1985.

58c. **Weir, D. and Scaiano, J. C.,** Study of the excited states of Nafion incorporated xanthone and benzophenone, *Tetrahedron,* 43, 1617, 1987.

59. **Nosaka, Y., Kuwabara, A., and Miyama, H.,** Effect of ionic polymer environment on the photoinduced electron transfer from zinc porphyrin to viologen, *J. Phys. Chem.,* 90, 1465, 1986.

60. **Kojima, K., Nakahira, T., Kosuge, Y., Saito, M., and Iwabuchi, S.,** Studies on polymeric sensitizers. II. Photocatalytic activities of porphyrins trapped in polymer matrices, *J. Polym. Sci., Polym. Lett. Ed.,* 19, 193, 1981.

61. **Guha, S. N., Naik, D. B., and Moorthy, P. N.,** Photoredox reactions in thionine-ferrous system in macromolecular environments, *Ind. J. Chem.,* 26A, 99, 1987.

62. **Staerk, H., Mitzkus, R., Kühnle, W., and Weller, A.,** Picosecond studies of intramolecular charge transfer processes in excited A-D molecules, in *Picosecond Phenomena, Part III,* (Springer Ser. Chem. Phys., Vol. 23), Eisenthal, K. B., Hochstrasser, R. M., Kaiser, M., and Laugereau, Eds., Springer-Verlag, New York, 1982, 205.

63. **Hiratsuka, H., Sekiguchi, K., Hatano, Y., Tanizaki, Y., and Mori, Y.,** Polarized absorption spectra of radical ions of some azanaphthalenes and biphenyls in stretched polymer films, *Can. J. Chem.,* 65, 1185, 1987.

64a. **Faulkner, L. R.,** Chemical microstructures on electrodes, *Chem. Eng. News,* 62, 28, February 27, 1984.

64b. **Murray, R. W.,** Chemically modified electrodes, *Acc. Chem. Res.,* 13, 35, 1980.

64c. **Krishnan, M., Zhang, X., and Bard, A. J.,** Polymer films on electrodes. XIV. Spectral sensitization of n-type SnO_2 and voltammetry at electrodes modified with Nafion films containing $Ru(bpy)_3^{2+}$, *J. Am. Chem. Soc.,* 106, 7371, 1984.

64d. **Kamat, P. V. and Fox, M. A.,** Dye loaded polymers on semiconductor electrodes. I. The electrochemical behavior of n-SnO_2 modified by adsorption of poly(4-vinyl-pyridine) films containing an anionic dye, *J. Electroanal. Chem.,* 159, 49, 1983.

64e. **Kamat, P. V., Basheer, R., and Fox, M. A.,** Polymer modified electrodes. Electrochemical and photoelectrochemical polymerization of 1-vinylpyrene, *Macromolecules,* 18, 1366, 1985.

65. **Kamat, P. V. and Chauvet, J.-P.,** *Photochem. Photobiol.,* submitted.

66. **Kamat, P. V. and Fox, M. A.,** Dye loaded polymer electrodes. III. Photogalvanic effect in poly(4-vinylpyridine) films coated on SnO_2 electrodes and incorporated with rose bengal, *J. Electrochem. Soc.,* 131, 1032, 1984.

67. **Tamilarasan, R. and Natarajan, P.,** Photovoltaic conversion by macromolecular thionine films, *Nature,* 292, 224, 1981.

68. **Shigehara, K., Sano, H., and Tsuchida, E.,** Photogalvanic effect of thin-layer photocells composed of thionine-polymer/Fe(II) systems, *Makromol. Chem.,* 179, 1531, 1978.

69. Improved photoresists for integrated circuit chips devised, *Chem. Eng. News,* 63, October 7, 1985, p. 27.

70. **Williams, J. L. R., Farid, S. Y., Doty, J. C., Daly, R. C., Specht, D. P., Searle, R., Borden, D. G., Chang, H. J., and Martic, P. A.,** The design of photoreactive polymer systems for imaging processes, *Pure Appl. Chem.,* 49, 533, 1977.

71. **Grassie, N. and Scott, G.,** *Polymer Degradation and Stabilization,* Cambridge University Press, Cambridge, 1985, 78.

72. **Ranby, B. and Rabek, J. F.,** *Photodegradation, Photo-Oxidation and Photostabilization of Polymers,* John Wiley & Sons, New York, 1975, 165.

73. **Geuskens, G., Delaunois, G., Lu-Vinh, Q., Piret, W., and David, C.,** Photooxidation of polymers. VIII. The photooxidation of polystyrene containing aromatic ketones, *Eur. Polym. J.,* 18, 387, 1982.

74. **Ikeda, T., Yamaoka, H., Matsuyama, T., and Okamura, S.,** Transient species during photosensitized degradation of poly(α-methylstyrene) by benzophenone, *J. Phys. Chem.,* 82, 2329, 1978.

75. **Todesco, R. V. and Kamat, P. V.,** Excited state behavior of poly[dimethyl-comethyl-(1-naphthyl)silylene], *Macromolecules,* 19, 196, 1986.

76. **Todesco, R. V., Basheer, R. A., and Kamat, P. V.,** Photophysical and photochemical properties of poly(1-vinylpyrene). Evidence for dual excimer fluorescence emission, *Macromolecules,* 19, 2390, 1986.

77. **Scott, G.,** Photodegradation and photostabilization of polymers, *J. Photochem.,* 25, 83, 1984.

78. **Zewail, A. H. and Batchelder, J. S.,** Luminescent solar concentrators: an overview, in *Polymers in Solar Energy Utilization,* Gebelein, C. G., Williams, D. J., and Deanin, R. D., Eds., (ACS Symposium Series, No. 220), American Chemical Society, Washington, D.C., 1983, 331.

79. **Batchelder, J. S., Zewail, A. H., and Cole, T.,** Luminescent solar concentrators. II. Experimental and theoretical analysis of their possible efficiencies, *Appl. Optics,* 20, 3733, 1981.

80. **Goetzberger, A. and Wittwer, V.,** Fluorescent planner collector-concentrators for solar energy conversion, in *Festkörperprobleme,* (Advances in Solid State Physics Ser., Vol. 19), Treusch, J., Ed., Vieweg, Braunschweig, 1979, 427.

81. **Weber, W. H. and Lambe, J.,** Luminescent greenhouse collector for solar radiation, *Appl. Optics,* 15, 2299, 1976.

82. **Meseguer, F. J., Cusso, F., Jaque, F., and Sanchez, C.** Temperature effects on the efficiency of luminescent solar concentrator (LSC) for photovoltaic systems, *J. Luminescence,* 24/25, 865, 1981.

Chapter 9

DIFFUSE REFLECTANCE LASER FLASH PHOTOLYSIS OF DYED FABRICS AND POLYMERS

Francis Wilkinson and Charles John Willsher

TABLE OF CONTENTS

I. INTRODUCTION

The need to understand primary photochemical processes occurring within irradiated polymers in the presence and absence of dissolved or chemically attached additives is of the upmost importance to an increasing number of scientists working on technological applications with plastics, rubbers, fabrics, coatings, etc. There is considerable interest in photopolymerization, photodegradation, photosensitized degradation, photostabilization, as well as in the optical and luminescence properties of irradiated polymers in the presence and absence of additives. A great deal of progress has been made recently in this area as witnessed by the appearance of several books dealing with various aspects of this subject.[1-3]

One of the most powerful methods for studying primary photochemical processes is the technique of flash photolysis. This technique combined with steady-state and time-dependent luminescence measurements has been applied to many systems and allowed an insight into the mechanism of several primary photochemical processes. However, as normally applied, flash photolysis is restricted to homogeneous transparent samples since the time-dependent spectral information is obtained using transmission spectroscopy as the analyzing technique. Thus, many polymer studies have been made using model systems, for example, using dilute fluid solutions and the relevance of such studies to the primary processes which occur in the solid polymer needs to be established. This chapter discusses the method of diffuse reflectance laser flash photolysis (DRLFP) which extends to opaque, heterogeneous, and often highly scattering samples, the advantages of being able to subject them to flash photolysis investigations by using diffusely reflected light in place of transmitted light as the analyzing source on time scale extending from several seconds[4] to picoseconds.[5]

In our first experiments using this technique in 1981 we were able to record triplet-triplet absorption spectra of fractions of monolayers of aromatic hydrocarbons adsorbed on γ-alumina powder with lifetimes greater than 10 ms.[4] Since then we have made many studies using a nanosecond laser as the excitation source, and obtained photoinduced transient spectra and decay kinetics from a wide variety of opaque samples, including fractions of monolayers of organic molecules adsorbed on catalytic oxide surfaces[4-6] and included within the hydrophobic zeolite silicalite,[7] semiconductor powders and sintered porous electrodes doped and undoped,[8,9] organic microcrystals,[10,11] powdered inorganic phosphors,[12] and dyes adsorbed on fabrics and chemically bound to polymers.[13,14] We have also successfully demonstrated using pump and probe techniques at the Rutherford Appleton Laboratory that DRLFP can be extended to picosecond time scales.[5]

In this review we discuss some of our recent results obtained using nanosecond laser excitation of dyed fabrics and various polymers. The technique is particularly useful for mechanistic investigation of opaque polymer samples where transmission flash photolysis is not possible. It needs to be stressed that the data obtained with this arrangement for opaque samples are often comparable to those obtained using transmission flash photolysis for transparent samples. The same apparatus allows time-dependent luminescence studies which can often be of considerable help in the assignment of the transient species observed.

II. PRINCIPLES OF THE METHOD

A. DIFFUSE REFLECTANCE

When a ray of light is incident on a plane transparent surface it is well known that part of the light is reflected with the angle of reflection equal to the angle of the incidence. This is termed specular or mirror reflection, while the rest of the light is refracted or transmitted at an angle given by Snell's Law such that $\sin\Theta_i/\sin\Theta_r = n$ where n is the relative refractive index of the media and Θ_i and Θ_r are the angles of incidence and refraction, respectively.

However, when light is incident on a matt surface composed, for example, of a fine powder, much of the reflected light and the transmitted light which penetrates below the surface is diffuse in nature, i.e., unpolarized and distributed symmetrically in the plane parallel to the surface and independent of the angle of incidence or the polarization of the incident light. This diffuse light arises from multiple scattering of reflected and refracted light from the individual particles. The diffuse light effectively penetrates into the interior of the particles and can be attenuated by absorption within the particles. It is interesting to note that the specular and the diffuse reflected light give opposite behavior in the region of an absorption band, that is, an increase in absorption coefficient leads to more specular but less diffuse reflection.

In the case of weakly absorbing finely divided powders and for dyes on fabrics, etc., where the light flux penetration of the sample rapidly becomes diffuse, the contribution from specular reflection is usually small, in which case the Kubelka Munk treatment can be applied. In this theory two light fluxes I and J are considered moving in opposite directions, perpendicular to the irradiated surface at x = 0. At the beginning of the irradiation the absorbers are assumed to be randomly distributed, and so it follows that the attenuation of the incident flux is given by Equation 1 and the generated flux J is given by Equation 2 where K and S are the absorption and scattering coefficients, respectively, with dimension, distance^{-1}. Note the change of sign in Equation 2 since this flux is moving in the opposite, i.e., the $(-x)$, direction.

$$dI = -I(K + S)dx + JSdx \qquad (1)$$

$$dJ = J(K + S)dx - ISdx \qquad (2)$$

The absorption coefficient K_A of an absorber A is related to the concentrations C_A and the Naperian extinction coefficient ϵ_A as follows:

$$K_A = 2\epsilon_A C_A \qquad (3)$$

If the incident intensity is I_0 at the surface and the intensity of the reflected light is J_0 the diffuse reflectance R is given by

$$R = J_0/I_0 \qquad (4)$$

If the layer is so thick that any further increase in thickness of the sample does not affect R it is said to be "optically thick" and for such samples the light flux falls off exponentially as it penetrates below the surface, i.e.,

$$I = I_0 exp(-bSx) \qquad (5)$$

and

$$J = RI_0 exp(-bSx) \qquad (6)$$

where

$$bS = \sqrt{(K^2 + 2KS)} \qquad (7)$$

In addition it follows that

$$(1 - R)^2/2R = K/S \qquad (8)$$

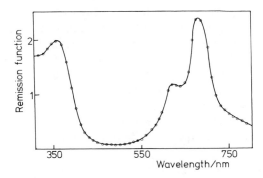

FIGURE 1.   Plot of remission function, $(1 - R)^2/2R$, against wavelength for cotton fabric dyed with aluminum(III) sulfonated phthalocyanine.

The expression on the left hand side of Equation 8 is known as the remission function. When S is independent of wavelength a plot of remission function against wavelength represents an absorption spectrum for a diffuse reflector (see Figure 1).

B. TRANSIENT ABSORPTION

Equations 5 to 8 are appropriate for randomly distributed absorbers in a diffuse reflector. However, in laser flash photolysis experiments the light absorbed from the laser causes photoconversion of the initial absorber, that is, a transient T is populated from the ground state species A, and then the distribution of A and T is not necessarily random. At wavelengths where ϵ_A is less than ϵ_T a transient reduction in diffuse reflectance is observed — this represents transient absorption — while in regions where ϵ_T is less than ϵ_A, enhanced reflection is observed due to ground state depletion. When $\epsilon_T = \epsilon_A$ the transient difference spectra show isosbestic points, where $\Delta R = 0$. If the yield of conversion to the transient is Ø and the incident intensity at the excitation wavelength is I_0^e, then for low percentage conversions it follows from Equations 5 and 6 that

$$\frac{dC_T}{dt} = K^e \emptyset I_0^e (1 + R^e) \exp(-b^e S x) \tag{9}$$

where the superscript e represents the value of the various parameters at the excitation wavelength. Equation 9 demonstrates that initially the transient concentration falls off exponentially with increasing penetration depth. For optically thick samples where a small percentage conversion of ground state molecules to transients takes place, a solution to Equations 1 and 2 is required in which K varies exponentially with the penetration depth x at analyzing wavelengths where only the transient absorbs. These equations have already been solved by Lin and Kan[16] who give

$$R^a = R_B^a \frac{1 + \dfrac{\gamma}{\delta} u + \dfrac{\gamma(\gamma + 1)u^2}{\delta(\delta + 1)2!} + \cdots}{1 + \dfrac{\gamma + 1}{\delta} u + \dfrac{(\gamma + 1)(\gamma + 2)u^2}{\delta(\delta + 1)2!} + \cdots} \tag{10}$$

where the superscript a represents the value of the various parameters at the analyzing wavelength and R_b^a represents the background reflectance and $\gamma = (b^e R_B^a)^{-1}$, $\delta = \gamma + 1 - R_B^a/b^e$ and $u = 2K^a(0)/b^e S$. This series converges for all values of u. Thus, kinetic analysis is possible for low surface percentage conversion using Equation 10 or for low percentage

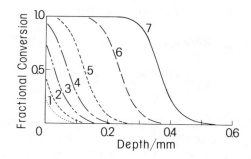

FIGURE 2. Calculated curves showing the effect of increasing laser fluence on the fractional conversion below an irradiated surface of a dye A adsorbed on a white powder with $S = 50$ cm^{-1}. For these calculations the initial concentration, c_A, was assumed to be 5×10^{-3} mol dm^{-3}, $\epsilon_A{}^e = 2.3 \times 10^4$ dm^3 mol^{-1} cm^{-1}, and ø, the quantum yield of conversion, equal to unity. The laser fluences for curves 1 to 7 are 5×10^{-10}, 10^{-9}, 2.5×10^{-9}, 5×10^{-9}, 10^{-8}, 2.5×10^{-8}, and 5×10^{-8} einsteins cm^{-2}, respectively.

absorption by using $1 - R_T$ as a function of time, since we have shown that for $(1 - R_T) < 0.1$, $(1 - R_T)$ is itself a linear function total of transient concentration.[17]

Another limiting condition is where the concentration of A is small compared with the laser fluence. In this case complete conversion occurs from A to T at the surface and for a considerable depth beneath the irradiated surface. Figure 2 shows typical depth profiles for various laser fluences used to excite a dye absorbed on a white powder. Thus, a single nanosecond laser pulse often establishes an essentially homogeneous layer of transient on top of a layer of unconverted ground state. In this limiting case kinetic analysis and spectra can be analyzed using Equation 8 or well-known modifications for two layers of homogeneous absorbing scatterers (see Reference 18). Between and including these limits the concentration profile can be calculated by a numerical iterative multithin-slice approach using Equations 11 and 12. The reflectance and transmission of a pair of layers i and j with light incident on layer i can be expressed in terms of the individual reflectances and transmittance of the layers, i.e.,

$$T_{i,j} = \frac{T_i T_j}{(1 - R_i R_j)} \tag{11}$$

and

$$R_{i,j} = R_i + \frac{T_i^2 R_j}{(1 - R_i R_j)} \tag{12}$$

where $(1 - R_i R_j)^{-1}$ is the sum of the infinite geometric series which accounts for the multiple scattering between the layers. Such calculations are discussed in Reference 19.

C. TRANSIENT DECAY

For a simple first-order decay

$$C_T = C_T(0) \exp(-k_1 t) \tag{13}$$

If either limiting case holds then it follows that for low percentage conversion

$$C_T \propto (1 - R_m/R_b) \tag{14}$$

while for an optically dense totally converted plug

$$C_T \propto \frac{(1 - R_m)^2}{2R_m} - \frac{(1 - R_b)^2}{2R_b} \tag{15}$$

where R_m and R_b are the reflectance of the transient at any time and of the background at the analyzing wavelength, respectively, and the first-order constant can be obtained by substitution of either Equations 14 or 15 as appropriate into Equation 13.

Second-order decay with

$$\frac{C_T(0)}{C_T} = 1 + k_2 C_T(0)t \tag{16}$$

would only be expected in the absence of concentration gradients in any direction. This might arise if energy, charge, or even material diffusion is fast relative to the transient decay, and in this case the transient would be randomly distributed throughout the irradiated sample; for optically thick samples Equation 15 can be substituted into Equation 16 for kinetic analysis. In all other cases there is no straightforward analysis method available for separating out contributions due to bimolecular reactions in the presence of concentration gradients in the irradiated sample.

D. TEMPERATURE JUMPS

For strongly absorbing samples where the laser penetration depth is only a few micrometers, temperature rises of several hundred degrees are expected and have been found experimentally in strongly absorbing crystals — Reference 20 gives a detailed discussion of thermal effects. The possibility that laser excitation can lead to large temperature rises needs to be borne in mind, but, provided that the penetration depth at the excitation wavelength is greater than 0.1 mm, the temperature rise at the irradiated surface calculated for fluences of 100 mJ cm^{-2} is not greater than 10 K even if all the laser energy is converted to heat. Thus, for most work on dyed fabrics large temperature jumps are not a problem.

III. EXPERIMENTAL

In the application of laser flash photolysis to opaque materials, virtually no analyzing light is transmitted through the sample, and this fact must be taken into account when designing apparatus with which to carry out experiments. As mentioned previously, an opaque substance generally gives both a specular and a diffuse reflection; the latter contains most of the transient information and can be differentiated from the former by its angular dependence. This must also be considered in the design of instrumentation. The apparatus to perform DRLPF studies is essentially the same as that required for transmission flash photolysis — indeed the apparatus in the authors' laboratory can be converted easily and quickly for transmission studies — but with modifications to account for the above characteristics of an opaque sample.

A. APPARATUS

The principle components of a laser flash photolysis apparatus comprise exciting and analyzing sources, a detector, and a storage device. In our case, the exciting source is a Nd-YAG laser (Lumonics Ltd.), of which the second and third harmonics are used in the experiments described later. The second harmonic gives intensities at 532 nm up to 200 mJ

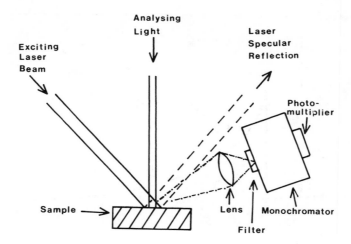

FIGURE 3.   Schematic diagram of sample, exciting and analyzing beams, and detector. The laser and monitoring beams are typically 0.7 and 0.5 cm in diameter, respectively.

in a pulse of about 20-ns duration, and the third harmonic at 354 nm provides intensities up to 70 mJ per pulse. In practice the lowest exciting intensity is used, such that a reasonable size of transient absorption can be obtained without causing significant thermal or photochemical damage to the sample. The analyzing source is a 250-W Xe arc lamp (Applied Photophysics Ltd.), which has the facility to give a high-intensity output pulse of 0.5-ms duration. However, many of the samples reported later have transient absorptions whose decay time is longer than the pulse width, and so the unpulsed output of the arc lamp must be used; in doing this a lower quality of signal is generally obtained, but this can often be overcome by signal averaging, provided no thermal or photochemical damage occurs to the sample on repeated excitation. The detector is an R928 photomultiplier (Hanamatsu Ltd.), with a rise time of about 2 ns, and the data are captured by a 7912AD programmable digitizer (Tektronix Ltd.). Digitized data are then transferred from the digitizer to floppy discs in a PDP/11 minicomputer (Digital Equipment Co. Ltd.) for storage and subsequent analysis. The operation of the apparatus is managed by the minicomputer which coordinates laser firing, arc lamp pulsing, shutter control, setting and reading the digitizer, etc.

B. THE DISPOSITION OF LIGHT BEAMS, SAMPLE, AND DETECTOR

Figure 3 shows detail of the experimental arrangement of light beams, sample, and detector. It must be stressed that this is not a unique arrangement, but one that has been adopted for practical purposes. There is no reason why, for example, colinear excitation and analysis should not be performed. The sample is held in a holder that is suited to the nature of the material under investigation. A powder is usually contained in a teflon holder and held under slight pressure against a quartz window. If it is necessary to deoxygenate the material, the sample is contained in a 1- or 2-mm quartz cuvette which is provided with gas entrance and exit needles. A cuvette is also used to study suspensions. A sample such as a dyed fabric can be contained in one of the above holders or, if it is rigid enough, no holder is necessary. The sample is positioned such that it receives the exciting laser beam at about 50° to the normal, while the analyzing light impinges on the sample at 90° — see Figure 3. The width of these beams and their positioning on the sample is arranged such that the excited area of the sample is larger than that receiving analyzing light. This is to ensure that all the reflected monitoring light will have probed the excited sample. The specular reflection of the exciting laser beam occurs at the angle of incidence, namely about

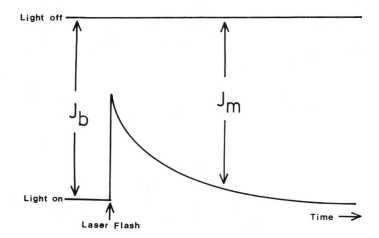

FIGURE 4.   Diagram showing a transient absorption and the measurements of diffusely reflected monitoring light which are necessary for spectral and kinetic analysis.

50° to the normal, and the monochromator is placed so that this reflection just misses the entrance slit. The filter in front of the monochromator is necessary to remove the diffuse reflection of the laser; for 354 nm an aqueous solution of sodium nitrite is used, and for 532 nm a solution of cobalt(II) sulfate is employed. If other materials or commercial filters are used, it is crucial that they are nonluminescent, since emission generated by the laser diffuse reflection can be strong enough to cause spurious signals resulting from fatigue and recovery of the photomultiplier. The specular reflection of the analyzing light occurs along the path of the input beam, and, as shown in Figure 3, a lens is positioned in order to focus a portion of the diffuse reflection onto the monochromator entrance slit. The use of an integrating sphere to gather all the diffusely reflected light failed to improve either the quantity of light detected or its quality (i.e., the signal-to-noise ratio). This is because (1) it is not possible to focus light from the sphere exit port onto the monochromator entrance slit, and (2) if the monochromator is placed between the arc lamp and the sphere, then scattered laser light and any luminescence from the sample overload the photomultiplier.

C. INTERPRETATION OF DIFFUSELY REFLECTED MONITORING LIGHT

The extraction of kinetic and spectral data from diffusely reflected monitoring light will now be considered. Figure 4 shows the measurements of reflected light which are made. J_m is the level of diffusely reflected monitoring light at a given time after excitation and J_b is the difference between light on and light off. If I_0 is the intensity of monitoring light received by the sample, then $I_0 = J_b$ only if it is a perfect scatterer, i.e., $R_b = 1$ (R_b is the background reflectance of the sample at the wavelength in question). In all other cases $I_0 = J_b/R_b$. The two limiting cases of transient population beneath the surface of an opaque sample, namely an exponentially falling off concentration and a homogeneous distribution (see earlier), need different approaches to relate transient concentration to J_m and J_b. The former case is the simpler one, in which calculations[17] show that in some cases the transient concentration is directly proportional to the size of the transient signal; this corresponds to $(J_b - J_m)$ in Figure 4. For spectral interpretation, the extinction coefficient of the transient species is proportional to $(J_b - J_m)$.

In the case of a homogeneous transient distribution, it is necessary to use the Kubelka-Munk function

$$(1 - R_m)^2/2R_m = (1 - R_t)^2/2R_t + (1 - R_b)^2/2R_b = (K_t + K_b)/S \qquad (17)$$

where R_m is the measured reflectance, R_t is the reflectance of the transient species, K_t and K_b are the absorption coefficients of the transient and background, respectively, and S is the scattering coefficient of the sample. Since $R_m = J_m/I_0$ and $I_0 = J_b/R_b$ an expression for R_m is obtainable in terms of the measured parameters, i.e.,

$$R_m = \frac{J_m R_b}{J_b} \qquad (18)$$

Inspection of Equation 18 shows that the background reflectance R_b (obtained from ground state spectroscopic measurements) is included in the expression which relates R_m to J_m and J_b. The concentration of transients, c_t, is related to the absorption coefficient through the equation $K_t = 2\epsilon_t c_t$. Thus, Equation 17 can now be written as

$$c_t = \frac{K_t}{2\epsilon_t} = \frac{S}{4\epsilon_t} \left[\frac{(1 - R_m)^2}{R_m} - \frac{(1 - R_b)^2}{R_b} \right] \qquad (19)$$

where ϵ_t is the extinction coefficient of the transient. When substitution for R_m from Equation 18 is made in Equation 19, then Equation 20 is obtained through which c_t is related to the measured levels of reflected light shown in Figure 4 and the background reflectance:

$$c_t = \frac{K_t}{2\epsilon_t} = \frac{S}{4\epsilon_t} \left[\frac{[1 - (J_m R_b)/J_b]^2}{(J_m R_b)/J_b} - \frac{(1 - R_b)^2}{R_b} \right] \qquad (20)$$

The constant $S/4\epsilon_t$ need not be known if the above expression is divided by the value of c_t when the laser fires, corresponding to time zero. Thus, first-order analysis is performed by plotting $\ell n\{c_t(\text{time} = t)/c_t(\text{time} = 0)\}$ vs. time, and second-order analysis by plotting $\{c_t(\text{time} = 0)/c_t(\text{time} = t)\}$ vs. time. It can be seen from Equations 19 and 20 that the extinction coefficient of the transient has the same proportionality to the measured parameters as the concentration, i.e.,

$$\epsilon_t \propto \left[\frac{[1 - (J_m R_b)/J_b]^2}{(J_m R_b)/J_b} - \frac{(1 - R_b)^2}{R_b} \right] \qquad (21)$$

IV. RESULTS

A. DYED FABRICS

It is very important to be able to understand the photochemical and photophysical behavior of the compounds employed to color natural and synthetic fibers, especially from a commercial point of view. Although it is possible to obtain useful information from ground state reflectance spectroscopy and luminescence studies of a dyed fabric, a complete investigation of the above properties has not been possible hitherto. It was necessary to undertake model studies of the dye in solution simply because transmission flash photolysis cannot be applied to a piece of cloth. The data obtained from model studies had then to be extrapolated to the environment within a woven fabric, which is quite different to a fluid solution. It is clearly more useful to be able to apply flash photolysis directly to a dyed fabric, and this is now possible through DRLFP. In this section, some results obtained from cotton fabric dyed with phthalocyanine and fluorescein derivatives will be discussed.

1. Cotton Fabric Dyed with Aluminum(III) Phthalocyanine (Sulfonated) (AlPcS)

Nonfluorescent woven cotton fabric dyed with AlPcS shows both transient absorption and ground state depletion following excitation at 354 nm (intensity = 15 mJ per pulse).[14] The spectral position and λ_{max} around 500 nm of the positive signal indicates that the transient absorption arises from the lowest triplet state of AlPcS, and the negative signal around 650 nm corresponds to depletion of the ground state. The decay of the absorption follows a first-order rate law (τ is in the region of 0.75 ms), in a water-free fabric, in the presence or absence of oxygen. Quenching by oxygen occurs only if the fabric is water saturated, but the overall quenching process is inefficient, since the decay of the absorption cannot be fitted as a pseudo-first-order process. A similar effect is observed if other potential quenchers are present in the fabric. It must be concluded that the rate of diffusion of molecules through the fibers is very much slower than in solution; this will be an important factor in determining the rate of reactions which may occur in the fabric between the dye and other molecules, such as brighteners or detergents.

2. Cotton Fabric Dyed with Halogenated Fluorescein Dyes

The triplet-triplet (T-T) absorption of halogenated fluorescein dyes (eosin, erythrosin, and rose bengal) contained in cotton fabric can be readily detected following excitation at either 354 or 532 nm. Figure 5A shows the transient difference spectrum for rose bengal in nonfluorescent cotton cloth. The positive part of the spectrum from 650 to 900 nm corresponds to absorption from the first triplet state of the dye, while the negative portion represents depletion of the ground state. The majority of the decay of the T-T absorption follows a first-order rate law (the half-life is in the region of 80 µs), but the initial portion of the decay can be attributed to a second-order process. Similar spectral and kinetic behavior is found for eosin and erythrosin dyed into cotton. As in the case of AlPcS cotton, water is necessary to observe partial quenching by oxygen of the triplet state of the fluorescein dyes. Figure 5B gives the laser-induced emission of rose bengal in cotton fabric; the feature at 765 nm corresponds to phosphorescence of the dye, and it has the same decay parameters as the T-T absorption. The smaller feature at 650 nm disappears more rapidly than the phosphorescence, and occurs at wavelengths where fluorescence from rose bengal would be expected. Its lifetime is too long for this feature to be prompt fluorescence, and it is more likely to be delayed fluorescence which results from T-T annihilation. This process would account for the fact that the beginning of the T-T absorption decay can be fitted to a second-order process. It is interesting to note that for rose bengal adsorbed onto different substrates, the relative sizes of the signals at 650- and 765-nm emission varies. When adsorbed on nylon or polyacrylamide, the emission spectrum appears as in Figure 5B, and the majority of the decay of the T-T absorption obeys a first-order rate law. With polystyrene as the substrate, the delayed fluorescence at 650 nm predominates and virtually no phosphorescence is detected. For this sample the T-T absorption is much shorter lived and decays almost entirely by a second-order rate law. It seems that polystyrene is a substrate which allows intermolecular T-T energy transfer to take place more readily than on the other adsorbents.

B. FUNCTIONALIZED POLYMERS

1. Benzoylated Polystyrene

2% divinylbenzene cross-linked polystyrene beads can be functionalized by Friedal-Crafts benzoylation to produce a polymer in which a given number of phenyl groups have been converted to the benzophenone moiety.[21] This insoluble photosensitizer has been successfully employed to bring about reactions such as 2 + 2-photocycloadditions and photooxidations.[21] The opaque, highly scattering nature of functionalized polystyrene beads make them ideal to study by DRLFP to investigate their photophysical behavior. Excitation at 354 nm of a sample where 1% of available phenyl groups are functionalized results in transient absorption, the time-resolved spectrum of which is given in Figure 6. The maximum at 530 nm could be assigned as either absorption from the lowest triplet state of benzophenone,

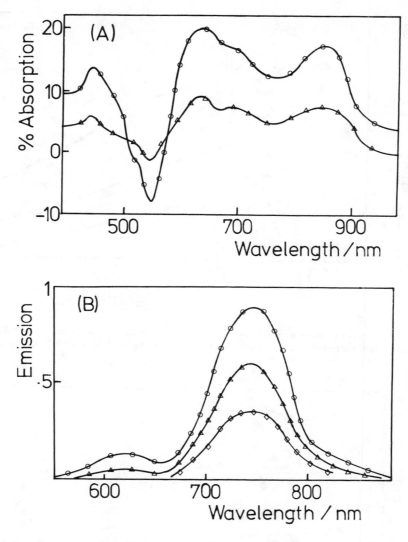

FIGURE 5. Rose bengal dyed into cotton fabric: (A) transient difference spectrum following pulsed excitation at 354 nm (40 mJ per pulse); -⊙- at flash, -△- after 50 μs. (B) Spectrum of emission generated by laser excitation at 354 nm; -⊙- at flash, -△- after 15 μs, -◇- after 50 μs.

or from the benzophenone ketyl radical. Laser-induced emission from this sample has a spectrum showing the three peaks characteristic of the phosphorescence of benzophenone, which confirms the triplet state to be one of the transient species produced on excitation at 354 nm. The decay kinetics of both the absorption and emission are identical, although somewhat complicated. However, if the percent functionalization is increased, the decay of the transient absorption becomes slower, while that of the emission is unchanged. It is believed that abstraction of benzilic hydrogens from the polystyrene support by the benzophenone triplet takes place at higher loadings, and this results in the formation of a radical that has a much longer lifetime than the triplet state.[22] Triplet energy transfer can be demonstrated if a diene quencher is included in the polymer chain. For example, in a sample that is 29% benzoylated and contains several units of the diene, the phosphorescence is quenched very efficiently but there is only a small effect on the transient absorption — the

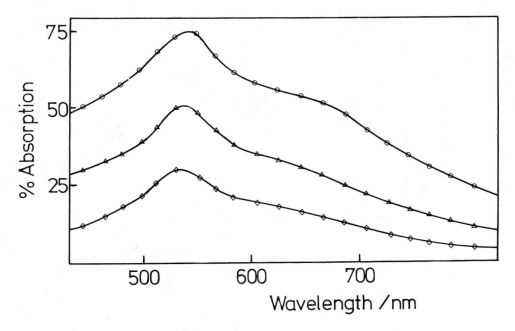

FIGURE 6.   Spectrum of transient absorption within deoxygenated benzoylated polystyrene beads (1% functionalization) after pulsed laser excitation at 354 nm (40 mJ per pulse); $\ominus$ at flash, $\triangle$ after 0.25 ms, $\diamond$ after 0.45 ms.

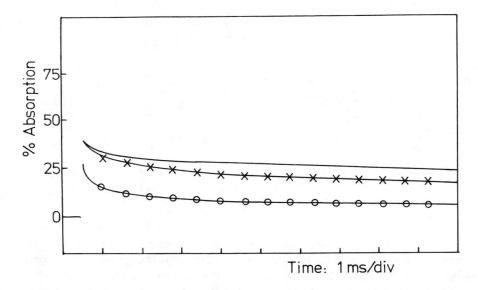

FIGURE 7.   Decay traces for deoxygenated benzoylated polystyrene beads (29% functionalization) after excitation at 354 nm (40 mJ per pulse), monitored at 510 nm; ----- no diene quencher attached, $\times$-5 available phenyl units functionalized with diene; $\ominus$ 20 units functionalized with diene.

latter phenomenon is illustrated in Figure 7. This is because the radical is the major transient species in this particular sample, and the low concentration of triplet therefore accounts for the very small quenching effect on the transient absorption. An important conclusion from photophysical studies of benzoylated polystyrene is that substantial radical formation takes place at high percentage functionalization, whereas the benzophenone triplet predominates

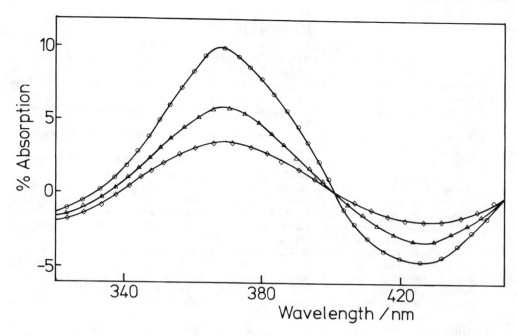

FIGURE 8.   Transient difference spectrum for deoxygenated $Ru(bpy)_3^{2+}$ covalently bound to polystyrene, following excitation at 532 nm (150 mJ per pulse); ⊖ after 100 ns, ▵ after 200 ns, ⋄ after 400 ns.

at low loadings. This is borne out by the experimental behavior of benzoylated polystyrene as a triplet photosensitizer, where a higher efficiency is observed with samples of lower percent functionalization.

2. Polymer-Bound Ruthenium(II) Tris-Bipyridyl, $Ru(bpy)_3^{2+}$

An enormous amount of research effort has been directed towards $Ru(bpy)_3^{2+}$, especially in its role as a photosensitizer which responds in the visible region of the spectrum. The complex has been employed both in homogeneous and heterogeneous systems, and we are interested in its photophysical behavior when linked to a polymeric support. The sample in question contains one of the three bipyridyl ligands covalently bound to polystyrene, and when suspended in a dioxan-water mixture transient absorption and depletion are recorded following laser excitation at 532 nm (see Figure 8). The absorption around 370 nm corresponds to $^*Ru(bpy)_3^{2+}$, and the negative part represents depletion of the MLCT ground state absorption. The sample also shows strong emission of the MLCT complex at 650 nm. If excitation is carried out at 354 nm, a very weak transient absorption is observed, the spectrum of which may be assigned as $Ru(bpy)_3^+$. The samples of polymer-bound $Ru(bpy)_3^{2+}$ are very dark in color, and illustrate that DRLFP is not suited to materials which have a very small background reflectance. With such samples signals of poor quality are often obtained and, in some cases, no signal at all can be recorded, especially at wavelengths where the detecter is not very sensitive.

ACKNOWLEDGMENTS

We wish to thank Drs. J. L. Bourdelande and J. Font of the Universitat Autònoma, Barcelona, Spain, for undertaking collaborative research work on the functionalized polymers, and C. J. Willsher wishes to thank the European Office of the U.S. Army for support under Contract No. DAJA 45-85-C-0010.

REFERENCES

1. **Guillet, J.**, *Polymer Photophysics and Photochemistry: An Introduction to the Study of Photoprocesses in Macromolecules*, Cambridge University Press, Cambridge, England, 1985, 391.
2. **Grassie, N. and Scott, G.**, *Polymer Degradation and Stabilisation*, Cambridge University Press, Cambridge, England, 1985, 222.
3. **Ranby, B. and Rabek, J. F.**, *Photodegradation, Photo-Oxidation and Photo-Stabilization of Polymers*, John Wiley & Sons, New York, 1975.
4. **Kessler, R. W. and Wilkinson, F.**, Diffuse reflectance triplet-triplet absorption spectroscopy of aromatic hydrocarbons chemisorbed on γ-alumina, *J. Chem. Soc., Faraday Trans. 1*, 77, 309, 1981.
5. **Wilkinson, F., Willsher, C. J., Leicester, P. A., Barr, J. R. M., and Smith, M. J. C.**, Picosecond diffuse reflectance laser flash photolysis, *J. Chem. Soc., Chem. Commun.*, 1216, 1986.
6. **Kessler, R. W., Oelkrug, D., and Wilkinson, F.**, Detection of transient spectra within polycrystalline samples using the new technique of diffuse reflectance flash photolysis, *Appl. Spectrosc.*, 36, 673, 1982.
7. **Wilkinson, F., Willsher, C. J., Casal, H. L., Johnston, L. J., and Scaiano, J. C.**, Intrazeolite photochemistry. IV. Studies of carbonyl photochemistry on the hydrophobic zeolite Silicalite using time-resolved diffuse reflectance techniques, *Can. J. Chem.*, 64, 539, 1986.
8. **Pouliquen, F., Fichou, D., Kossanyi, J., Willsher, C. J., and Wilkinson, F.**, Excited state emission and absorption within sintered doped zinc oxide, *Proc. 10th IUPAC Symp. Photochemistry*, Interlaken, Presses Polytechniques Romandes, Lausanne, Switzerland, 1984, 339.
9. **Wilkinson, F., Willsher, C. J., Pouliquen, J., Fichou, D., Valet, P., and Kossanyi, J.**, The use of time-resolved diffuse reflectance spectroscopy to observe transient absorption in transition metal-ion doped zinc oxide powders, *J. Photochem.*, 35, 381, 1986.
10. **Wilkinson, F. and Willsher, C. J.**, Detection of triplet-triplet absorption in microcrystalline benzophenone by diffuse-reflectance laser flash photolysis, *Chem. Phys. Lett.*, 104, 272, 1984.
11. **Wilkinson, F. and Willsher, C. J.**, Triplet-triplet absorption in microcrystalline benzil detected by diffuse reflectance laser flash photolysis, *Appl. Spectrosc.*, 38, 897, 1984.
12. **Wilkinson, F. and Willsher, C. J.**, Photoinduced transient absorption within zinc sulphide phosphors in powder form detected by diffuse reflectance laser flash photolysis, *J. Lumin.*, 33, 187, 1985.
13. **Wilkinson, F., Willsher, C. J., and Pritchard, R. B.**, Laser flash photolysis of dyed fabrics and polymers. I, *Eur. Polym. J.*, 21, 333, 1985.
14. **Wilkinson, F. and Willsher, C. J.**, Detection of transient absorption in a dyed cotton fabric and in semiconductor powders by diffuse reflectance laser flash photolysis, *J. Chem. Soc. Chem. Commun.*, 142, 1985.
15. **Kubelka, P.**, New contributions to the optics of intensely light-scattering materials. I, *J. Opt. Soc. Am.*, 38, 448, 1948.
16. **Lin, T. and Kan, H. K. A.**, Calculations of reflectance of a light diffuser with non uniform absorption, *J. Opt. Soc. Am.*, 60, 1252, 1970.
17. **Kessler, R. W., Krabichler, G., Uhl, S., Oelkrug, D., Hagen, W. P., Hyslop, J., and Wilkinson, F.**, Transient decay following pulse excitation of diffuse scattering samples, *Opt. Acta*, 30, 1099, 1983.
18. **Wendlandt, W. and Hecht, H. G.**, *Reflectance Spectroscopy*, John Wiley & Sons, New York, 1966.
19. **Wilkinson, F., Willsher, C. J., Oelkrug, D., and Honnen, W.**, Modelling of transient production and decay following laser excitation of opaque materials, *J. Chem. Soc. Faraday Trans 2*, 83, 2081, 1987.
20. **Wilkinson, F., Willsher, C. J., Uhl, S., Honnen, W., and Oelkrug, D.**, Optical detection of a photoinduced thermal transient in titanium dioxide powder by diffuse reflectance laser flash photolysis, *J. Photochem.*, 33, 273, 1986.
21. **Bourdelande, J., Font, J., and Sánchez-Ferrando, F.**, The use of insoluble benzoylated polystyrene beads (polymeric benzophenone) in photochemical reactions, *Can. J. Chem.*, 61, 1007, 1984.
22. **Wilkinson, F., Willsher, C. J., Bourdelande, F., Font, J., and Greuges, J.**, Photophysical evidence for intramolecular energy transfer in insoluble polymeric benzophenone (benzoylated polystyrene beads), *J. Photochem.*, 38, 381, 1987.

Chapter 10

TIME-RESOLVED TOTAL INTERNAL REFLECTION FLUORESCENCE SPECTROSCOPY FOR DYNAMIC STUDIES ON SURFACE

Hiroshi Masuhara and Akira Itaya

TABLE OF CONTENTS

I. INTRODUCTION

Structure and dynamic behavior of solid surface is generally considered to be different from bulk characteristics. Molecular motion, physical, as well as chemical functions of polymers and organic solids may be changed from the boundary surface to the bulk. These problems are one of the most attractive topics in chemistry, physics, and material science. Powerful tools for surface investigations are various kinds of electron spectroscopy, attenuated total reflection infrared (ATR IR) absorption spectroscopy, and Raman scattering measurement under total internal reflection (TIR) condition. These have been applied not only to metals and semiconductors but also to polymers and organic solids, which gives detailed data on surface structures. However, all these methods cannot provide direct information on dynamic processes in the nano- and picosecond time regions.

Recently, importance of electronic properties has been emphasized in the studies on organic solids. Carrier generation and its transport, excitation energy relaxation, and electron transfer depend, of course, upon the structure, so that they may change from the surface to the bulk. Furthermore, these electronic properties are dynamic in nature and should be followed in real time. From this viewpoint we proposed a time-resolved TIR fluorescence spectroscopy as a new methodology, and demonstrated it for the first time.

In this chapter, principle, experimental details, and an analysis method of the time-resolved TIR fluorescence spectroscopy are summarized and discussed. Factors determining the dynamic range of this spectroscopy and affecting the depth-resolution are described. Its applications to vacuum-deposited films, silk fabrics, and cast polymer films are reported and considered. Finally, a TIR fluorescence measurement under a fluorescence microscope, which enables us to obtain time- and space-resolved data on fluorescence characteristics, is given as an extension of the present new methodology.

II. CONCEPT OF TIR FLUORESCENCE SPECTROSCOPY[1-6]

Under the TIR condition, a laser beam penetrates into material with lower refractive index from a glass plate with higher value. This is called an evanescent wave and is used as an excitation beam for surface studies. In the case of s-polarization, an amplitude of this wave is given by the following equations:

$$E = E_0 \exp(-rz) \tag{1}$$

$$r = (2\pi n_1/\lambda)(\sin^2\theta - (n_2/n_1)^2)^{0.5} \tag{2}$$

Here, z is the depth from the interface between both materials, E_0 and E are intensities of the evanescent wave at the interface and at the depth z, respectively, θ is an incident angle of the beam, λ is its wavelength, and n_1 and n_2 are the refractive indices of the denser and rarer materials, respectively. A surface area whose depth is determined by experimental parameters in these equations can be elucidated by analyzing fluorescence emitted from this area.

III. EXPERIMENTAL

A. OPTICAL SET FOR THE TIME-RESOLVED TIR FLUORESCENCE SPECTROSCOPY[1,2]

Sapphire was selected as an internal reflection element with a high value of refractive index ($n_1 = 1.81$ at 313 nm). Its dimension was $30 \times 10 \times 1$ mm^3, and the longest edge was along the c-axis. The contact surface (30×10 mm^2) with the sample was the (1010) plane.

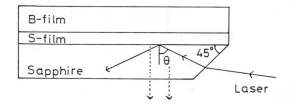

Fluorescence

FIGURE 1. Optical set for total internal reflection fluorescence spectroscopy. θ: incident angle of an excitation laser beam, S-film: thin surface film, and B-film: thick bulk film.

Samples were firmly contacted on the sapphire plate. As an example, an optical set for a model system of layered polymer films (described later) is shown in Figure 1. This combination was fixed on a goniometer by which an incident angle was adjusted with the precision of less than 0.1°.

B. MEASURING SYSTEMS

A pulsed laser is indispensable for the time-resolved fluorescence measurement. A time-correlated single photon counting method is the most preferable as a detection system, because this has a high sensitivity, a wide dynamic range, and a high time resolution. Measuring systems used in the present experiment follow.

1. System I[1,2]

A synchronously pumped, cavity-dumped dye laser (Spectra Physics 375 and 344S) combined with a mode-locked Ar^+ laser (Spectra Physics 171-18) was used as a picosecond excitation light source. The laser was operated with a repetition rate of 800 kHz. The pulse duration was 6 ps (full-width of half maximum, fwhm). The excitation wavelength was 315 nm which was obtained by using a KDP crystal. Its beam divergence was 1.7 ± 0.4°. A Nikon monochromator was used, and the fluorescence was detected by a HTV R1294U microchannel plate photomultiplier. Time-resolved fluorescence spectra and rise as well as decay curves were measured with time-correlated single photon counting method. This system gave an instrumental response function with 60 ps (fwhm), details of which were published elsewhere.[7] Polarizers and filters were set in front of the monochromator, and a quartz lens was used to collect the fluorescence efficiently. Excitation polarization was usually adjusted to be perpendicular to the plane of incidence which includes the c-axis of the sapphire and, to be normal to the contact surface, using Babinet-Soleit plate. In this case of s-polarization, the polarization of the evanescent wave is independent of the incident angle.

2. System II[8]

Lumonics TE430T-2 excimer laser (308 nm, 6 ns, 10 mJ) was used as an excitation light source, and its effective diameter was reduced to 1 mm with an aperture. Time-resolved fluorescence spectra at a fixed delay time and time variation of fluorescence intensity were measured by using Jovin Yvon H-10 monochromator, HTV 1P28 photomultiplier, and Iwatsu SAS 601B sampling oscilloscope. Since this detection system has a low sensitivity compared to the single photon counting one and need a high excitation intensity, special caution should be taken in analysis in order to exclude complicated behavior due to S_1-S_1 annihilation.

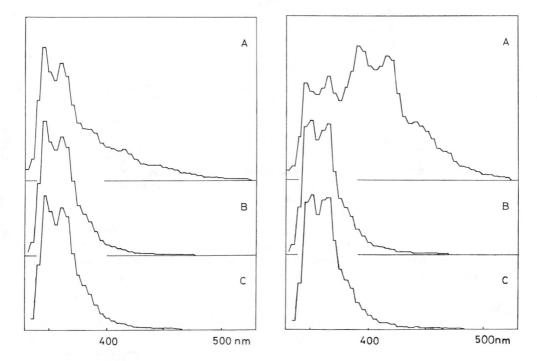

FIGURE 2.   Time-resolved fluorescence spectra of a bilayer system of the 0.01-μm S-film doped with 1×10^{-2} mol dm^{-3} POPOP and the thick B-film doped with 1.3×10^{-2} mol dm^{-3} N-ethylcarbazole. θ is 69.6° (left) and 71.9° (right). Gated times are (A) 0.4 to 1.4 ns, (B) 10.4 to 12.4 ns, and (C) 24.4 to 44.4 ns.

IV. TIME-RESOLVED TIR FLUORESCENCE SPECTROSCOPY[1,2]

A model system of bilayered polystyrene films was adopted for demonstrating a high potential of this spectroscopy and for proposing an analysis method. Thin surface (S) films with 0.01-, 0.09-, and 0.4-μm thickness was doped with p-bis(2-(5-phenyleneoxazo-lyl)benzene) (POPOP). Thick bulk (B) films with about 30 μm were doped with N-ethyl-carbazole. Absorbance of both films at the excitation wavelength was smaller than 0.1. The S-film was formed on the sapphire, while the B-film was pressed firmly to the S-film in order to obtain an intimate contact. The refractive index of polystyrene (n_2) is 1.68 at 313 nm, so that the critical angle (θ_c) given by sin $\theta_c = n_2/n_1$ was calculated to be 68.15°. The present investigation was carried out with the System I.

A. TIME-RESOLVED TIR FLUORESCENCE SPECTRA

Both spectra measured under the TIR condition and by the conventional illumination are similar to fluorescence spectrum of N-ethylcarbazole in the B-film. This is because the latter was far thicker than the S-film. The time-resolved fluorescence spectra of 0.01-μm S-film measured with an incident angle of $\theta_c + 1.45°$ is shown in Figure 2 (left). The structured bands of N-ethylcarbazole below 380 nm were observed as a main band inde-pendent of the gated time. Only at the early gated interval was POPOP fluorescence above 380 nm slightly observed. On the other hand, when $\theta = \theta_c + 3.75°$ (Figure 2, right), POPOP fluorescence intensity is larger than the N-ethylcarbazole one of the B-film at the early gated time interval. As the fluorescence lifetime of POPOP (1.2 ns) is shorter than that of N-ethylcarbazole (13 ns), the spectrum at the late gated time was independent of θ. This result demonstrated a possibility that a surface information is emphasized to a great extent by applying time-resolved measurement.

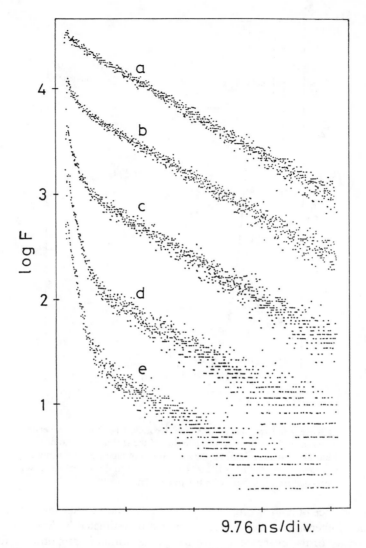

9.76 ns/div.

FIGURE 3.   Fluorescence decay curves of a bilayer system of the 0.09-μm S-film doped with 0.1 mol dm^{-3} POPOP and the thick B-film doped with 1.4 × 10^{-2} mol dm^{-3} N-ethylcarbazole. Observation wavelength was 385 nm. Incident angles are (a) θ_c + 0.27°, (b) θ_c + 0.68°, (c) θ_c + 0.98°, (d) θ_c + 1.49°, and (e) θ_c + 1.89°.

B. FLUORESCENCE DECAY CURVES AS A FUNCTION OF INCIDENT ANGLE

An analysis of fluorescence decay curves gives more quantitative information on the TIR phenomena compared to the time-resolved spectra. The results on the present bilayer system with the 0.09-μm S-film at 385 nm are shown in Figure 3. At $\theta = \theta_c + 0.27°$, a contribution of the shorter component POPOP is slightly detected, while it increased sharply upon a little increase of θ by about 1°. All the curves were reproduced by a sum of two exponentials; $F(t) = F_S\exp(-t/\tau_S) + F_B\exp(-t/\tau_B)$ where τ_S and τ_B are fluorescence lifetimes of POPOP and N-ethylcarbazole, respectively. Although preexponential factors F_S and F_B vary depending on absorbance, spectrum, yield, and lifetime, the values of F_B/F_S are related to a ratio of the number of fluorescent molecules in each layer. Therefore, a relation between F_B/F_S and θ can be used as a probe for surface analysis. One of the examples is shown in Figure 4.

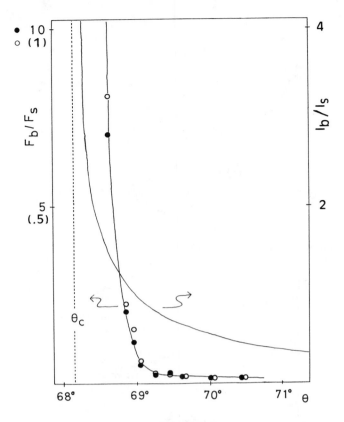

FIGURE 4.   Experimental (○,●) and theoretical (——) relations between θ and the ratio of fluorescence intensity of the B-film over that of the S-film. The system examined is the same as that of Figure 3. The monitoring wavelength are 385 nm (○) and 420 nm (●). The theoretical result was calculated for the system with 0.1-μm S-film.

   Since absorbance of both fluorophores is small, the fluorescence intensity is proportional to the excitation intensity of each layer, which can be estimated by Equations 1 and 2. An effective intensity of the evanescent wave for s-polarization is proportional to $\cos^2 \theta$. Assuming the B-film has an infinite depth, the excitation intensity ratio of the B-film to the S-film is theoretically calculated by using Equation 3.

$$I_B/I_S = \exp(-2rd)/(1 - \exp(-2rd)) \tag{3}$$

where d is the thickness of the S-film. The value of I_B/I_S of the 0.1-μm S-film system was calculated and is shown in Figure 4 for comparison.

   Differences between the theoretical and experimental curves are observed in the angle region showing a large decrease of the F_B/F_S and above 69°. Various experimental factors such as a beam divergence of the excitation pulse, fluctuation of beam direction, optical quality of polymer films, flatness of the contact face and the 45° edge of the sapphire, contact condition of both films, and an error of angle setting of goniometer led to the difference in the former region. The other was ascribed to the fact that the F_B/F_S value approaches to the S/N value. This problem is clearly related to the dynamic range of the present method.

   On the basis of the above results, we proposed how to analyze the θ-dependence of fluorescence decay curves. The F_B/F_S values are plotted against θ by normalizing them to the value obtained by excitation with $\theta < \theta_c$. This makes it easier to compare fluorescence characteristics observed under different conditions. The normalized value can be calculated theoretically by the following equation.

$$f_B/f_S = C(\exp(-2rd) - \exp(-2rl))/(1 - \exp(-2rd)) \qquad (4)$$

where l is the thickness of the B-film. The constant C includes all the experimental parameters such as concentration of both fluorophores, molar extinction coefficient at the excitation wavelength, and so on. In the case of $\theta < \theta_c$, this equation is approximated to be proportional to l/d. The experimental value F_B/F_S can be correlated to f_B/f_S. Considering impurity fluorescence, the maximum count decided by the machine time, and an accuracy of two-exponential analysis, the present analysis method was confirmed to have a practical dynamic range of F_B/F_S less than 2 orders of magnitude.

C. EFFECT OF EXCITATION POLARIZATION

In the case of p-polarization, the polarization of the evanescent wave depends upon the incident angle, which makes it complicated to analyze the data. However, it is experimentally important to examine the relation between F_B/F_S and θ with this polarization excitation. We found that the F_B/F_S values with this polarization were larger than those with s-one, while their dependencies upon the incident angle were identical with each other within experimental error (0.13°).

D. EFFECT OF FLUOROPHORE CONCENTRATION

As the complex refractive index is a fucntion of absorbance, the θ-dependence of the F_B/F_S upon fluorophore concentration was investigated by using the bilayer models of 0.4- and 0.09-μm S-film. It is confirmed that this factor is practically neglected under the present experimental conditions. For example, the θ-dependence of F_B/F_S value of the 10^{-2}-10^{-2} mol dm^{-3} concentration pair for POPOP in the S-film and N-ethylcarbazole in the B-film, respectively, was confirmed to be identical to the 10^{-3}-10^{-3} one.

On the other hand, the θ-dependence of F_B/F_S is affected by the concentration ratio of both fluorophores. This is observed as a shift of the F_B/F_S-θ curve to the large or small θ region according to the present analysis method, which is shown in Figure 5. It was concluded that a concentration difference of 1 order of magnitude of S- or B-film is detected by analyzing the F_B/F_S-θ curve.

E. DEPTH RESOLUTION

In order to demonstrate a depth resolution of the present spectroscopy, relations between F_B/F_S and θ for three model systems with 0.4-, 0.09-, and 0.01-μm S-films were investigated by fixing the ratio of both chromophore concentrations. As shown in Figure 6, an angle region where the value of F_B/F_S sharply decreases with an increase of θ is shifted to the larger θ region as the S-film becomes thinner. This indicates that a selective excitation of the S-film is possible as the incident angle is $\theta > \theta_c + 3°$.

V. APPLICATIONS

A. ANALYSIS OF VACUUM-DEPOSITED, MULTILAYERED FILMS OF ORGANIC MOLECULES[9]

This layered model is schematically shown in Figure 7. Deposited molecules were 12-(1-pyrenyl)dodecanoic acid (15 nm), stearic acid (150 nm), and perylene (100 nm) for A-, B-, and C-layers, respectively. Steady-state fluorescence spectra were measured by using the System I, which is shown in Figure 8. Fluorescence below 500 nm was ascribed to

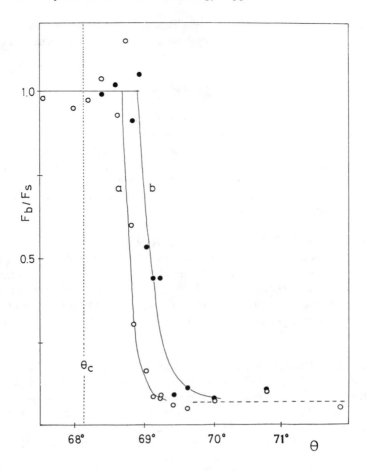

FIGURE 5.   Concentration effect upon a relation between θ and the ratio of fluorescence intensity of the B-film over that of the S-film. The model systems are (a) the 0.4-μm S-film doped with 2×10^{-2} mol dm^{-3} POPOP and the thick B-film doped with 1.3×10^{-2} mol dm^{-3} N-ethylcarbazole, and (b) the 0.4-μm S-film doped with 3×10^{-3} mol dm^{-3} POPOP and the thick B-film doped with 1.3×10^{-2} mol dm^{-3} N-ethylcarbazole. Observation wavelength is 420 nm.

pyrenyl chromophor (A-layer),[10] with a broad emission above 500 nm to perylene excimer (C-layer). The spectrum at θ = 55.9° corresponds to that measured under the normal optical condition. It is reasonable that the contribution of perylene is larger than that of 12-(1-pyrenyl)dodecanoic acid in this spectrum. As being expected, the intensity ratio of the A-layer to the C-layer increases with an increase of the incident angle.

Time-resolved fluorescence spectra made the θ-dependence more clearly. As measured in the spectra gated at 0 to 400 ps after excitation (Figure 9), perylene monomer fluorescence gave a peak at 520 nm. This intensity decreases with an increase of θ from 55.9 to 70.0°, which indicates that the time-resolved fluorescence measurement gives more distinct information on the depth profile.

It is demonstrated that the time-resolved TIR fluorescence spectroscopy makes it possible to obtain the information on the 15-nm region from the interface of the 300-nm thickness film and has a potential to elucidate photoprocesses and molecular motions in the nano- and picosecond time domains.

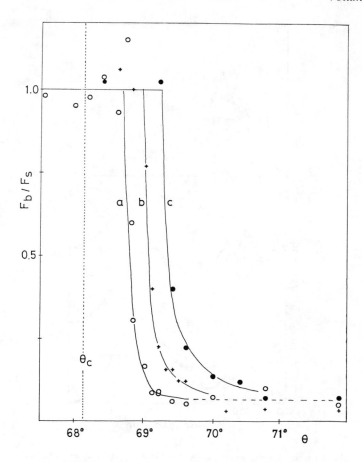

FIGURE 6.   Relations between θ and the ratio of fluorescence intensity of the B-film over that of the S-film. Observation wavelength is 420 nm. The model systems are (a) the 0.4-μm S-film doped with 2×10^{-2} mol dm^{-3} POPOP and the thick B-film doped with 1.3×10^{-2} N-ethylcarbazole, (b) the 0.09-μm S-film with 2.3×10^{-3} mol dm^{-3} POPOP and the B-film doped with 1.2×10^{-3} mol dm^{-3} N-ethylcarbazole, and (c) the 0.01-μm S-film doped with $\sim\!10^{-2}$ mol dm^{-3} POPOP and the B-film doped with 1.3×10^{-2} mol dm^{-3} N-ethylcarbazole.

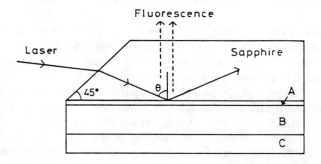

FIGURE 7.   Optical set for the multilayered film. Vacuum-deposited molecules are (A) 12-(1-pyrenyl)dodecanoic acid (15 nm), (B) stearic acid (150 nm), and (C) perylene (100 nm).

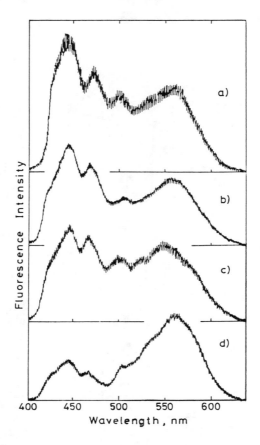

FIGURE 8.   Fluorescence spectra of the multilayered film as a function of an incident angle. (a) 70.0°, (b) 65.8°, (c) 61.8°, and (d) 55.9°.

B. A DEPTH DISTRIBUTION OF FLUORESCENT DYES IN SILK FABRICS[8]

Silk woven fabrics were dyed with carthamin according to the literature and pressed firmly to the sapphire plate whose optical alignment is schematically shown in Figure 10. The fabric consists of many fibers, so that some of them are excited under the TIR condition and others under the normal one. Their relative contributions are changed with the incident angle. Therefore, the fluorescence spectra under both conditions were simply compared with each other.

As shown in Figure 11 where the System II was used, a fluorescence spectrum of silk fabrics consisted of a peak at 340 nm, a shoulder below 400 nm, and a descending tail in the long wavelength region. This spectral shape was independent of the excitation condition, indicating that the silk fabrics have a homogeneous structure along the depth of the yarn. Fluorescence spectra of a dyed silk showed an additional band at 585 nm due to carthamin. The relative intensity of the silk and the dye bands was sensitive to excitation condition; that is, the ratio of the dye fluorescence intensity to the silk under the TIR condition was smaller than that under the normal condition. This means that the carthamin concentration near the surface is lower than that in the bulk. This suggests a depth distribution of chemical reactions such as dyeing, oxidation, photodegradation, etc., which have an important role in determining physical and chemical properties of silk fabrics.

Although the contact between the sapphire plate and silk fabrics is not so good, the difference in fluorescence spectra under TIR and normal conditions was observed. This promises that the present spectroscopy is useful even for materials with optically anisotropic or scattering properties.

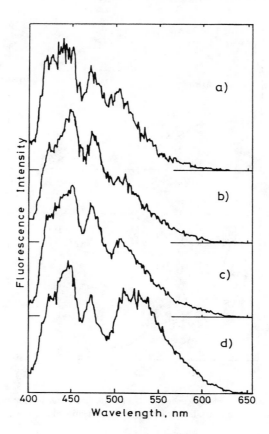

FIGURE 9. Normalized time-resolved fluorescence spectra of the multilayered film as a function of an incident angle. (a) 70.0°, (b) 65.8°, (c) 61.8°, and (d) 55.9°. The gate time interval is 0—400 ps.

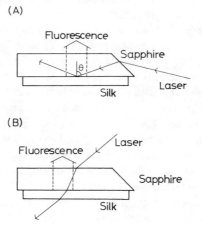

FIGURE 10. Optical set for silk fabrics. (A) TIR condition and (B) normal excitation condition.

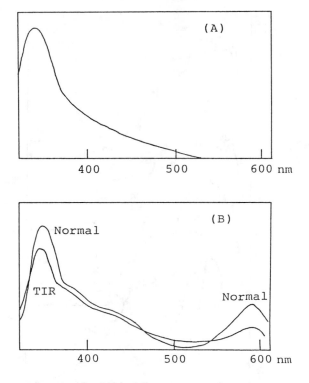

FIGURE 11.   Fluorescence spectra of a silk fabric under the normal condition (A) and of a silk fabric dyed with carthamin under both conditions
(B).

C. A DEPTH DISTRIBUTION OF FLUORESCENT DOPANTS IN CAST POLYMER FILMS[11]

Poly(*N*-vinylcarbazole) (PVCz) films doped with perylene and PMMA films doped with 1-ethylpyrene were formed on a sapphire plate by casting anisole and 1-chlorobenzene solutions, respectively. Their thickness was a few micrometers. Fluorescence spectra were measured by using a picosecond laser pulse as an excitation light.

As given in Figure 12, fluorescence spectra of PVCz film doped with perylene consists of broad structureless excimer bands of the polymer with a shoulder at 375 nm and a peak at 420 nm, and the perylene band with a vibrational structure above 450 nm. The excitation energy migrates over carbazolyl chromophores and is trapped in the doped perylene efficiently in spite of its low concentration. Therefore, the present result was examined as a function of the dopant concentration. It is worth noting that the perylene fluorescence intensity under the TIR condition is relatively weaker than that under the normal one. Since the boundary surface is selectively excited under the former condition, the doped perylene is concluded to be less dissolved in the interface layer than in the bulk.

A similar experiment was performed for the PMMA film doped with 1-ethylpyrene. As shown in Figure 13, fluorescence spectra were composed of a structured monomer and a red-shifted broad excimer bands. As molecular diffusion in film is negligible during fluorescence lifetime, the latter band is due to the ground state dimer of pyrene which is easily formed under its high concentration. It should be notified that the fluorescence intensity ratio of monomer to excimer emissions under the TIR condition is larger than that under the normal one. This is a direct indication that the pyrenyl concentration in the interface layer is lower than that in the bulk.

The thickness of the interface layer in problems where the dopant concentration is

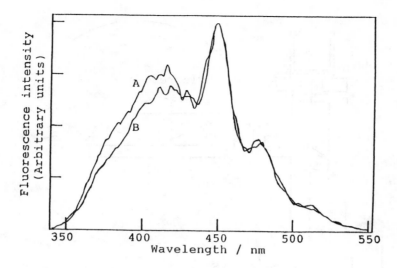

FIGURE 12. Fluorescence spectra of cast poly(*N*-vinylcarbazole) film doped with 6.2×10^{-5} mol perylene per mole carbazole unit. (A) TIR condition with $\theta = 73.6°$. (B) Normal condition. Both spectra are normalized at the 0-0 band of perylene fluorescence.

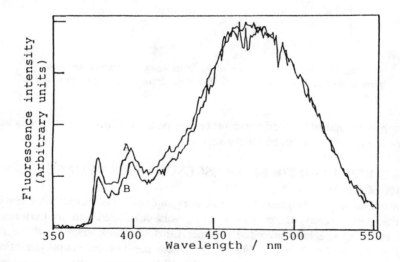

FIGURE 13. Fluorescence spectra of cast PMMA film doped with 7.7×10^{-2} mol 1-ethylpyrene per mole MMA unit. (A) TIR condition. (B) Normal condition. Both spectra are normalized at the excimer band.

different from that of the bulk is estimated as follows. Since an absorption coefficient of PVCz at 295 nm is 1.54×10^5 cm^{-1}, the depth where the excitation intensity is l/e of the initial value is calculated to be 0.065 μm under the normal condition. On the other hand, the penetration depth of the evanescent wave is a function of the incident angle, and it is difficult to calculate it here because the complex refractive index cannot be estimated correctly from the large absorbance at the laser wavelength. At present we can say that the TIR phenomenon was really observed and that the effective thickness under the TIR condition is thinner than 0.065 μm. According to the similar consideration, the concentration gradient of the dopant in PMMA is in the depth region less than 1.4 μm. The present results indicate

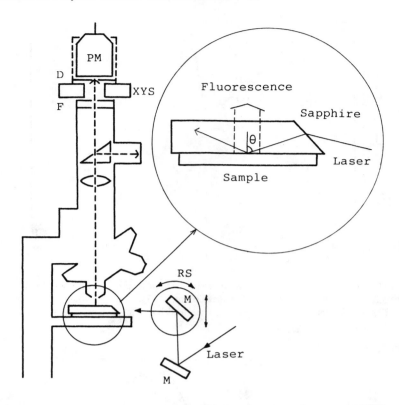

FIGURE 14.   Schematic diagram of ps fluorescence microprobe apparatus. PM: photomultiplier, D: diaphragm, XYS: X-Y stage, F: filter, M: laser mirror, and RS: rotating stage.

that photophysical and photoconductive properties of these films should be elucidated as a function of the depth from the boundary surface.

D. A TIME-RESOLVED TIR FLUORESCENCE MEASUREMENT UNDER A MICROSCOPE[12]

As an extension of the present methodology, an optical setup for time-resolved TIR fluorescence spectroscopy under a microscope was developed and its performance was examined by using specially prepared polymer films. A block diagram of the developed system is schematically shown in Figure 14, where the TIR excitation condition is also illustrated. Fluorescence microscope Nikon XF-EFD equipped with an adaptor PFX or Olympus BHS-RFK-A was chosen, while their internal optics for excitation beam was not used, since it absorbs a UV laser beam to some extent. We selected a long-distance-working objective lens Nikon ELWD M Plan 40 or Olympus ULWDCDPL40X which gives a more space between the lens and the sample compared to the conventional lens system, and makes it possible to set the coated sapphire system. Incident angle θ was adjusted by inclining the laser mirrors and by sliding the rotating stage on which the mirror is mounted. The beam diameter was reduced to 1 mm with an aperture. On the top of the microscope, an X-Y stage with a 1-mm diaphragm was set in order to choose the microsection whose fluorescence dynamics should be probed. A HTV R2809U-01 microchannel plate photomultiplier was used as a fast-response detector. All these optical instruments were designed and constructed in the present work. When two-dimensional resolution was examined, the film sample on the quartz plate was set instead of the sapphire sample and was illuminated from the upper-right side. An excitation laser and a detection system are due to the System I.

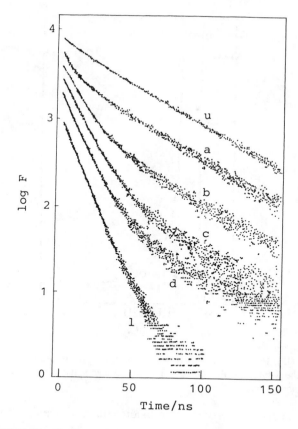

FIGURE 15.   Fluorescence decay curves of a bilayer system of 0.4-μm S-film doped with 4.8 × 10⁻³ mol *N*-ethylcarbazole per mole phenyl group and the thick B-film doped with 5 × 10⁻³ mol 9-ethylphenanthrene per mole phenyl group. Incident angles are (a) θ_c + 0.16°, (b) θ_c + 1.09°, (c) θ_c + 1.49°, and (d) θ_c + 2.28°. The upper (u) and lower (l) curves are due to B- and S-films, respectively.

First we describe the depth information which was confirmed by examining the similar bilayer model systems as that described in IV. Thin S- and thick bulk B-films had the thickness of 0.4 and a few μm, and were doped with *N*-ethylcarbazole and 9-ethylphenanthrene, respectively. The fluorescence decay curves of the model film were measured by using a UV-33 filter whose results are shown in Figure 15. The fluorescence lifetime of *N*-ethylcarbazole and 9-ethylphenanthrene in polystyrene matrix was obtained to be 12 and 40 ns, respectively. Just above θ_c, the short decay component of the S-film was slightly detected, while it was greatly enhanced by a little increase of θ. These decay curves were treated according to the analysis method, which is shown in Figure 16. Since the obtained relation between F_B/F_S and θ is similar to that obtained without a microscope, we concluded that the present optical system works well and the time-resolved TIR measurement combined with the fluorescence microscope is practically possible.

The two-dimensional resolution of the present optical system was examined by using a polymer film microfabricated with the 308-nm excimer laser. A cast PMMA film containing 1-ethylpyrene was ablated with a photolithographically prepared mesh mask in contact mode. In the ablated area, concentration as well as distribution of the dopant and its microenvironment were affected to a great extent which can be monitored by fluorescence measurement. Here the fluorescence decay curves were used for reproducing an etched pattern, although

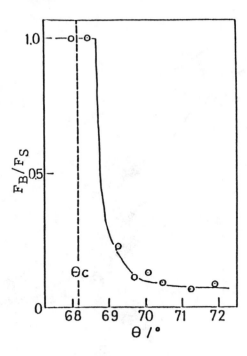

FIGURE 16.   A relation between θ and the ratio of fluorescence intensity of the B-film over that of the S-film. The system is the same as that of Figure 15.

the molecular mechanism how laser ablation modify fluorescence dynamics of the film are being studied.

In Figure 17, typical decay curves of l-ethylpyrene aggregates in the film are given, where fluorescence was measured with Toshiba UV 33 and Hoya U 330 filters. It is clear that a difference was induced by laser ablation. By shifting the diaphragm of the microscope, fluorescence decay curves were measured as a function of the horizontal position. We plotted the time when the intensity became one twentieth of the initial one against the position in Figure 4 (C). The values about 80 and around 100 correspond to the masked and the ablated areas, respectively. Therefore, it is concluded that the present result represents the cross-section of the masked area. The width of nonablated area was obtained to be 37 μm, which is in agreement with the mask dimension.

The practical performance of the present method is summarized as follows: (1) depth resolution, 0.1 μm, (2) two-dimensional resolution, 5 μm, (3) time-resolution, 10 ps, and (4) wavelength resolution, 10 nm by using an interference filter. Each resolution is easily improved if a measuring time longer than a few tens of minutes is permitted. We believe that the present new methodology will be fruitful for dynamic studies on microfabricated materials, biological tissues, cells, fabrics, and so on.

ACKNOWLEDGMENTS

The authors express their sincere thanks to Prof. S. Tazuke, Prof. I. Yamazaki, Dr. Y. Taniguchi, and Dr. N. Tamai for their fruitful discussions and help. The present work is supported partly by Grant-in-Aid from the Japanese Ministry of Education, Science, and Culture (59850146) and by the Joint Studies Program of the Institute for Molecular Science.

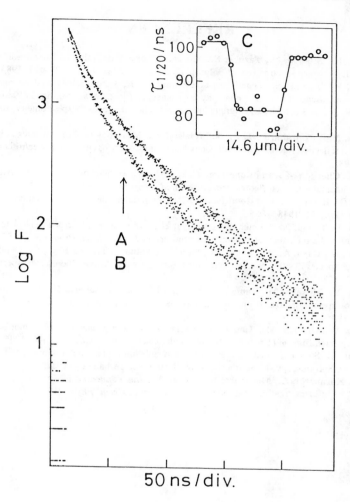

FIGURE 17.   Fluorescence decay curves of ablated (A) and nonablated (B) areas of PMMA film doped with 7.75×10^{-2} mol 1-ethylpyrene per mole MMA unit. (C) A cross-section of an ablated PMMA film revealed by analyzing fluorescence data.

REFERENCES

1. **Masuhara, H., Mataga, N., Tazuke, S., Murao, T., and Yamazaki, I.,** Time-resolved total internal reflection fluorescence spectroscopy of polymer films, *Chem. Phys. Lett.,* 100, 415, 1983.
2. **Masuhara, H., Tazuke, S., Tamai, N., and Yamazaki, I.,** Time-resolved total internal reflection fluorescence spectroscopy for surface photophysics studies, *J. Phys. Chem.,* 90, 5830, 1986.
3. **Harrick, N. J. and Loeb, G. I.,** Multiple internal reflection fluorescence spectrometry, *Anal. Chem.,* 45, 687, 1973.
4. **Thompson, N. L., Burghardt, T. P., and Axelrod, D.,** Measuring surface dynamics of biomolecules by total internal reflection fluorescence with photobleaching recovery or correlation spectroscopy, *Biophys. J.,* 33, 435, 1981.
5. **Lok, B. K., Cheng, Y.-L., and Robertson, C. R.,** Protein adsorption on crosslinked polydimethylsiloxane using total internal reflection fluorescence, *J. Colloid Interface Sci.,* 91, 104, 1983.
6. **Ausserre, D., Hervet, H., and Rondelez, F.,** Concentration profile of polymer solution near a solid wall, *Phys. Rev. Lett.,* 54, 1948, 1985.
7. **Yamazaki, I., Tamai, N., Kume, H., Tsuchiya, H., and Oba, K.,** Microchannel-plate photomultiplier applicability to the time-correlated photon-counting method, *Rev. Sci. Instrum.,* 56, 1187, 1985.
8. **Kurahashi, A., Itaya, A., Masuhara, H., Sato, M., Yamada, T., and Koto, C.,** Depth-distribution of fluorescence species in silk fabrics as revealed by total internal reflection fluorescence spectroscopy, *Chem. Lett.,* 1413, 1986.
9. **Taniguchi, Y., Mitsuya, M., Tamai, N., Yamazaki, I., and Masuhara, H.,** Time- and depth-resolved fluorescence spectra of layered organic films prepared by vacuum deposition, *J. Colloid Interface Sci.,* 104, 596, 1985.
10. **Taniguchi, Y., Mitsuya, M., Tamai, N., Yamazaki, I., and Masuhara, H.,** Fluorescence spectra of vacuum-deposited films ω-(1-pyrenyl)alkanoic acids, *Chem. Phys. Lett.,* 132, 516, 1986.
11. **Kurahashi, A.,** Structural Analysis along the Depth and Non-Linear Photochemistry of Polymer Films by Fluorescence Spectroscopy, M.Sc. thesis, Kyoto Institute of Technology, 1987.
12. **Itaya, A., Kurahashi, A., Masuhara, H., Tamai, N., and Yamazaki, I.,** Dynamic fluorescence microprobe method utilizing total internal reflection phenomenon, *Chem. Phys.,* 1079, 1987.

Chapter 11

APPLICATION OF LASERS TO TRANSIENT ABSORPTION SPECTROSCOPY AND NONLINEAR PHOTOCHEMICAL BEHAVIOR OF POLYMER SYSTEMS

Hiroshi Masuhara

TABLE OF CONTENTS

I. INTRODUCTION

Transient UV absorption spectroscopy was proposed by Norrish and Porter and has made a great contribution to the studies on physical, chemical, and biological processes.[1] The time-resolution of flash photolysis measurement has been improved from microsecond to tens of femtosecond by introducing various kinds of pulsed lasers as excitation and monitoring light sources. The dynamic behavior of unstable radicals and molecular triplet states was the main subject of concern for studies in the microsecond time range. Nanosecond laser flash photolysis made it possible to observe dynamics of excited singlet (S_1) states and n-π^* triplet states. The most active areas of investigation in the nanosecond time range are given for electron and charge transfer phenomena in chemistry,[2] and S_1-S_1 annihilation processes in solution, molecular crystals, and semiconductors.[3] Laser photolysis method in the picosecond and femtosecond time domains is now available.[4] Thus, transient UV absorption spectroscopy is recognized as an indispensable technique in chemistry, physics, biology, and engineering.

In most cases, transmittance optical alignment has been adopted and optically clear, transparent samples such as gaseous and solution systems have been studied. The present transient absorption spectroscopy is now being extended to opaque as well as scattering systems and to solid surface. One is a diffuse reflectance laser photolysis method which analyzes laser-induced change of diffuse reflected light from scattering materials instead of transmitted monitoring light. This powerful technique was first developed by Wilkinson[5] and applied to organic microcrystals, semiconductor, as well as insoluble polymer powders, dyed fabrics, molecules adsorbed on silica gels, etc.[6,7] The surface region of optically clear solid can be elucidated by applying attenuated reflection (ATR) UV absorption spectroscopy proposed by us.[8] We consider that various kinds of materials are now fruitful target systems for photophysical as well as photochemical studies by means of transient absorption spectroscopy. All the types mentioned here are schematically shown in Figure 1.

II. EXPERIMENTAL DETAILS

In these measurements, dynamic processes have been analyzed primarily by probing a transient absorption at one wavelength. In general, however, absorption spectra of excited states and chemical intermediates overlap with each other. Furthermore, conformational change and orientational relaxation of the surrounding solvent molecules result in a time dependence of the spectral band shape. Intramolecular exciplex systems give an absorption spectrum, the band shape of which is a function of solvent properties and delay times.[9] Examples of phenomena which have been studied by analyzing absorption spectral changes in the picosecond time region are isomerization of *trans*-stilbene from the S_1 state[10] and vibrational relaxation of the triplet benzophenone[11] as well as the excited singlet anthracene.[12] It is indispensable in these kinds of experiments to measure absorption spectra over a wide range of wavelengths for elucidating the photophysical and photochemical processes accurately.

From this viewpoint, we have constructed several laser photolysis systems all of which are controlled and processed by a microcomputer. Nanosecond excimer and picosecond Nd^{3+}:YAG transmittance laser photolysis systems are shown in Figures 2 and 3, respectively.[13,14] The monitoring lamp is a 150-W Xe lamp (Wacom) which is operated in the DC and pulsed modes. This is synchronized to an excimer laser (Lumonics, 430T2 or Hyper400) through a timing circuit (homemade or Leonix). A time-profile of the transmitted monitoring light is digitized by a transient memory (Kawasaki Electronica M-50E) or a storagescope (Iwatsu TS-8123), transferred to a microcomputer (NEC PC9801) and processed. In the picosecond photolysis the monitoring beam is a picosecond continuum produced by focusing

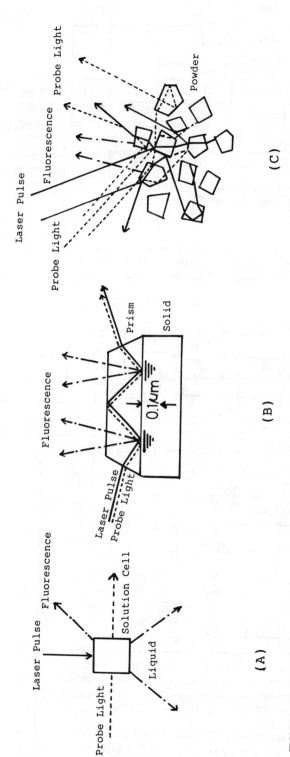

FIGURE 1. Classification of transient absorption spectroscopy. (A) Transmittance laser photolysis, (B) time-resolved attenuated reflection (ATR) UV absorption spectral measurement, and (C) diffuse reflectance laser photolysis.

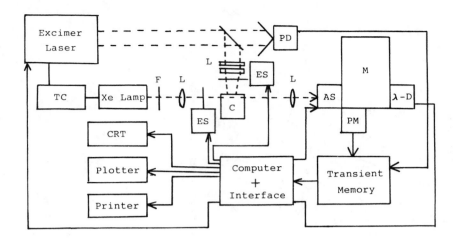

FIGURE 2.   Schematic diagram of the microcomputer-controlled nanosecond excimer laser photolysis system. PD, photodiode; PM, photomultiplier; ES, electromagnetic shutter; AS, automatic slit; λ-D, wavelength driver; L, lens; C, sample cell; F, filter; TC, timing circuit; and M, monochromator.

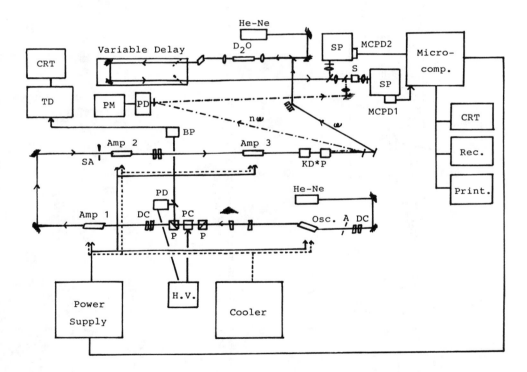

FIGURE 3.   Schematic diagram of the microcomputer-controlled picosecond Nd^{3+}:YAG laser photolysis system. DC, dye cell; A, aperture; P, polarizer; PC, Pockels cell; PD, photodiode; SA, soft aperture; BP, biplanar photodiode; SP, spectrograph; S, sample; Rec., recorder; PM, power meter; and TD, transient digitizer.

the fundamental pulse (1064 nm) into a quartz cell of 10-cm length with a lens of 15-cm focal length. The solution for generating this continuum is D_2O (Merk UVASOL. 99.95%) which makes it possible to measure the full range of the visible region. The excitation pulse is either the second (532 nm), the third (355 nm), or the fourth (266 nm) harmonic of the Nd^{3+}:YAG laser. The spectrum of the picosecond continuum is detected by a multichannel photodiode array (MCPD). Two diode arrays are attached to each spectrograph (f = 25 cm, Jovin Yvon grating 300 lines per millimeter), covering a wavelength range of 380 nm. The output of each MCPD is sent to a microcomputer (SORD M223 Mark II), whose processing principle is the double-beam method. This one-shot measurement is also possible in nanosecond photolysis by using a gated photodiode array or a streak camera. These optical and electronic arrangements are considered to be standard at the present stage of investigation.

Transient absorption spectroscopy, of course, can be applied to nonluminescent species, which makes it possible to investigate a wider range of systems compared to fluorescence techniques. This spectroscopy probes several electronic transitions, while the fluorescence measurement follows only one. This indicates that more detailed information on electronic structure of the excited state and the factors affecting it are available in absorption spectroscopy. However, two disadvantages have to be pointed out. The first is its small dynamic range. The second is a difficulty in obtaining accurate transient absorption spectra and their rise as well as decay curves. Since a concentration of 10^{-4} to 10^{-5} M is required to detect the excited states or intermediates, a high intensity laser has to be used as an excitation light source. This sometimes induces nonlinear behavior which affects absorption spectroscopy. We have performed transmittance laser photolysis studies on exciplexes, charge transfer complexes, polymers, aromatic liquids, and molecular aggregates such as micelles, organic films, and vesicles, and have studied their high intensity effects. On the basis of these results, we summarize the important phenomena affecting the transient absorption spectral measurements in solution, as follows:[15]

1. A spatial stray light effect. The monitoring beam which passes the unexcited volume acts as a stray light for transient absorption spectroscopy.[16]
2. A depletion effect on the ground state molecules. If the excitation condition is such that most of the molecules in the ground state are excited, the concentration of the excited species no longer increases in proportion to the excitation intensity.
3. A transient inner filter effect. The excitation pulse may be absorbed by the excited singlet state or by other photoinduced chemical intermediates, suppressing the effective number of excitation photons.[17,18]
4. A nonlinear refractive index change. Since the excitation photon density per unit area per unit time is very high, a change of the refractive index of the solvent may be induced.[19] At its worst, an excitation pulse produces a nonuniform distribution of the refractive index which leads to scattering of the monitoring light, resulting in an apparent transient absorption.
5. A thermal lensing effect. A similar refractive index change is also induced by local heating around the absorbing molecule.[20]
6. Resonance Raman scattering due to solvents. In the case of aliphatic solvents, the Stokes-Raman scattering induced by a pulsed laser disturbs the absorption spectral measurements.[20]
7. Two-photon excitation of solvents. It has been demonstrated that the picosecond 355-nm photolysis of aromatic liquids and solvents containing double bonds gives $S_n \leftarrow S_1$ absorption spectra with appreciable intensity,[21,22] which may modify the absorption spectral shape.

All these phenomena are considered to induce an apparent spectral change and to affect the absorption rise and decay curves. Therefore, it is indispensable to examine experimentally or to evaluate numerically these effects in order to obtain reliable and accurate data. Among the above-described phenomena, effects 6 and 7 can be excluded by selecting an appropriate solvent and increasing the solute concentration. The effect of spatial stray light 1 can also be decreased to a great extent by adjusting the optical alignment carefully. The contribution

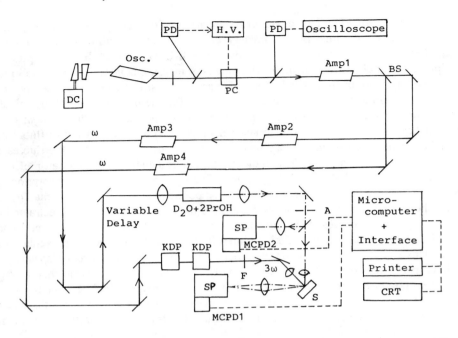

FIGURE 4.   Schematic diagram of the picosecond diffuse reflectance laser photolysis system. DC, dye cell; H.V., high-voltage power supply; PC, Pockels cell; PD, photodiode; A, aperture; SP, spectrograph; S, sample; BS, beam splitter; and F, filter.

of effects 2 and 3 in transient absorbance can be estimated numerically if molar extinction coefficients of the ground and the excited states at the excitation wavelength are known. These are experimental conditions under which we have studied photophysical and photochemical processes of molecules in solution.

The details of diffuse reflectance laser photolysis method are described in this book by Wilkinson.[23] Here we mention our picosecond system which enables for the first time transient absorption spectral and kinetic study on powder samples in tens of picoseconds.[24] A microcomputer-controlled picosecond diffuse reflectance laser photolysis system with a repetitive mode-locked Nd^{3+}:YAG laser is shown in Figure 4 where a double beam optical arrangement is adopted. The excitation and monitoring lights are frequency doubled or tripled pulse and picosecond continuum, respectively, as in the transmittance photolysis. As proposed by Kessler et al.,[5] the transient absorption intensity is displayed as percent absorption which is given $100 \times (1-R/R_0)$. Here R and R_0 represent the intensity of the reflected picosecond continuum with and without excitation, respectively. If we block the excitation pulse, this percent absorption should give zero. In this case the obtained spectrum is called a baseline which is used as a standard for judging the optical alignment. As shown in Figure 5, the baseline is reasonably flat. As the reflected intensity of the monitoring pulse is usually weak and in the same order of magnitude as that of emission, the latter contribution should be subtracted from the spectrum. The corrected result and the absorption spectrum of the triplet benzophenone in 2-propanol solution are compared with each other in Figure 5. Although a small red shift (about 10 nm) of the peak position and a broadening were observed, they are similar to each other. The spectrum of Figure 5c is also in agreement with the spectra obtained by nanosecond diffuse reflectance laser photolysis of the powder samples[5,25] and nanosecond transmittance laser photolysis of its single crystal.[26]

A microcomputer-controlled system of a nanosecond ATR UV spectroscopy is schematically shown in Figure 6.[27] A laser beam excites the sample which is firmly contacted to a sapphire plate and produces excited states and chemical transients. A monitoring light

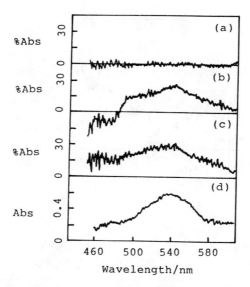

FIGURE 5. Transient absorption spectra of benzophenone microcrystals at 80 ps under various conditions: (a) baseline without excitation, (b) without correction for emission, (c) with correction, and (d) the triplet absorption spectrum of $3 \times 10^{-3} M$ benzophenone in 2-propanol at 450 ps, obtained by the transmittance laser photolysis method.

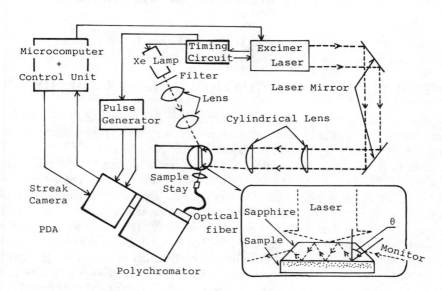

FIGURE 6. Schematic diagram of the nanosecond ATR UV absorption spectroscopic system. PDA, photodiode array and θ, incident angle.

is lead through the sapphire plate with an incident angle larger than the critical one. Under this condition the light penetrates into the sample with lower refractive index from the sapphire with higher value. This is called an evanescent wave and is used as a monitoring beam for the surface area. Although the penetration depth is given as a function of the wavelength, both refractive indices, and the incident angle as described in Chapter 10,[28] it is almost in the order of the wavelength. Therefore, transient absorption spectral and kinetic study is now possible even for the solid surface with the depth of the wavelength. One of

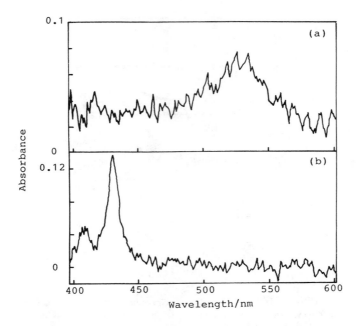

FIGURE 7. Transient ATR UV absorption spectra ascribed to the triplet state of the dopants in polymer films. Delay time, 1 μs; gate time, 167 ns; and incident angle, 60°. (a) 15 w% benzophenone in PMMA cast film and (b) 2 w% anthracene in PMMA cast film.

the results of demonstration experiments is given in Figure 7. When picosecond excitation and monitoring pulses are used, the time resolution is improved up to tens of picoseconds.[27]

III. GENERAL CLASSIFICATION OF LASER CHEMISTRY

In order to get a general insight into laser chemistry and to correlate polymer dynamics with gaseous as well as dilute solution photochemistry, we propose here a schematic diagram in Figure 8. The x-axis represents the photon number per unit area per laser shot, y-axis the number of molecules in aggregation, and z-axis reaction quantity. The origin of these axes is responsible to reaction of one-molecule-one-photon system. The xz-plane shows dependencies of one-molecule (unimolecular) reaction upon photon density. One-photon chemistry is proportional to photon density and then saturated in the high-density region, while multiphoton chemistry increases quickly in the latter. The chemistry of gas phase and dilute solution just corresponds to this case, and has been studied as state-to-state chemistry and ultrafast phenomena by physical chemists. Photochemistry of molecular aggregates such as concentrated solutions, colloid ones, powders, membranes, biological tissues, and materials by low excitation intensity light is given on the yz-plane. The increased reaction efficiency accompanied by an increase of molecular number is, for example, observed for reaction center of photosynthetic unit of green plants.

Until now physical and chemical phenomena in xz- and yz-planes have been mainly elucidated as chemistry; however, a new field is opened in the region with large x- and y-values. Namely, multimolecule-multiphoton chemistry seems to be promising. We suppose that this chemistry is nonequilibrium, just like a plasma state, and cannot be simply interpreted in terms of chemistry in xz- and yz-planes. Excited states, transient species, hot sites, and detects formed interact with each other, absorb the successive photons, and produce new molecular species. Therefore, it can be said that multimolecule-multiphoton chemistry is nonlinear with respect to molecular and photon numbers.

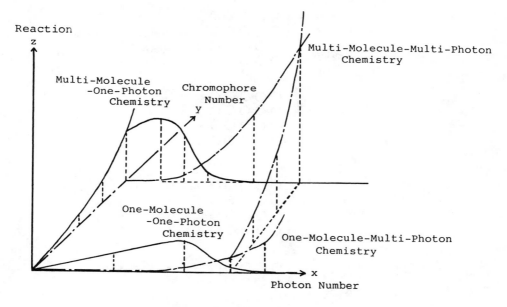

FIGURE 8.   General classification of photochemistry.

Molecular and electronic processes of the present nonlinear photochemistry are quite diversified and depend upon molecular nature, aggregation structure, laser wavelength, its output energy, and pulse width. Using photon density h (1/cm² pulse) and absorption cross-section σ (cm²), where σ is related to molar extinction coefficient ϵ as $\sigma = 3.83 \times 10^{-21} \times \epsilon$, we classify absorption process in order to get a practically important viewpoint for studies on laser photochemistry. The condition $h\sigma > 1$ means that absorption of photon has a low probability, while the reverse inequality indicates a high one. In the latter case deplection of the absorbing state is sometimes induced. Considering the case that absorption by transient species is enough competitive to absorption by the ground state, we include both absorption cross-sections of the ground (σ_g) and the transient (σ_e) states, and classify various absorption processes. The result is illustrated in Figure 9.

$$\text{Case (i)} \quad h\sigma_g < 1, \quad h\sigma_e < 1$$

A small number of excited states are formed. Therefore, a one-photon photochemistry is observed even under high density excitation.

$$\text{Case (ii)} \quad h\sigma_g < 1, \quad h\sigma_e > 1$$

Although the number of the formed excited state is low, the latter state absorbs the laser photon more efficiently than the ground state. The higher excited state is formed and interacts with the surrounding ground state molecules.

$$\text{Case (iii)} \quad h\sigma_g > 1, \quad h\sigma_e < 1$$

The excited singlet states are densely produced, interact with each other, and undergo into new processes.

$$\text{Case (iv)} \quad h\sigma_g > 1, \quad h\sigma_e > 1$$

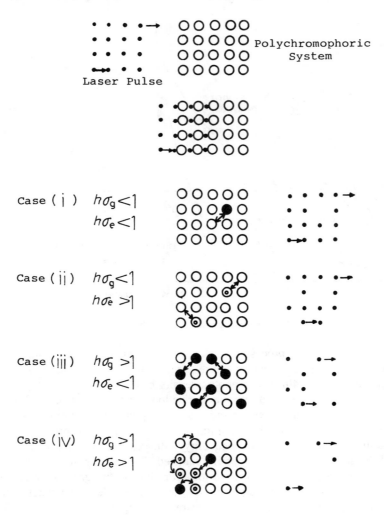

FIGURE 9.   Classification of absorption process of laser photon by polychromophoric system. (○), Ground state, (●), excited state, and (◎), higher excited state.

Both excited and higher excited states are produced and their mutual interactions result in a quite new dynamics.

In this chapter, we summarize our results on polymer systems according to the above classification. In the studies on laser photochemistry of polymers, we have examined the effects of the number of molecules in addition to photon density effects by using polymers with different degrees of polymerization. This is the reason why laser photochemistry of polymers is so interesting even from basic viewpoints. In Case (i), photochemical behavior is just like one-photon photochemistry. We have already summarized the dynamics of poly(N-vinylcarbazole) in solution which was revealed by transient absorption spectroscopy under this condition.[29] Case (ii) was realized by exciting polymers having pyrenyl chromophores with the third harmonic of the Nd^{3+}:YAG laser. The higher excited state interacts with the ground state pyrene and leads to electron transfer. Namely, an intrapolymer multiphoton charge separation is induced. Case (iii) was examined in most detail and their mechanism was established as intrapolymer interactions between excited states and transient polyelectrolyte formation. Case (iv) corresponds to excimer laser excitation of polymer solids which has been studied as laser ablation. All these dynamics are described in the following sections.

-Py; 1-Pyrenyl,

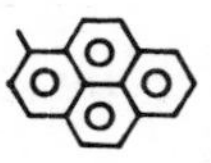

Poly(1-pyrenylalanine):P(L-PyA)-n, P(DL-PyA)-n
  -(NHCH(CH$_2$-Py)co)$_n$-(NHCH(C$_2$H$_4$CO$_2$CH$_2$)CO)$_{150}$-NH(CH$_2$)$_5$CH$_3$

N-Acetyl-L-1-pyrenylalaninemethylester:L-PyA
  CH$_3$CONHCH(CH$_2$-Py)COOCH$_3$

Monomer ester model
  CH$_3$COOCH$_2$CH(CH$_2$-Py)CH$_2$OCOCH$_3$

Dimer ester model
  CH$_3$COOCH$_2$CH(CH$_2$-Py)CH$_2$OCO(CH$_2$)$_m$COOCH$_2$CH(CH$_2$-Py)CH$_2$OCOCH$_3$

Polyester
  -(OCH$_2$CH(CH$_2$-Py)CH$_2$OCO(CH$_2$)$_m$CO)$_n$-

Poly(1-vinylpyrene)
  -(CH$_2$CH-Py)$_n$-

Dipeptide model
  CH$_3$CONHCH(CH$_2$-Py)CONHCH(CH$_2$-Py)COOCH$_3$

FIGURE 10.   Molecular structures and abbreviations of polymers studied and their reference compounds. The notation n is the mean degree of polymerization.

IV. INTRAPOLYMER MULTIPHOTON CHARGE SEPARATION[30,31]

Case (ii) in Figure 9 corresponds to successive two-photon absorptions which are experimentally identical with inner filter effects in transient absorption measurements. This leads to the formation of the higher excited states with the singlet or triplet spin manifolds which are deactivated rapidly to the lowest excited state or eject electron, depending upon solvent polarity. In addition to these processes, a novel behavior characteristic of polymers with specific structures was confirmed recently. Geometrical structures of the polymers studied are listed in Figure 10. The same 1-pyrenyl group is connected to polypeptide, polyester, and polyvinyl chains, so that effects of higher-order structure, mean chromophore distance, and/or local concentration upon the photochemical processes can be elucidated. It is worth noting that the polymer, P(L-PyA)-n, has a helical structure.[32]

The picosecond 355-nm photolysis was performed in *N,N*-dimethylformamide (DMF) solution. In Figure 11 the absorption spectra of P(L-PyA)-20 at 33 picosecond and P(DL-PyA)-20 at 100 picosecond consist of three bands at 420, 460, and 500 nm. These are quite different from the $S_n \leftarrow S_1$ absorption spectra of the monomer model compound, L-PyA, measured under the same experimental conditions. In order to assign these three peaks, transient absorption spectra of L-PyA quenched with O_2, an electron acceptor, and a donor were measured. Upon quenching, intersystem crossing and electron transfer processes were enhanced, giving the triplet, cationic, and anionic states of L-PyA. On the basis of these reference data, we concluded that the triplet and ionic states are formed in this polypeptide immediately after picosecond 355-nm excitation. All these states are formed by multiphoton processes which is supported by examining an excitation intensity effect upon the transient absorption spectra. As shown in Figure 12, the spectrum of P(DL-PyA)-20 at low excitation intensity loses its structure and can be interpreted in terms of a monomer-excimer equilibrium. Upon lowering the excitation intensity to one tenth of that of normal laser photolysis, the ionic absorptions were still observed whereas the triplet band was suppressed. This detailed examination shows that the dependence of the triplet band upon the excitation intensity is

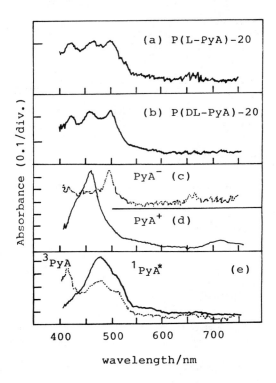

FIGURE 11.   Transient absorption spectra of the polypeptide and reference spectra of the model compound. (a) P(L-PyA)-20 at 33 ps, (b) P(DL-PyA)-20 at 100 ps, (c) L-PyA anion, (d) L-PyA cation, and (e) the excited singlet and triplet PyA.

larger than that of the ionic bands. We concluded that the intrapolymer charge separation resulting in the pyrenyl cation and anion is induced by a two-photon excitation, and that the triplet may be formed via a three-photon process. In other words, the triplet state may be formed from a higher excited state of the charge-separated state.

There are two possible interpretations for the present intrapolymer charge separation. One interpretation is that an electron ejection occurs from the higher excited state of pyrene which is formed by a successive two-photon absorption. The energy level of such a state can be estimated to be 6.8 eV, and higher than the ionization potential of pyrene in DMF. The ejected electron could be captured by the neighboring pyrenyl chromophore, leading to intrapolymer charge separation. However, picosecond excitation of the monomer model, L-PyA, did not result in the two-photon ionization, which may be due to the fact that the solvent DMF is an electron donor. The other interpretation is to consider the role of the Rydberg state. This state of pyrene in solution can be estimated to be around 5 eV above the ground state according to the rule that the lowest Rydberg state has an energy level lower than the gas-phase ionization potential by 2.5 eV.[33] The electron density of the molecule in this state is more expanded compared to other electronic configurations, so interchromophoric interaction is considered to be more easily induced. Even if the Rydberg state is not reached directly by two-photon excitation, a mixing of this state with the higher excited states results in a similar kind of interaction. In the case that the neighboring chromophore, in its ground state, is within the long interaction radius of the Rydberg state, electron transfer from the latter to the former may be brought about. This idea is similar to the discussion on photoionization of some amines[34] and can explain the present characteristic behavior of polypeptides.

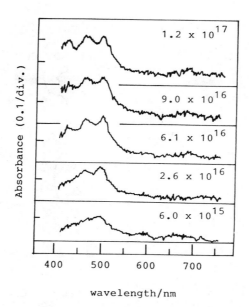

FIGURE 12. Excitation intensity dependence of the transient spectrum of P(DL-PyA)-20 in DMF at 100 ps. Excitation intensity in photons per square centimeter are given in the figure.

Although the history of photophysics of π-electronic molecules is very long, a charge separation between identical molecules is a rare phenomenon. An exceptional case was reported for some dyes and for octaethylporphyrins,[35] and no report has been given for aromatic hydrocarbons. This suggested that the higher-order structure characteristic of polypeptides has an important role in the present primary process; however, polyesters whose conformation are flexible showed the similar dynamics. In the case of poly(1-vinylpyrene) normal monomer-excimer dynamics was observed under the same experimental condition.[36] We conclude that a loose structure of relevant chromophores is an important factor for inducing electron transfer between identical moieties and that their strict mutual orientation is not required. This process is included in a schematic diagram for laser photochemistry of polymers in Figure 13.

V. INTRAPOLYMER INTERACTIONS BETWEEN EXCITED SINGLET STATES

High density of excited states in one polymer chain in solution can be easily realized in Case (iii) of Figure 9. Formed excited states migrate along the polymer chain and interact with each other, leading to an efficient deactivation. This intrapolymer S_1-S_1 annihilation process was first confirmed in the carbazole polymers shown in Figure 14, although interaction between excited states are quite familiar phenomena for concentrated solutions of molecules, semiconductors, and molecular crystals.[37] Here we summarize the experimental results confirming the intrapolymer S_1-S_1 annihilation as follows:

1. Excitation intensity dependence of fluorescence spectra.[17] In the case of PVCz, two kinds of excimers are observed, and their relative intensity shows an interesting dependence upon excitation intensity. The so-called second excimer with a partial overlap structure increases relatively with increasing excitation intensity, which is more pronounced in the case of PVCz with higher degree of polymerization. This behavior can

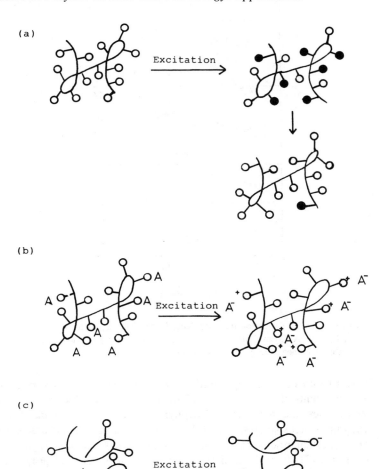

FIGURE 13.   Schematic diagram of laser photochemistry of polymers. (a) Intra-polymer S_1-S_1 annihilation, (b) transient polyelectrolyte formation, and (c) intra-polymer multiphoton charge separation. ●, Excited state; ○⁺, cationic state; ○⁻, anionic state; A, electron acceptor; and A⁻, acceptor anion.

be ascribed to the fact that the sandwich excimer is produced from the monomer as well as the second excimer states and all these fluorescent states interact with each other.

2.   Excitation intensity dependence of fluorescence intensity.[38] We compared excitation intensity dependences of the total fluorescence intensity and of the transmitted intensity of N_2 laser pulse for some polymers. As shown in Figure 15, the fluorescence intensity increases with the excitation intensity, but a deviation from a linear relation is observed for all the systems. The observed curves approach to a saturation rather rapidly at high excitation intensity, and its degree is in the order of EtCz < Pu-I-44 = PU-II-25 < PCzEVE-33 < PVCz-42. This tendency corresponds to a decreasing order of the mean distance between neighboring chromophores. On the other hand, the deviation of the transmitted intensity of the excitation pulse from a linear relation is in the order of EtCz > PU-I-44 = PU-II-25 > PCzEVE-33 > PVCz-42. This is opposite to that of

FIGURE 14. Molecular structures and abbreviations of polymers studied and their reference compounds. The notation n is the mean degree of polymerization.

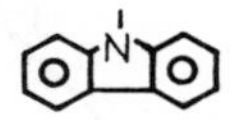

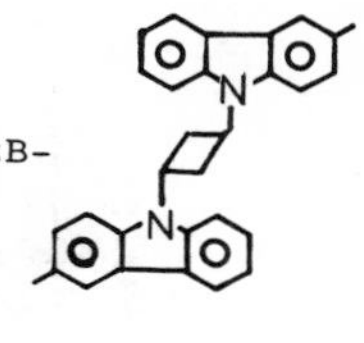

fluorescence intensity, suggesting that the ground state molecules are recovered less efficiently in monomer than in polymers.

3. Excitation intensity dependence of fluorescence quenching efficiency.[17,39] In the systems containing electron donors or acceptors, the S_1-S_1 annihilation competes with electron transfer quenching, which means that a simple Stern-Volmer relation does not hold. Actually we confirmed that quenching efficiencies measured by steady light and intense laser pulse excitation are different from each other.

4. Excitation intensity dependence of fluorescence decay curve. Direct experimental proof for the interaction between excited states is due to a change in fluorescence decay curves. In the nanosecond time region, the sandwich excimer emission of PVCz-400 shows a shortening from 35 to 14 nanosecond under intense laser excitation.[17] A more rapid decay in the picosecond time region was also observed by using a picosecond Nd^{3+}:YAG pulse (355 nm) and a streak camera system.[38] As demonstrated in Figure 16, the present systems give a rapid decay in addition to the slow one.

On the basis of these results, we concluded that an intrapolymer S_1-S_1 annihilation is inevitably involved in the polymers, which corresponds to Case (iii) in Figure 9. Actually, this process is confirmed for polymers having pyrenyl groups excited by the second harmonics of Ruby laser[40] and for poly(methyl methacrylate) with pendant anthryl moieties.[41] It was experimentally confirmed that the ground state molecule or the triplet state is formed by the S_1-S_1 annihilation process:

$$S_1 + S_1 \longrightarrow S_n + S_0 \longrightarrow S_1 + S_0$$
$$\searrow T_1 + T_1$$

In film, not only intrapolymer but also interpolymer processes operate efficiently, be-

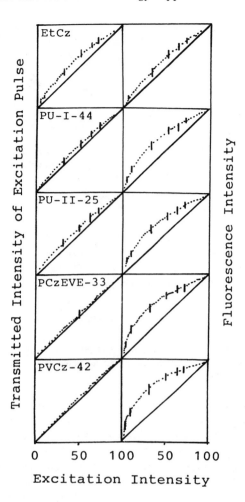

FIGURE 15. Variations of the fluorescence intensity (420 nm) and the transmitted intensity of the excitation pulse (337 nm) with the latter excitation intensity in DMF. The excitation intensity 100 corresponds to about 10^{16} photons per square centimeter. The lines connecting the origin and the value measured at intensity 100 are given to have a slope of 45° as a measure for linearity.

cause unimolecular decay kinetics is hardly measured by a pulsed laser excitation.[42] The present S_1-S_1 annihilation process is schematically shown in Figure 13. Intrapolymer interactions between excited states are, of course, extended to singlet-triplet and triplet-triplet pairs which was confirmed by Webber et al.[43]

VI. FORMATION OF TRANSIENT POLYELECTROLYTE

It is well established that electron transfer quenching of the excited singlet state in polar solvents results in the formation of a donor cation and an acceptor anion.[2,44] In laser photolysis of polymer-quencher systems, this electron transfer competes with the S_1-S_1 annihilation. By increasing the quencher concentration, the quenching is able to overcome the latter process, and some of the excited states in one polymer chain are converted to the corresponding ion radicals. The net effect of this process is that polymers having a number of ionic chromophores are formed by one pulse excitation. For example, a ruby laser excitation

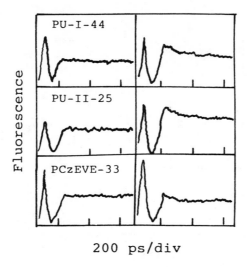

FIGURE 16.   Fluorescence decay curves observed by low (left) and high (right) excitation. The first sharp pulse is a reference pulse used for calibration and data processing.

of PVCz-400 resulted in a polymer with pendant about 70 cationic chromophores.[17,45] Energy migration and interaction between the excited states are determined by dipole-dipole interactions, while electron migration along the polymer is due to an exchange interaction. Since the latter interaction is a short-range one and electron or hole transfer requires solvent reorientation and segment motion, the charge may migrate slowly. Recombination of the polymer and quencher ion radicals determines their decay time. In consequence, ionic states in the polymer are much longer lived than the excited singlet states. We named this process "laser-induced formation of a transient polyelectrolyte".[17] This process is also shown schematically in Figure 13.

Examining various systems, the conditions favoring the laser-induced formation of transient polyelectrolytes are summarized as follows.

1.   The absorption cross-section of the chromophore bound to the polymer is large at the excitation wavelength.
2.   Nonradiative processes other than the formation of free ions occur with small rate constants.
3.   The mean distance between chromophores is long.

Condition 1 is determined by the electronic structure of the chromophore, and 2 is characteristic of the donor-acceptor pair. The present pair of carbazole-dimethyl terephthalate satisfies these conditions to some extent, and PU-I and PU-II are the polymers having a suitable structure. Finally, it has to be mentioned that the residence time of the polyelectrolyte will be lengthened by preventing its recombination with the counter ion radical.

VII. LASER ABLATION OF POLYMER FILMS

All the phenomena described here for polymers having *N*-carbazolyl and 1-pyrenyl groups have been performed mainly with laser intensity around 10 mJ/cm². What happens, if the excitation intensity is increased furthermore, namely, if case (iv) of Figure 9 is conditioned, is the subject of the present section. We investigated the results on poly(*N*-vinyl-carbazole) films[46] because their laser photochemistry has been elucidated in most detail. An

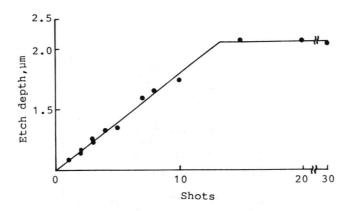

FIGURE 17.   Etch depth of poly(N-vinylcarbazole) film with a thickness of about 2.1 μm as a function of the number of laser shots. The irradiation intensity was 160 mJ/cm^2.

irradiation with a 248-nm excimer laser leads to etching of films. It was clear that the surface of this film is homogeneous and very smooth, while laser irradiation leads to morphological changes. The irradiated area was removed to some extent and a hole was left, indicating laser-induced etching. A periodic pattern extended from the masked area to the inner irradiated one, which covers over 20 μm. This behavior was observed over most of the irradiation intensity range examined.

In order to confirm the present etching phenomenon, a relation between etch depth and the number of laser shots was examined. One of the examples, obtained by fixing the irradiation intensity to 160 mJ/cm^2, is given in Figure 17. The etch depth increased in proportion to the number of laser shots and the film with thickness of 2.1 μm was completely etched with less than 15 shots. This means that optical and chemical properties of the etched surface are almost the same to those of the original one. The etch rate, which is a depth brought about with one shot of irradiation, is plotted against irradiation intensity in Figure 18. The etch rate increased with the latter intensity; however, linear relations were not obtained. We estimated 30 mJ/cm^2 to be the threshold of laser-induced etching. It is worth noting that the latter laser intensity and the molar extinction coefficient at 248 nm satisfies the condition of Case (iv) in Figure 9.

Under the same experimental conditions, the etching behavior of poly(N-vinylcarbazole) films doped with photophysically interesting molecules, perylene and dimethyl terephthalate (DMTP), was investigated. The concentration of additives was around 3% mol per mole carbazole unit, so that an initial absorption process of laser photons still occurred in the carbazole chromophore. The result on the perylene-doped film is included in Figure 18. The relation between etch rate and irradiation intensity was identical with that for the neat film within an experimental error. This result was supported by morphological studies, using SEM and depth profile measurement. On the other hand, etched profiles of polymer films containing DMTP were different from those of the neat film described above. In some irradiated areas an efficient removal of the polymer occurred similarly as in the neat film, while large residuals were left randomly. Namely, no smooth etched profile was obtained, whose typical example is shown in Figure 19. Such a distinct difference was independent of laser irradiation intensity.

The present ablation phenomena are considered here in terms of laser chemistry. Since this ablation was not observed under weak irradiation intensity and has a threshold value, multiphoton absorption processes should be involved. As the formation of the sandwich as well as the partial overlap excimers is almost completed around 1 ns,[47] the excimer states

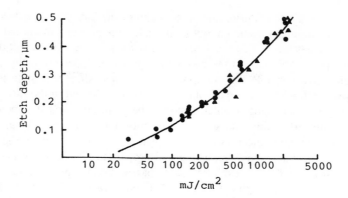

FIGURE 18. Variation of etch rate (etch depth/shot) with irradiation intensity. Poly(*N*-vinylcarbazole) films doped with 2.6×10^{-2} mol perylene per mole carbazole unit (▲) and without dopant (●).

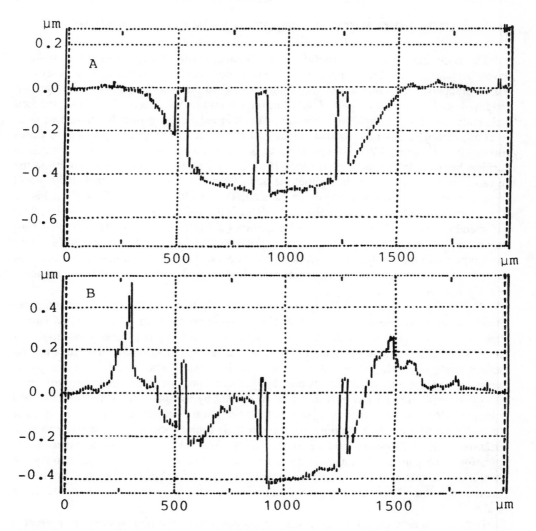

FIGURE 19. Depth profiles of poly(*N*-vinylcarbazole) films (A) without dopant and (B) with 3.3×10^{-2} mol dimethyl terephthalate per mole carbazole unit. The irradiation intensity was 1.29 J/cm². The effective width of the mask used was different in (A) and (B).

absorb the successive photons during the pulse width of the 248-nm laser, leading to higher excited states. This high electronic energy is dissipated to various vibrational modes which may induce bond cleavage. Since the latter local density is very high, the decomposition reaction "grew up" to ablation. In the case of the perylene-doped film, the excited singlet state of perylene is responsible for multiphoton processes, since this state is also formed rapidly during irradiation. The similar ablation behavior observed in the neat and perylene-doped films suggests that higher excited states of perylene may transfer excitation energy to the carbazole unit efficiently and the same reaction induced. This explanation is based on the fact that the doping concentration of perylene is too low to change the morphology of this film and its mechanical as well as thermal properties. On the other hand, successive multiphoton absorptions by the exciplex of carbazole-DMTP pair produced the excited exciplex. This means that the high concentration of active ion-pair-like state is brought about by laser irradiation, which is characterized by strong mutual interactions due to Coulomb force. We consider that this particular state, formed by multimolecule-multiphoton interaction, may induce different energy dissipation processes, different chemical reactions, and morphological changes, leading to the different etching profiles.

VIII. TWO-PHOTON PHOTOLYSIS OF POLYMERS

The wide applicability and usefulness of transient absorption spectroscopy in solution photochemistry are well recognized, but reports on polymer films are rare. In the latter case the chromophore concentration is very high, leading to the fact that the excited states are produced only in the surface part of the film. Interactions between excited states are induced efficiently, resulting in rapid local heating. This thermal energy cannot be dissipated as in solution, causing cracking or optical damage of the solid. This is just responsible to Cases (iii) or (iv) of Figure 9 and is considered to be the main factor for making it difficult to perform laser photolysis studies on films. If excited states were formed, not densely in the surface, but homogeneously throughout the bulk, interactions between excited states would be suppressed and no cracking would be induced. This condition, namely Case (i) of Figure 9, is attained simply when the molar extinction coefficient at the excitation wavelength is sufficiently small. Actually, the $T_n \leftarrow T_1$ absorption spectrum of a single crystal of benzophenone was measured by N_2 gas laser photolysis, where the excitation photon (337 nm) is absorbed into the weak n-π* transition. We came to the conclusion that a simultaneous two-photon excitation is a useful excitation method to produce the excited states homogeneously in bulk.

We[21] and Hamanoue et al.[22] reported independently that simultaneous two-photon excitation of neat aromatics so efficient that the absorption spectrum of the excited singlet state can be measured under the normal picosecond photolysis condition. This nonlinear effect, summarized as 7 in the Section II, provides a new spectroscopic method for studying various kinds of solid. In the following we introduce briefly the experimental results on two-photon photolysis and discuss its possibility. Neat liquids of several benzene derivatives in a quartz cell with a 1-cm path length were excited by the 347- and 351-nm picosecond pulses.[21,48] It should be noted that one-photon absorption is completely neglected at these wavelengths. Measured absorption spectra and their time-dependence were interpreted in terms of excimer formation and decay dynamics. After examining various experimental conditions, we concluded that these excimers are formed via a simultaneous two-photon absorption. If this solution sample is replaced by polymer films, solid-state photolysis becomes possible. This idea was recently demonstrated for polystyrene films by Miyasaka et al.[49]

Here our recent experimental result on poly(N-vinylcarbazole) systems is presented briefly. We examined a very viscous, glass-like solution of this polymer in THF. This sample

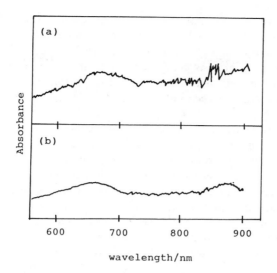

FIGURE 20. Transient absorption spectra of poly(*N*-vinylcarbazole) at 100 ps: (a) 532-nm two-photon photolysis of the glass-like solution and (b) 355-nm one-photon photolysis of the dilute solution.

was used instead of a block of solid polymer because the latter was very difficult to prepare. Its picosecond 532-nm photolysis gave the absorption spectrum shown in Figure 20.[50] Since the carbazolyl chromophore has no one-photon absorption at this wavelength, the present result is due to simultaneous two-photon excitation. A comparison with the spectrum of a dilute solution of the same polymer, measured by the normal 355-nm photolysis, is also given in Figure 20. It is said that the electronic structure of the polymer glass seems to be very similar to that of the isolated polymer in dilute solution. This experimental demonstration indicates that two-photon photolysis is a general method for photophysical and photochemical studies on polymer solids.

IX. SUMMARY

In this chapter, we have summarized the present stage of transient absorption spectroscopy, classified laser chemistry, and described the characteristic behavior of polymers excited by an intense laser pulse. Nonlinear photochemical behavior is quite general for polymer systems and commonly observed in their transient absorption spectroscopy. An exceptional case where one-photon photochemistry occurs mainly is a laser photolysis of polymers having carbonyl groups which was reported by Schnabel et al.[51] No distinct contribution of intra-polymer S_1-S_1 annihilation was observed, which was ascribed to a small value of the extinction coefficient at the laser wavelength and to fast intersystem crossing. On the other hand, studies on primary photoprocesses of polymers with pendant aromatic groups can be performed only by examining excitation intensity effects in detail. This has been demonstrated clearly by us and Webber et al., as explained above in detail.

Some important questions on polymer dynamics, which have nothing to do with high-intensity effects, can be answered by studying nonlinear photochemical behavior. By analyzing the rapid fluorescence decay curves due to the S_1-S_1 annihilation, a rate of excitation energy migration along the polymer chain was estimated.[38] As this annihilation process is a function of the mean interchromophoric distance, different polymer conformations in poor and rich solvents were confirmed by analyzing excitation intensity dependences of fluorescence intensity.[39] Concerning the dynamic aspects of dilute polyelectrolyte solutions, we believe that the study of transient polyelectrolytes will yield fruitful information. Furthermore, laser photolysis studies on polymers have made interesting contributions to photochemistry and spectroscopy, themselves. Electron transfer between identical chromophores is a rare process; however, this was confirmed in polymer systems and a new role of the higher excited states was pointed out. Simultaneous two-photon photolysis makes it possible to measure transient absorption spectra of solids and should now open a new field of solid photochemistry.

Before closing this chapter, we emphasize again that nonlinear photochemical behavior is quite general, and occurs during any interactions between an intense laser pulse and a condensed organic material. Not only polymer systems but also any molecularly associated ones such as crystals, amorphous solids, films, powders, and so on, will show similar phenomena upon excitation with an intense laser pulse. Molecular and electronic aspects of nonlinear multimolecule-multiphoton chemistry will be correlated with one-molecule-one-photon chemistry, which will be a beginning of the present new field. As an example, photodynamics of poly(N-vinylcarbazole) is summarized in Figure 21. We believe that this type of approach is indispensable and strongly required at the present stage of investigation.

ACKNOWLEDGMENTS

The author wishes to express his sincere thanks to Prof. A. Itaya and Dr. N. Ikeda for their discussion and help. The present work is partly supported by Grant-in-Aid (61470006), one for Special Project Research (62113007), and one for Scientific Research on Priority Area of "Dynamic Interactions and Electronic Processes of Macromolecular Complexes" (62612507) from the Japanese Ministry of Education, Science, and Culture.

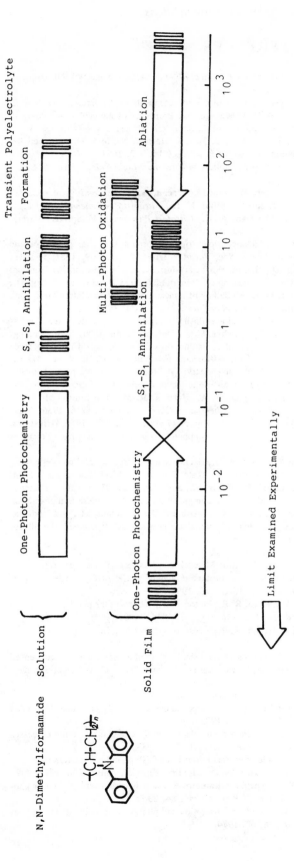

FIGURE 21.   Dependences of poly(*N*-vinylcarbazole) photophysics and photochemistry upon excitation intensity (mJ/cm²).

REFERENCES

1. **Claesson, S.,** *Fast Reactions and Primary Processes in Chemical Kinetics,* Almqvist and Wiksell, Uppsala, 1967.
2. **Mataga, N. and Ottolenghi, M.,** Photophysical aspects of exciplexes, in *Molecular Association,* Vol. 2, Forster, R., Ed., Academic Press, London, 1979, 1; **Mataga, N.,** Properties of molecular complexes in the electronic excited states, in *Molecular Interactions,* Vol. 2, Ratajczak, H. and Orville-Thomas, W. J., Eds., John Wiley & Sons, London, 1981, 509; **Masuhara, H. and Mataga, N.,** Ionic photodissociation of electron donor-acceptor systems in solution, *Accounts Chem. Res.,* 14, 312, 1981.
3. **Von der Linde,** Picosecond interactions in liquids and solids, in *Ultrashort Light Pulses,* Shapiro, S. L., Ed., Springer-Verlag, Berlin, 1977, 203.
4. **Eisenthal, K. B., Hochstrasser, R. M., Kaiser, W., and Laubereau, A.,** *Picosecond Phenomena III,* Springer-Verlag, Berlin, 1982; **Auston, D. H. and Eisenthal, K. B.,** *Ultrafast Phenomena IV,* Springer-Verlag, Berlin, 1984; **Fleming, G. R. and Siegman, A. E.,** *Ultrafast Phenomena V,* Springer-Verlag, Berlin, 1986.
5. **Kessler, R. W. and Wilkinson, F.,** Diffuse reflectance triplet-triplet absorption spectroscopy of aromatic hydrocarbons chemisorbed on γ-alumina, *J. Chem. Soc. Faraday Trans. 1,* 77, 309, 1981; **Kessler, R. W., Krabichler, G., Uhl, S., Oelkrug, D., Hogan, W. P., Hyslop, J., and Wilkinson, F.,** Transient decay following pulse excitation of diffuse scattering samples, *Optica Acta,* 30, 1099, 1983; **Oelkrug, D., Honnen, W., Wilkinson, F., and Willsher, C. J.,** Modelling of transient production and decay following laser excitation of opaque material, *J. Chem. Soc. Faraday Trans. 2,* 83, 2081, 1987.
6. **Wilkinson, F., Willsher, C. J., and Pritchard, R. B.,** Laser flash photolysis of dyed fabrics and polymers-1, *Eur. Polym. J.,* 21, 333, 1985; **Wilkinson, F., Willsher, C. J., Uhl, S., Honnen, W., and Oelkrug, D.,** Optical detection of a photoinduced thermal transient in titanium dioxide powder by diffuse reflectance laser flash photolysis, *J. Photochem.,* 33, 273, 1986; **Wilkinson, F., Willsher, C. J., Casal, H. L., Johnston, L. J., and Scaiano, J. C.,** Intrazeolite photochemistry. IV. Studies of carbonyl photochemistry on the hydrophobic zeolite silicalite using time-resolved diffuse reflectance techniques, *Can. J. Chem.,* 64, 539, 1986; **Wilkinson, F., Willsher, C. J., Leicester, P. A., Barr, J. R. M., and Smith, M. J. C.,** Picosecond diffuse reflectance laser flash photolysis, *J. Chem. Soc. Chem. Commun.,* 1216, 1986; **Wilkinson, F.,** Diffuse reflectance flash photolysis, *J. Chem. Soc. Faraday Trans. 2,* 82, 2073, 1986; **Wilkinson, F. and Willsher, C. J.,** The use of diffuse reflectance laser flash photolysis to study primary photoprocesses in anisotropic media, *Tetrahedron,* 43, 1197, 1987.
7. **Ikeda, N., Imagi, K., Hayabuchi, M., and Masuhara, H.,** Excited states in organic solid powders; time-resolved diffuse reflectance spectroscopic measurements, in *Abstr. 52nd Annu. Meet. Chem. Soc. Japan,* Vol. 1, Tokyo, 1986, 290; **Koshioka, M., Imagi, K., Ikeda, K., and Masuhara, H.,** Diffuse reflectance laser photolysis study on adsorbed molecules and inclusion complexes, in *Abstr. Japanese Symposium on Photochemistry,* Sendai, 1987, 165; **Imagi, K., Ikeda, N., Masuhara, H., Nishigaki, M., and Isogawa, M.,** Photochemical transient species of poly(ethylene terephthalate) powders as revealed by the diffuse reflectance laser photolysis method, *Polym. J.,* 19, 999, 1987.
8. **Kuroda, T., Ikeda, N., and Masuhara, H.,** to be published.
9. **Okada, T., Migita, M., Mataga, N., Sakata, Y., and Misumi, S.,** Picosecond laser spectroscopy of intramolecular heteroexcimer systems; time-resolved absorption studies of p-$(CH_3)_2NC_6H_4$-$(CH_2)_n$-(1-pyrenyl) and -(9-anthryl) systems, *J. Am. Chem. Soc.,* 103, 4715, 1981.
10. **Greene, B. I., Hochstrasser, R. M., and Weisman, R. B.,** Picosecond dynamics of the photoisomerization of trans-stilbene under collision-free conditions, *J. Chem. Phys.,* 71, 544, 1979.
11. **Greene, B. I., Hochstrasser, R. M., and Weisman, R. B.,** Picosecond transient spectroscopy of molecules in solution, *J. Chem. Phys.,* 70, 1247, 1979.
12. **Anderson, R. W.,** Vibrational relaxation in the lowest excited singlet state of anthracene in the condensed phase, in *Picosecond Phenomena II,* Hochstrasser, R. M., Kaiser, W., and Shank, C. V., Eds., Springer-Verlag, Berlin, 1980, 163.
13. **Imagi, K., Ikeda, N., and Masuhara, H.,** to be published.
14. **Masuhara, H., Ikeda, N., Miyasaka, H., and Mataga, N.,** Microcomputer-controlled picosecond Nd^{3+}:YAG laser photolysis system, *J. Spectrosc. Soc. Jpn.,* 31, 19, 1982.
15. **Miyasaka, H., Masuhara, H., and Mataga, N.,** Picosecond absorption spectra and relaxation processes of the excited singlet state of pyrene in solution, *Laser Chem.,* 1, 357, 1983.
16. **Goldschmidt, C. R.,** Some considerations in the interpretation of laser flash photolysis measurements, in *Lasers in Physical Chemistry and Biophysics,* Joussot-Dubien, J., Ed., Elsevier, Amsterdam, 1975, 499.
17. **Masuhara, H., Ohwada, S., Mataga, N., Itaya, A., Okamoto, K., and Kusabayashi, S.,** Laser photochemistry of poly(N-vinylcarbazole) in solution, *J. Phys. Chem.,* 84, 2363, 1980.
18. **Fischer, M. M., Veyret, B., and Weiss, K.,** Nonlinear absorption and photoionization in the pulsed laser photolysis of anthracene, *Chem. Phys. Lett.,* 28, 60, 1974.

19. **Auston, D. H.,** Picosecond nonlinear optics, in *Ultrashort Light Pulses,* Shapiro, S. L., Ed., Springer-Verlag, Berlin, 1977, chap. 4.
20. **Kliger, D. S.,** Thermal lensing; a new spectroscopic tool, *Accounts Chem. Res.,* 13, 129, 1980; **Fuke, K.,** private communication, 1982.
21. **Masuhara, H., Miyasaka, H., Ikeda, N., and Mataga, N.,** Picosecond two-photon photolysis of neat liquids, *Chem. Phys. Lett.,* 82, 59, 1981.
22. **Hamanoue, K., Hidaka, T., Nakayama, T., and Teranishi, H.,** Excimer formation of neat benzene derivatives studied by picosecond spectroscopy, *Chem. Phys. Lett.,* 82, 55, 1981.
23. **Wilkinson, F. and Willsher, C. J.,** Diffuse reflectance laser flash photolysis of dyed fabrics and polymers, in *Lasers in Polymer Science and Technology: Applications,* Vol. 2, Fouassier, J. P. and Rabek, J. F., Eds., CRC Press, Boca Raton, FL, 1989, chap. 9.
24. **Ikeda, N., Imagi, K., Masuhara, H., Nakashima, N., and Yoshihara, K.,** Picosecond transient absorption spectral and kinetic study on benzophenone microcrystals by diffuse reflectance laser photolysis method, *Chem. Phys. Lett.,* 140, 281, 1987.
25. **Ikeda, N., Imagi, K., Hayabuchi, M., and Masuhara, H.,** Excited states in organic solid powders; time-resolved diffuse reflectance spectroscopic measurement, *Abstr. 52nd Annu. Meet. Chemical Society of Japan,* Vol. 1, Chemical Society of Japan, Tokyo, 1986, 290.
26. **Morris, J. M. and Yoshihara, K.,** Interband transitions in molecular crystals; triplet-triplet absorptions in ketone crystals, *Mol. Phys.,* 36, 993, 1977.
27. **Kuroda, T.,** M.Sc. thesis, Kyoto Institute of Technology, 1988.
28. **Masuhara, H. and Itaya, A.,** Time-resolved total internal reflection fluorescence spectroscopy for dynamic studies on surface, in *Lasers in Polymer Science and Technology: Applications,* Vol. 2, Fouassier, J. P. and Rabek, J. F., Eds., CRC Press, Boca Raton, FL, 1989, chap. 10.
29. **Masuhara, H.,** Laser photochemistry of polymers, *Makromol. Chem. Suppl.,* 13, 75, 1985.
30. **Masuhara, H., Tanaka, J. A., Mataga, N., Sisido, M., Egusa, S., and Imanishi, Y.,** Intrapolymer charge separation induced by picosecond multiphoton excitation; synthetic polypeptides with a pendant 1-pyrenyl group in *N,N*-dimethylformamide, *J. Phys. Chem.,* 90, 2791, 1986.
31. **Masuhara, H., Tanaka, J. A., Mataga, N., Higuchi, Y., and Tazuke, S.,** Intrapolymer charge separation induced by picosecond multiphoton excitation; polyesters with pendant 1-pyrenyl groups in DMF, *Chem. Phys. Lett.,* 125, 246, 1986.
32. **Egusa, S., Sisido, M., and Imanishi, Y.,** Synthesis and spectroscopic properties of poly(L-1-pyrenylalanine), *Chem. Lett.,* 1307, 1983; **Egusa, S., Sisido, M., and Imanishi, Y.,** One-dimensional aromatic crystals in solution. IV. Ground- and excited-state interactions of poly(L-1-pyrenylalanine) studied by chiroptical spectroscopy including circularly polarized fluorescence and fluorescence-detected circular dichroism, *Macromolecules,* 18, 882, 1985; **Egusa, S., Sisido, M., and Imanishi, Y.,** One-dimensional aromatic crystals in solution. V. Empirical energy and theoretical circular dichroism calculations on helical poly(L-pyrenylalanine), *Macromolecules,* 18, 890, 1985.
33. **Robin, M. B.,** *Higher Excited States of Polyatomic Molecules,* Vol. 1 and 2, Academic Press, New York, 1974 and 1975.
34. **Nakato, Y.,** Photoelectron ejection from tetraaminoethylenes in aromatic hydrocarbons; role of the molecular "Rydberg" state in excited charge-transfer states and long-range electron-transfer processes, *J. Am. Chem. Soc.,* 98, 7203, 1976.
35. **Koizumi, M., Kato, S., Mataga, N., Matsuura, T., and Usui, Y.,** *Photosensitized Reactions,* Kagaku Dojin, 1977, chap. 6; **Kurabayashi, Y., Kikuchi, K., and Kokubun, H.,** Relaxation processes of excited octaethylporphyrins, in *Abstr. 24th Jpn. Symp. Photochemistry,* Tsukuba, 1983, 43.
36. **Tanaka, J. A., Masuhara, H., and Mataga, N.,** Absorption spectra of poly(l-vinylpyrene) in the excited and ionic states, *Polym. J.,* 18, 181, 1986.
37. **Nakashima, N. and Mataga, N.,** Electronic excitation transfer between the same kind of excited molecules in rigid solvents under high-density excitation with lasers, *J. Phys. Chem.,* 79, 1788, 1975, and papers cited therein.
38. **Masuhara, H., Tamai, N., Inoue, K., and Mataga, N.,** Intrapolymer interactions between the excited singlet states in dilute solution, *Chem. Phys. Lett.,* 91, 109, 1982.
39. **Masuhara, H., Shioyama, H., Mataga, N., Inoue, T., Kitamura, N., Tanabe, T., and Tazuke, S.,** Laser photochemistry of polymers having 1,2-trans-dicarbazolylcyclobutane groups in solution, *Macromolecules,* 14, 1738, 1981.
40. **Masuhara, H., Ohwada, S., Seki, Y., Mataga, N., Sato, K., and Tazuke, S.,** Photophysics and ionic photodissociation of polyesters with pendant 1-pyrenyl groups in solution, *Photochem. Photobiol.,* 32, 9, 1980.
41. **Hargreaves, J. S. and Webber, S. E.,** Photophysics of anthracene polymers; fluorescence, singlet energy migration, and photodegradation, *Macromolecules,* 17, 235, 1984.
42. **Masuhara, H., Tamai, N., Ikeda, N., Mataga, N., Itaya, A., Okamoto, K., and Kusabayashi, S.,** Excimer dynamics in poly (*N*-vinylcarbazole) films, *Chem. Phys. Lett.,* 91, 113, 1982.

43. **Pasch, N. F. and Webber, S. E.,** The effect of molecular weight on triplet exciton processes. II. Poly(2-naphthyl methacrylate), *Macromolecules,* 11, 727, 1978; **Kim, N. and Webber, S. E.,** Effect of molecular weight on triplet excitation processes. IV. Delayed emission of solid poly(2-vinylnaphthalene), *Macromolecules,* 13, 1233, 1980; **Pratte, J. F. and Webber, S. E.,** Intracoil triplet-triplet annihilation in poly(4-vinylbiphenyl) in benzene solution, *Macromolecules,* 16, 1193, 1983; **Pratte, J. F. and Webber, S. E.,** Intracoil triplet-triplet annihilation in poly(2-vinylnaphthalene) in benzene solution, *J. Phys. Chem.,* 87, 449, 1983.

44. **Masuhara, H.,** Electron transfer dynamics in the excited polymer and related systems in solution, in *Photophysical and Photochemical Tools in Polymer Science,* Winnik, M., Ed., D. Reidel, Dordrecht, Netherlands, 1985, 65.

45. **Masuhara, H., Ohwada, S., Yamamoto, K., Mataga, N., Itaya, A., Okamoto, K., and Kusabayashi, S.,** Laser-induced formation of transient polyelectrolyte in solution, *Chem. Phys. Lett.,* 70, 276, 1980.

46. **Masuhara, H., Hiraoka, H., and Marinero, E. E.,** Non-linear photochemistry of polymer films; laser ablation of poly(*N*-vinylcarbazole), *Chem. Phys. Lett.,* 135, 103, 1987.

47. **Itaya, S., Sakai, H., and Masuhara, H.,** Excimer dynamics of poly(*N*-vinylcarbazole) films revealed by time-correlated single photon counting measurements, *Chem. Phys. Lett.,* 138, 231, 1987.

48. **Miyasaka, H., Masuhara, H., and Mataga, N.,** Picosecond ultraviolet multiphoton laser photolysis and transient absorption spectroscopy of liquid benzenes, *J. Phys. Chem.,* 89, 1631, 1985.

49. **Miyasaka, H., Ikejiri, F., and Mataga, N.,** Picosecond dynamics of ionized and excited states in pure solid polystyrene film, *J. Phys. Chem.,* 92, 249, 1988.

50. **Masuhara, H.,** unpublished results, 1982.

51. **Schnabel, W.,** Laser flash photolysis of polymers, in *Development in Polymer Photochemistry,* Vol. 3, Allen, N. S., Ed., Elsevier-Applied Science, New York, 1982, chap. 2.

INDEX